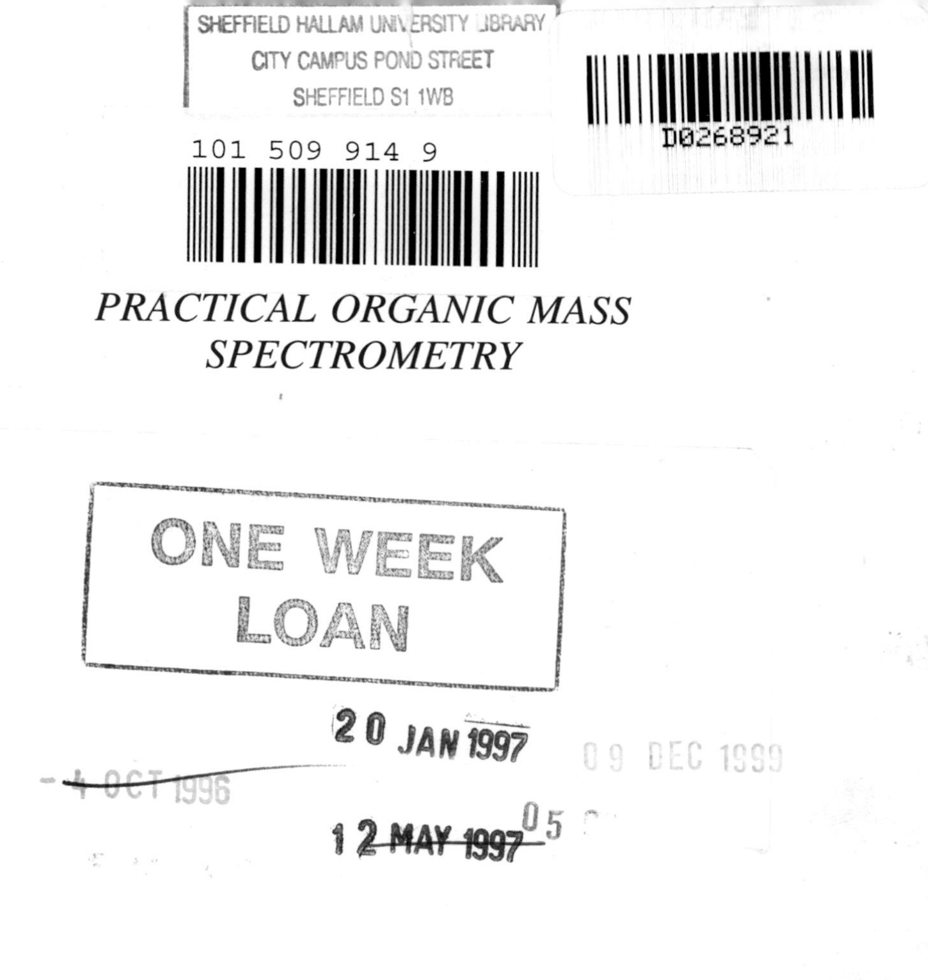

PRACTICAL ORGANIC MASS SPECTROMETRY

PRACTICAL ORGANIC MASS SPECTROMETRY

A Guide for Chemical and Biochemical Analysis

Second Edition

J.R. CHAPMAN
Kratos Analytical Instruments
Manchester, UK

JOHN WILEY & SONS
Chichester · New York · Brisbane · Toronto · Singapore

Baffins Lane, Chichester,
West Sussex PO19 1UD, England

National 01243 779777
International (+44) 1243 779777

Reprinted May 1994
Reprinted in paperback June 1995

Other Wiley Editorial Offices

John Wiley & Sons, Inc., 605 Third Avenue,
New York, NY 10158-0012, USA

Jacaranda Wiley Ltd, GPO Box 859, Brisbane,
Queensland 4001, Australia

John Wiley & Sons (Canada) Ltd, 22 Worcester Road,
Rexdale, Ontario M9W 1L1, Canada

John Wiley & Sons (SEA) Pte Ltd, 37 Jalan Pemimpin #05-04,
Block B, Union Industrial Building, Singapore 2057

British Library Cataloguing in Publication Data

A catalogue record for this book is available from the British Library

ISBN 0 471 95831 X

Typeset in 10/12pt Times from author's disks by Production Technology Department,
John Wiley & Sons Ltd, Chichester
Printed and bound in Great Britain by Biddles Ltd, Guildford and King's Lynn
This book is printed on acid-free paper responsibly manufactured from sustainable forestation,
for which at least two trees are planted for each one used for paper production.

Contents

Preface to First Edition

The aim of this book is to provide a practical guide to the use and utility of some of the alternative ionization and operating techniques now available in mass spectrometry. With this in mind, Chapters 3 to 6 in particular are each broadly divided into three sections—theory, practical requirements, and applications.

Theory is, of necessity, brief in a book of this size but hopfully presents those aspects that are most relevant for appreciation and application of each technique. Again, detailed practical requirements will vary from instrument to instrument but there is, nevertheless, a body of generally relevant knowledge that this book draws on.

Each section covering applications has been supplemented as far as possible with tabulations of references to published data in order to provide the intending user with a starting point for assessing the utility of a particular method. These tabulations are not completly up to date, mainly because of the labour involved, but I hope that both they and the more general references which follow each chapter will prove to be useful.

I am solely responsible for any opinions expressed within the book and for the emphasis given to any particular aspect of the subject. For example, field desorption, although now no longer a pre-eminent technique, is discussed at length since much of the work that has been published regarding the mechanisms that operate in field desorption provides invaluable background material for an intending user of any of the ionization methods designed for higher molecular weight, labile materials.

In conclusion, I would like to thank my wife for her assistance and encouragement provided throughout this enterprise and acknowledge the co-operation of the copyright owners, including my employers, in allowing me to incorporate many of their illustrations in the book.

J. R. Chapman

Preface to Second Edition

As with any new edition in a field that progresses as rapidly as mass spectrometry, a main objective has been to extend the coverage so as to include new techniques that have been developed in the intervening period. Most obviously, the new edition contains two chapters on the ionization of labile materials and a greatly extended section on LC–MS techniques. The introductory chapter on instrumentation now includes a much wider range of instrument types and a thorough updating of the chapter on tandem mass spectrometry has proved necessary, particularly in the areas of instrumentation and collision-induced dissociation.

As a result of these additions, I have chosen to abbreviate some of the existing text, but no techniques have been removed from the discussion altogether. Again, although lack of time and effort has made it impossible to update the tables of applications at the end of Chapters 3, 4, and 5, I felt that the information presented was useful and should be retained. Throughout all these changes, I have tried to retain the original chapter format—theory, practical requirements, and applications—wherever possible.

With the introduction of this paperback edition, I have tried to increase the general utility of the book with the addition of two appendices; a) a list of typical masses and neutral losses for use in the interpretation of electron impact spectra, and b) a list of acronyms and abbreviations, with explanations, used in this book and in mass spectrometry generally. I have also taken this opportunity to correct a few errors found in the hardback edition.

I would like to thank Professor P.J. Derrick (Warwick University) for his comments on the manuscript for Chapter 7 and would again thank my wife for her continued assistance and encouragement throughout this second enterprise. Finally, I would like to acknowledge the co-operation of copyright owners, including my employers, in allowing me to incorporate their illustrations in this second edition.

J. R. Chapman

Chapter 1

Instrumentation

1.1 Introduction

The first precision measurements of ionic masses and abundances were reported in 1918–19 by Aston [1] and Dempster [2] respectively. This work followed the pioneering demonstration of the existence of isotopic forms of stable elements by Thomson [3] using the first instrument built for positive ray analysis. From these beginnings, using magnetic and electrostatic fields for mass separation, steady improvements in the performance of mass analysers have been made and new forms of analyser, e.g. the quadrupole mass filter, have been introduced. The principles of mass analysers and of ion detection are discussed in this chapter, following a brief description of the standard electron impact ion source.

A considerable impetus to the development of organic mass spectrometry has also come from advances in sample handling and in methods of ion formation which have both enormously increased the range of samples amenable to mass spectrometric analysis. Sample handling techniques, particularly the combination of mass spectrometry with on-line chromatographic separation, are the subject of Chapter 2, while Chapters 3 to 6 cover alternative ionization techniques in some detail.

The observation of the further fragmentation of ions formed in the mass spectrometer has added another dimension to the measurement of ionic masses and abundances. It has proved to be particularly fruitful as a source of additional structural information and as a basis for methods for the analysis of specific compounds in complex mixtures. This so-called tandem mass spectrometry technique is described in Chapter 7. Quantitative aspects of mass spectrometry are then discussed in the final chapter, Chapter 8.

1.2 The Electron Impact Ion Source

In the electron impact (EI) source first used by Dempster [4], and subsequently developed by Nier [5], sample vapour at a reduced pressure flows through a region traversed by an electron beam. A schematic diagram of the ion source is shown in Fig. 1.1.

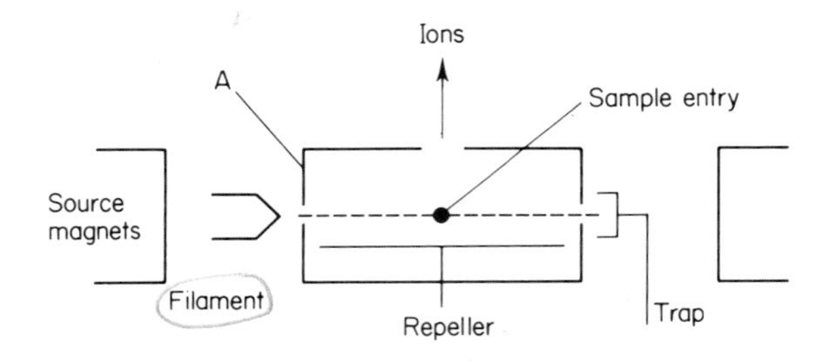

Fig. 1.1 Schematic diagram of electron impact ion source

Electrons formed by heating a tungsten or rhenium ribbon filament are accelerated by a voltage of from 5 to 100 volts towards plate A (Fig. 1.1). Some of these electrons pass through the slit in plate A and traverse the ionization region. The ionizing beam current may be controlled by monitoring total filament emission or, with more accuracy, by feedback control from the current reaching the trap plate in the EI mode. A magnetic field of the order of a few hundred gauss is maintained along the length of the electron beam to confine it to a narrow helical path.

Gas molecules entering the ionizing region interact with the electrons. Some of these molecules lose an electron to become positively ionized and often subsequently undergo fragmentation. Only about 1 in 10^3 of the molecules present in the source is ionized. The ionization probability differs among substances, but it is found that the cross-section for most molecules is a maximum for electron energies from approximately 50 to 100 eV. A typical plot of ion current versus electron energy is shown in Fig. 1.2. Note the dramatic fall in sensitivity at low electron energies.

The total ion current and the fragmentation pattern both depend on the electron energy. Figure 1.3 shows the variation in the mass spectrum of ethyl acetate obtained using nominal electron energies of 14, 20, and 60 eV. Most existing compilations of electron impact mass spectra are based on spectra recorded with approximately 70 eV electrons, since sensitivity is here close to a maximum and fragmentation is unaffected by small changes in electron energy about this value.

Ions formed in the ionization volume are extracted through the ion exit slit by a small voltage applied to the ion repeller plate (Fig. 1.1) and partly by penetration of the accelerating voltage field through the ion exit slit in the case of magnetic sector instruments.

Most electron bombardment sources are reasonably gas-tight apart from the small electron entry hole and ion exit slit necessary for their operation, so that a high sample pressure and consequently a high sensitivity is maintained. At the same time, fast pumping is maintained in the source housing and analyser regions (see Fig. 2.1 and section 2.1). A high pumping speed in the source housing reduces the sample pressure in the region of the hot filament so as to prolong filament lifetime and

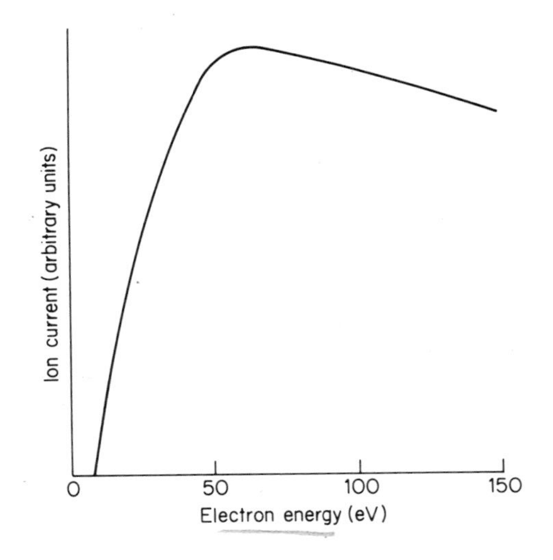

Fig. 1.2 Plot of ion current versus electron energy

minimizes the possibility of sample molecules re-entering the ionization chamber. A high pumping speed in the analyser and source housing regions also minimizes the possibility of unwanted ion–molecule collisions, although fragmentation of ions can be deliberately induced by ion–molecule collisions with a high pressure gas in a specially constructed collision cell (section 7.2). Too high a pressure in the analyser can have a number of other undesirable consequences, e.g. loss of sensitivity, loss of resolution, and a shortening of electron multiplier lifetime.

Despite its relative gas-tightness, the half-life for sample molecules which are continuously pumped from the small volume of the source and attached sample lines is only a fraction of a second. Thus the mass spectrometer as a detector has a very rapid response, entirely compatible with the rapid changes in sample concentration experienced in high resolution chromatography. The effective pumping speed is, however, dramatically reduced by any cooler surfaces which can trap less volatile materials (cf. section 2.2.4.2). Even heating of the source block and adjacent areas is therefore essential. Adjustment of the source block temperature by cooling as well as by heating is a useful facility. In this connection, it should be appreciated that the heat of the filament alone will maintain an ion source temperature of 100–150°C.

Although the electron bombardment source was the first source developed and used in organic mass spectrometry, it is still the most popular. Several factors contribute to its popularity. Thus, its stability, ease of operation and control of beam intensity, lack of contamination problems, and relatively high sensitivity are practical advantages. The lack of selectivity under electron impact conditions is a further advantage when a wide range of compounds is to be analysed.

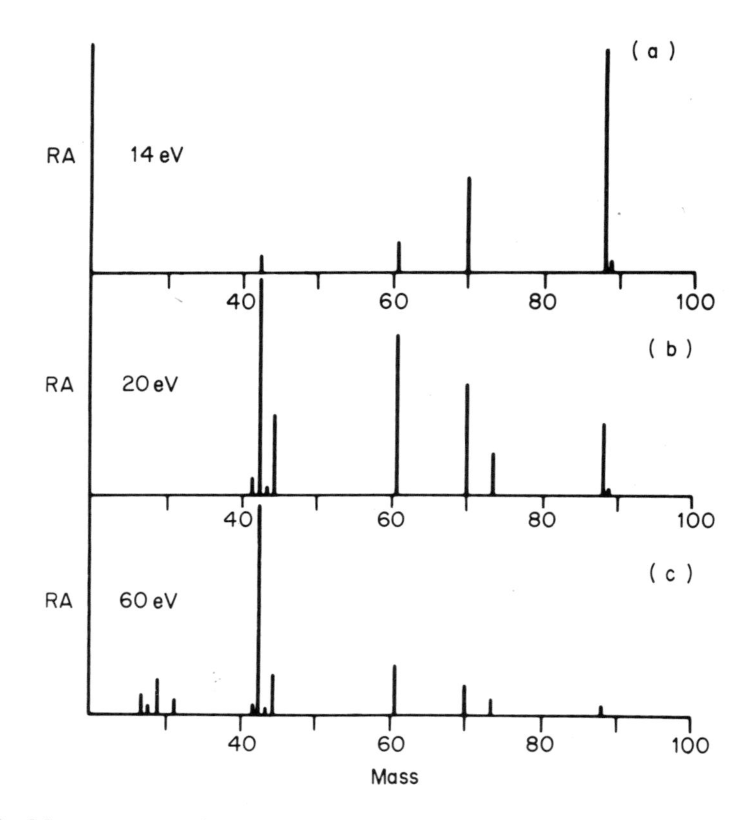

Fig. 1.3 Mass spectrum of ethyl acetate as a function of electron energy: (a) 14 eV; (b) 20 eV; (c) 60 eV. (Reproduced with permission from ref. 65)

Some groups of isomeric compounds give very similar mass spectra, but generally an electron impact mass spectrum is specific and characteristic of the chemical structure of the sample. Furthermore, the major compilations of mass spectral reference data presently available consist entirely of electron impact spectra.

1.3 Mass Analysers [6]

Several forms of mass analyser which separate ions according to their mass-to-charge ratio exist, e.g. magnetic sector analysers, quadrupole mass filters, quadrupole ion traps, time-of-flight analysers, and ion cyclotron resonance instruments. The first two types presently account for the great majority of instruments used in organic analysis. Each analyser is now discussed in more detail in the following sections.

1.3.1 Magnetic sector analyser [6,7]

The principle of the single-focusing magnetic deflection mass spectrometer is illustrated in Fig. 1.4. Ions formed in the source are accelerated through a voltage V (V = 2000–8000 V) towards the source slit which is at earth potential. The fall in potential energy for the ions is equal to their gain in kinetic energy, which is summarized as follows:

$$zeV = \frac{mv^2}{2} \tag{1.1}$$

where e is the charge on an electron, z is the number of such charges on the ion, m is the mass of the ion, V is the accelerating voltage, and v is the velocity of the ion.

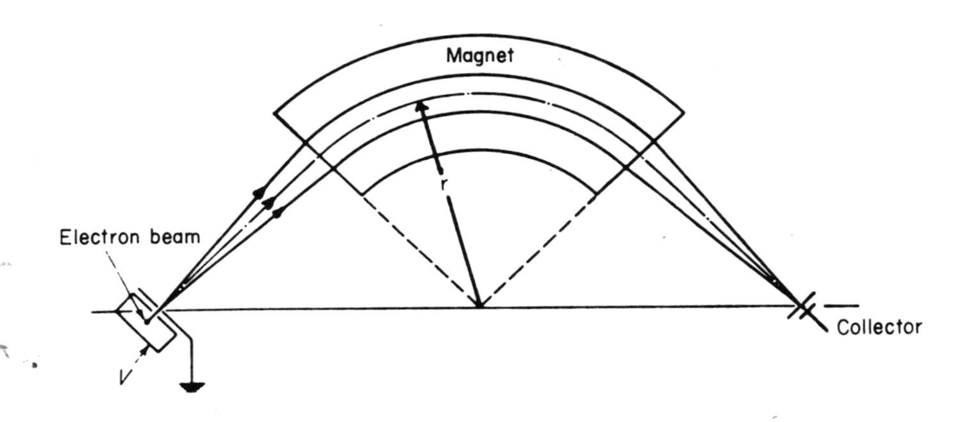

Fig. 1.4 Schematic diagram of single-focusing magnetic sector instrument

For an ion to reach the collector slit and be recorded, it must traverse a path of radius of curvature r through the magnetic field of strength B. The equation of motion of the ion

$$\frac{mv^2}{r} = Bzev \tag{1.2}$$

expresses the balance between the angular momentum and the centrifugal force caused by this field. Note that rearrangement of equation (1.2) in the form $mv = Bzer$ demonstrates the fact that a magnetic sector is a momentum analyser, rather than a mass analyser as is commonly assumed. Combining equations (1.1) and (1.2) gives the basic mass spectrometer equation:

$$\frac{m}{z} = \frac{B^2 r^2 e}{2V} \tag{1.3}$$

Expressed in practical units the atomic mass (M) of a singly charged ion is given by

$$M = 4.83 \times 10^3 \left(\frac{B^2 r^2}{V}\right) \tag{1.4}$$

where r is in centimetres, B is in tesla (1 tesla = 10^4 gauss), and V is in volts. For example, a maximum field strength of 2 tesla gives a maximum mass of just over 10 000 daltons (10 000 Da = 10 kDa) for an instrument of 65 cm radius operating with an accelerating voltage of 8000 V.

Equation (1.3) shows that by varying either B or V, ions of different m/z ratio, separated by the magnetic field, can be made to reach the collector. The most common form of mass scan is the exponential magnet scan, downward in mass. This scan has the advantage of producing mass spectral peaks of constant width. The equations appropriate to this form of scan are as follows:

$$m = m_0 \mathrm{e}^{-kt} \tag{1.5}$$

$$t_\mathrm{p} = \frac{t_{10}}{2.303R} \tag{1.6}$$

where m_0 is the starting mass at time $t = 0$, m is the mass registered at time t, t_p is the peak width between its 5% points, t_{10} is the time taken to scan one decade in mass (e.g. m/z 200 to 20 or m/z 500 to 50), and R is the resolving power measured by the 10% valley definition (see section 1.7).

Scanning of the accelerating voltage V would, at first sight, appear to be advantageous because of the ease of rapid scanning and the ease of scan control. Change of the accelerating voltage over even a twofold range, however, causes defocusing and loss of sensitivity and is consequently little used as a method of scanning.

1.3.2 Double focusing magnetic sector analyser [6,7]

Since a magnetic sector is a momentum rather than a mass analyser, ions of the same mass but of differing translational energy are not brought to a point focus in a single-focusing magnetic deflection instrument. The translational energy spread of the ions formed in an electron impact source will therefore limit the available resolution, and the existence of charging effects and contact potentials as source contamination builds up in use causes some worsening of this situation. Other ion sources, e.g. the field desorption source, produce ions with a much greater translational energy spread.

The inability of a single magnetic sector instrument to provide more than a limited resolution under practical conditions can be overcome by the addition of an electrostatic sector (Fig.1.5). The combination of the two sectors can then be designed to have velocity focusing properties.

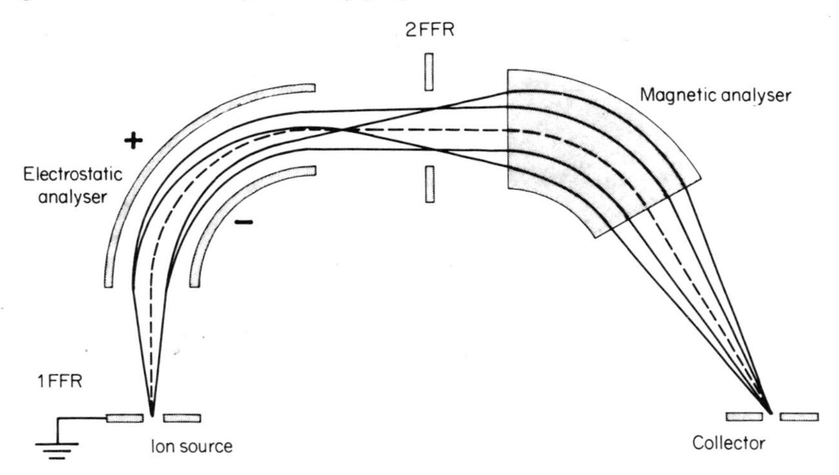

Fig. 1.5 Schematic diagram of double-focusing magnetic sector instrument of Nier–Johnson geometry

An ion entering the electrostatic field travels in a circular path of radius R such that the electrostatic force acting on it balances the centrifugal force. The equation of motion is

$$\frac{mv^2}{R} = ezE \tag{1.7}$$

where E is the electrostatic field strength. Hence the radius of curvature of the ion path in the electrostatic sector is dependent upon its energy, but not its mass.

A narrow slit placed in the image plane of an electrostatic sector could be used to transmit a narrow band of ion energies. If such an energy filter were used as the ion source for a magnetic sector instrument, the resolution could be improved but sensitivity would be lost since most ions would be rejected. This loss in sensitivity can be avoided by the choice of a suitable combination of electrostatic and magnetic sectors such that the velocity dispersion is equal and opposite in the two analysers. The Nier–Johnson geometry [8] illustrated in Fig. 1.5 is of this type and is one in which both mass and energy focusing occur at a single point. The collector slit is located at this focal point and the detector is placed immediately behind this beam-defining slit.

In Fig. 1.5 the ion following the optical axis has mass m and velocity v. A second ion of mass m may have a velocity $v(1+\beta)$ and also an angular divergence

a from the optical axis leaving the ion source. The expression giving the image displacement at the collector is a power series of the form

$$A_{10}\alpha + A_{01}\beta + A_{20}\alpha^2 + A_{11}\alpha\beta + A_{02}\beta^2 + \cdots \quad (1.8)$$

A system is said to be double focusing when the aberrations $A_{10} = A_{01} = 0$. In addition the second-order coefficients A_{20}, A_{11}, and A_{02} can be reduced by a suitable choice of sector geometry. Thus, a well-designed double focusing instrument can offer good resolution and sensitivity even when α and β are both relatively large.

In the Nier–Johnson geometry the electrostatic sector precedes the magnetic sector, but double focusing can be achieved whichever field is traversed first by the ion beam[9]. Most sector instruments intended for medium or high performance work in organic analysis are based on either conventional or reversed (Fig. 1.6) Nier–Johnson geometry. Although only one mass can be exactly in double focus with these geometries, there is a limited range of nearby masses that will fall on an approximate plane of double focus and can be detected simultaneously. Ion detection in this case uses an array detector which can provide improved detection limits (section 1.4.4).

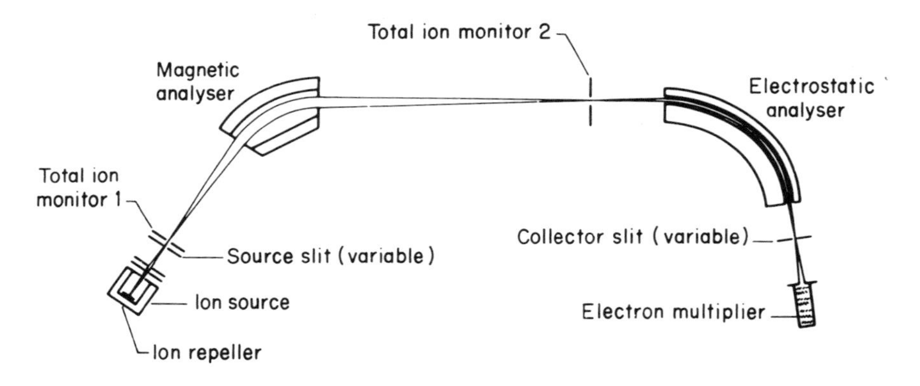

Fig. 1.6 Schematic diagram of double-focusing magnetic sector instrument of reversed geometry. (Reproduced with permission from ref. 66)

The resolution of a magnetic sector instrument is measured from a comparison of the mass dispersion effected by the magnet with the width of the image at the collector. This image width, and therefore the resolution for a given dispersion, is changed by changing the widths of the source and collector slits. Thus, an increase in resolution is obtained at the expense of sensitivity.

Magnetic sector instruments are more expensive and require a more sophisticated approach to operation than most quadrupole instruments (section 1.3.3). They do not therefore compete with quadrupole instruments in the market for unsophisticated routine instrumentation. They are also somewhat more difficult to interface to many inlet systems since the ion source of a sector instrument operates at a high voltage whereas the inlet system is usually at earth potential (Chapter 2).

Double-focusing sector instruments are, however, very widely used and owe their popularity to their versatility and high specifications for mass range and resolution. They are able to provide mass measurement data of ppm accuracy as well as access to tandem mass spectrometry experiments (section 7.3.1) and more specific forms of selected ion monitoring (e.g. section 8.3.1 and 8.3.2). The development of fast scanning, fully laminated magnets has introduced specifications for speed and convenience of operation in the scanning and selected ion monitoring modes to equal those of quadrupole instruments while offering mass ranges in excess of 10 000 daltons at full sensitivity.

1.3.3 Quadrupole analyser [10–12]

The quadrupole mass filter consists of four parallel rods of hyperbolic or circular cross-section arranged symmetrically to the z axis (Fig. 1.7). A voltage made up of a d.c. component U and a radio-frequency (r.f.) component $V_o \cos \omega t$ is applied between adjacent rods. Opposite rods are electrically connected. Ions injected into the filter with a very small accelerating voltage, typically 10–20 V, are made to oscillate in the x and y directions by this electric field.

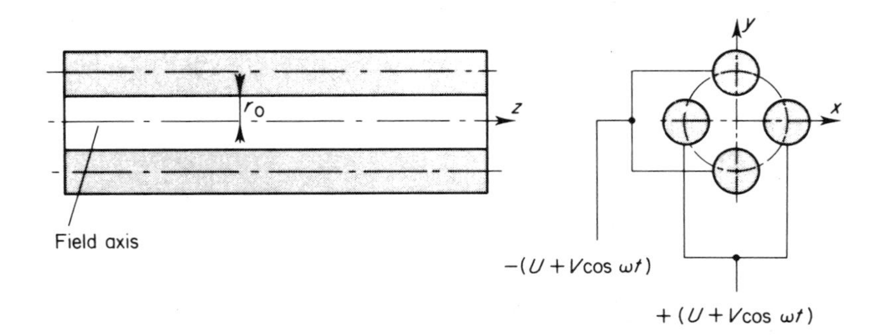

Fig. 1.7 Schematic diagram of a quadrupole analyser

The parameters a and q can be defined by

$$a = \frac{8eU}{mr_0^2\omega^2} \tag{1.9}$$

$$q = \frac{4eV_0}{mr_0^2\omega^2} \tag{1.10}$$

In these equations $2r_o$ is the rod spacing and ω is the frequency of the r.f. voltage. For certain values of a and q the oscillations performed by the ions are stable, i.e. their amplitude remains finite, but for other values of a and q the oscillations are unstable and the amplitude becomes infinite. The stability diagram, which is also known as a Mathieu diagram (Fig.1.8), shows the values of a and q for which these conditions apply.

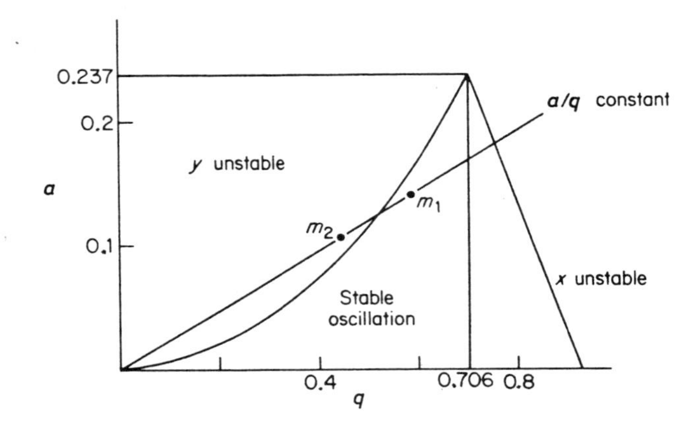

Fig. 1.8 Stability diagram for a quadrupole analyser

For masses such as m_1 where the point falls within the region of stable oscillation, ions may continue on an oscillatory path within the bounds of the rods to reach the detector, whereas for other masses such as m_2 the oscillations become unstable and the ions are lost on the rod assembly. Hence mass separation is achieved. The mass spectrum is scanned by varying U and V_o while maintaining the ratio U/V_o constant. The recorded mass m is proportional to V_o so that a linear increase of V_o provides an easily calibrated linear mass scale. Scanning by varying the frequency ω while keeping U and V_o constant is also possible but is less convenient.

The resolution required is selected by adjusting the d.c. voltage U. At high resolution, corresponding to high values of a/q, a significant proportion of the selected ions oscillate far enough to be lost so that transmission is reduced. At lower values of a/q the range of stabilization is extended and the resolution is decreased. When the d.c. voltage is switched off, all ions perform stable oscillations for a sufficiently small r.f. voltage. This feature has been used to great advantage to provide a collision chamber of high transmission in the triple quadrupole (section 7.3.2) and more recently in hybrid geometry instruments (section 7.3.4).

In theory the mass resolution of the quadrupole filter can be increased to a very high value by operating with an a/q value close to the apex of the stability triangle. In practice, however, the attainable resolution depends upon the initial ion velocities in the x and y directions and upon the positions at which the ions enter the filter; hence the quadrupole is essentially a low resolution mass filter. Unlike magnetic sector instruments, no narrow resolution defining apertures are required but the ion beam is usually restricted in cross-section and angular divergence at the entrance to the filter. Ion transmission through a simple quadrupole mass filter decreases with increasing mass-to-charge ratio and losses can be severe at higher mass values. High mass transmission may be improved by the use of a short r.f.-only pre-filter [13].

Quadrupole instruments are deservedly popular since they are compact, relatively inexpensive, and need little experience to operate. They have two particular advantages compared with magnetic sector instruments, viz. ease of data system control and ease of interfacing with a wide range of inlet systems (cf. section 3.2.2). In addition the basic quadrupole analyser is able to separate positive or negative ions without modification (section 4.3). All of these features have contributed to the conversion of a relatively unsophisticated mass filter into an analytical instrument which can be applied routinely to a great many problems.

Quadrupole analysers do not provide the extended m/z range and high resolution or the wide range of data types offered by magnetic sector instruments. They do, however, also constitute an integral part of more sophisticated instruments, such as triple quadrupole (section 7.3.2) and instruments of hybrid geometry (section 7.3.4), specifically designed for tandem mass spectrometric analyses. In addition, m/z range limitations may be offset by the use of ionization techniques, such as electrospray, which result in the formation of multiply charged ions (section 6.3.2.2).

1.3.4 Time-of-flight analyser [14,15]

For an ion that is accelerated through a voltage V, the resulting velocity v is characteristic of the mass-to-charge ratio. In the time-of-flight mass spectrometer (Fig. 1.9), ions are separated according to their velocity and the mass-to-charge ratio is determined from a measurement of the time taken to traverse a specified flight path (L) to the detector

$$t = \left(\frac{m}{2zeV}\right)^{1/2} L \qquad (1.11)$$

Ions of a very high mass-to-charge ratio may be recorded after an appropriate length of time. Thus, ions with an m/z value in excess of 500 000 [16] have recently been recorded in matrix-assisted laser desorption experiments.

An important advantage of a time-of-flight instrument is its potentially high sensitivity which results from two instrumental features: first, high transmission due to the absence of beam defining slits; second, temporal separation of ions which, unlike spatial separation, does not direct any ions away from the detector. Early

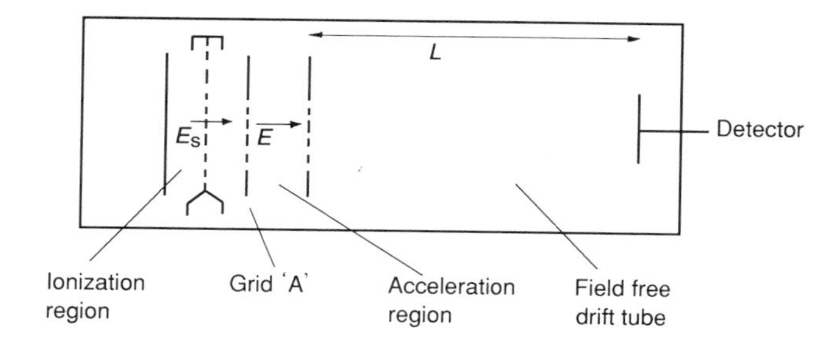

Fig. 1.9 Schematic diagram of time-of-flight instrument

time-of-flight instruments were unable to make full use of this latter advantage, but the recent development of high speed recording devices has made it possible to record ion currents (transients) that contain data corresponding to the entire mass range. Erickson *et al.* [17] have recently demonstrated the recording and summing of arrival-time transients to give up to 20 mass spectra per second in gas chromatography–mass spectrometry (GC–MS) operation.

Good mass resolution is less readily achieved in a time-of-flight instrument since it is affected by factors that increase the time width of the ion packet that reaches the detector. These factors are the length of the ion formation pulse, the finite size of the volume in which the ions are formed, and the kinetic energy spread of the ions. Thus, careful ion source design is essential for a time-of-flight mass spectrometer.

In the first commercial time-of-flight mass spectrometer, based on the design reported in 1955 by Wiley and McLaren [18], ions were formed in the gas phase by electron impact ionization. The fact that ions can be created at different positions within the electron beam is compensated for by the use of an ion source with two ion accelerating regions (Fig. 1.9). The provision of a relatively small extraction field (E_S) in the first region means that ions formed at the rear of the source acquire more energy than those formed close to the grid (A) and so may catch up with the latter. The second stage accelerates the ions to their final energy.

The electron beam is pulsed for approximately 1 μs to provide sufficient ions for analysis and time focusing is then achieved by pulsing the first accelerating voltage over a much shorter time period so that all the ions in the source are accelerated almost simultaneously. Kinetic energy spread is reduced by introducing a time lag between ion formation and acceleration. Ions that are initially moving away from the detector would normally arrive at the detector after other ions of the same mass. During the time lag, however, the former ions will move to positions where the application of the extraction voltage will cause them to acquire greater energy than the other ions and again catch up at the detector.

In contrast to continuous gas phase ionization processes, such as electron impact ionization, time-of-flight mass spectrometers are directly and simply compatible with ion formation by direct desorption from a surface, e.g. laser desorption and plasma desorption mass spectrometry. These pulsed ionization techniques provide a short, precisely defined ionization time and a small, precisely defined ionization region which is ideal for time-of-flight analysis.

Resolution in a time-of-flight instrument may be improved by the use of a reflectron (or ion mirror) [19]. The reflectron, which is used to compensate for the difference in flight times of ions with different kinetic energies, is shown later in Fig. 5.15. After traversing the flight tube, ions enter a retarding field, defined by a series of grids, and are turned around and sent back through the flight tube. The principle of the reflectron is that an ion with a higher energy will penetrate the retarding field more deeply, will spend more time turning around, and will just catch up with a slower ion (of the same mass) at the time they both reach the detector. The reflectron results in the loss of some ion signal. In part this is due to the introduction of additional grids, so that gridless reflectrons have been designed [20].

Time-of-flight methods are very suitable for measurements where the available ion current is limited by the method of ionization, e.g. ^{252}Cf fission fragments (section 5.3.3) or laser pulses (section 5.3.4), and for problems where sample is limited. Pulsed ionization techniques may also be used with a continuous flow of sample, but the process is, naturally, much less efficient with respect to sample consumption [21]. An interesting implementation in this case is to present the sample as a molecular beam (section 5.3.4) which provides internal energy cooling and reduces the kinetic energy spread [22]. Further correction of the energy spread following laser ionization is possible using a reflectron and under these conditions relatively high resolving powers may be recorded [23]. More recently, efforts to make continuous ionization processes more compatible with time-of-flight analysis have used ion storage techniques which offer the benefit of much higher duty cycles [24–27].

1.3.5 Ion trap [28,29]

Figure 1.10 shows a cross-section of an ion trap. Whereas a quadrupole mass filter is formed by linear extension of this cross-section, the three-dimensional ion trap is a solid of revolution produced by rotation of the cross-section about the z axis. The ion trap therefore comprises three cylindrically symmetrical electrodes: viz. two end-caps (A,B) and a ring (C).

Each electrode has accurately machined hyperbolic internal surfaces. The ring electrode is fed with an r.f. voltage (V), and sometimes an additional d.c. voltage (U), relative to the end-cap electrodes, which are held at earth potential. Operating parameters (a_z and q_z), which are analogous to those for the quadrupole mass filter (equations 1.9 and 1.10), may also be defined for the ion trap where r_0 is now the internal radius of the ring electrode. The ion trap is a small-scale device so that typically r_0 is 1 cm.

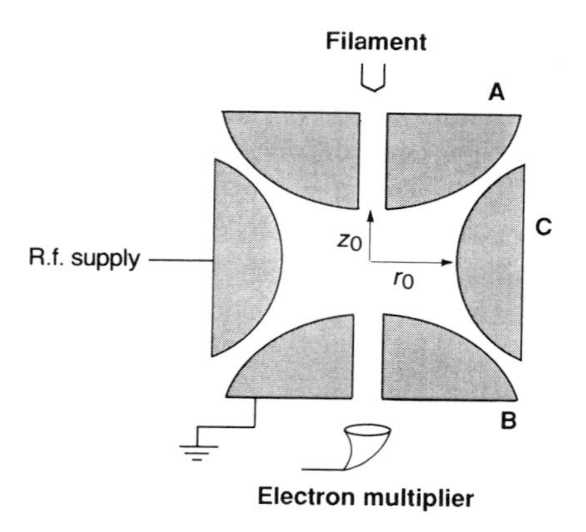

Fig. 1.10 Schematic diagram of ion trap

The use of an r.f. voltage causes rapid reversals of field direction so that ions are alternately accelerated and decelerated in the axial (z) direction and vice versa in the radial direction. Regions of stable motion are described by a Mathieu diagram. Stable orbits, in which ions are trapped in the cell, correspond to regions of stable motion in both the radial and axial directions. For example, Fig. 1.11 shows that ions of m/z 200, 300, and 400 may be held in stable orbits by the use of an r.f. voltage of 1000 volts. The same figure also shows that the range of m/z values trapped changes as the r.f. amplitude is changed.

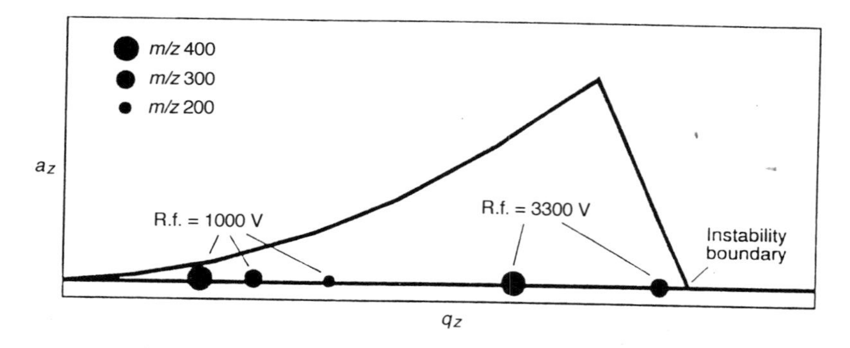

Fig. 1.11 The effect of change of r.f. voltage on ion stability in an ion trap. (Reproduced with permission from ref. 62)

Scanning mass analysis with an ion trap uses what is known as the mass-selective instability mode (Fig. 1.12). In this mode, the d.c. component is zero while the r.f. frequency (typically 1.1 MHz) and initial amplitude are chosen so that all ions

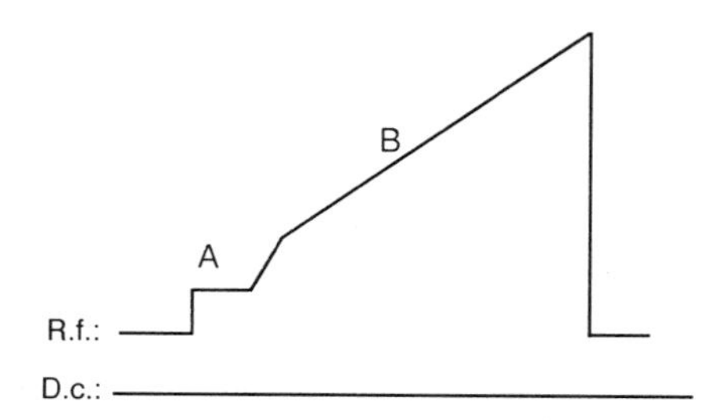

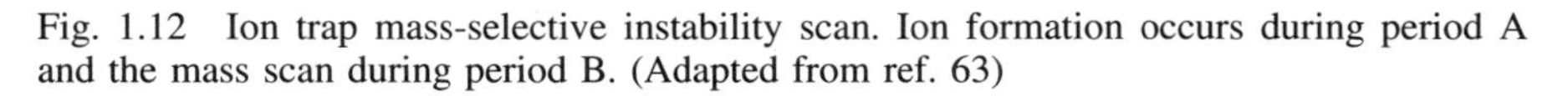

Fig. 1.12 Ion trap mass-selective instability scan. Ion formation occurs during period A and the mass scan during period B. (Adapted from ref. 63)

with an m/z value greater than a threshold value, e.g. $m/z = 4$, are stored. Thus, first the electron beam is used to create ions within the trapping field. Then, after *ca.* 1 ms, the electron beam is turned off and ions with an m/z value below the threshold value are allowed to escape.

Now, the amplitude of the r.f. voltage is increased and the motion of the ions becomes more energetic so that they eventually develop unstable trajectories along the z axis ($q_z = q_{ej} = 0.908$) [30]. As a result, and in order of increasing m/z value, ions leave the trap through holes in one of the end-caps and strike the detector. In this way, a mass spectrum is generated.

A number of developments have greatly increased the analytical utility of the ion trap. First, it was discovered during early work that the presence of a low pressure of helium (10^{-3} torr) greatly improves the quality of the spectra in terms of both resolution and sensitivity. This effect is due to collisions with the helium gas which reduce the kinetic energy of the ions and cause migration of trapped ions towards the centre of the trap. As a result of this discovery, the ion trap has become very popular as a compact, low cost instrument for GC–MS work.

Although the principal applications of the ion trap have involved the creation of ions within the trap, e.g. either by electron impact or by chemical ionization, the injection of ions from an external source is obviously preferable with ionization techniques such as liquid secondary ion mass spectrometry (LSIMS) or electrospray. Here again, the use of a helium buffer gas is the key to successful operation as collisions with this gas decrease the kinetic energy of injected ions and increase the probability of trapping [31].

A build-up of ion density can lead to space-charge effects which will modify the electric fields within the trap. This problem has been tackled by a second development, the use of an 'automatic gain control' (AGC). Here, an initial ionization period of fixed duration (e.g. 0.2 ms) is used, after which low mass ions formed from background gases are removed and the remaining analyte ions detected without mass separation. This 'total ion' signal is then used to calculate the optimum time that will avoid space-charge effects for a second, more extended, period of ionization. Following this second ionization period, the r.f. voltage is ramped to effect a mass scan as before. The main practical advantage of the AGC is a great increase in the usable dynamic range before the onset of space-charge effects [32].

The third development is the technique known as 'axial modulation'. In this technique, a supplementary r.f. voltage with a frequency that is lower than that of the main r.f. voltage fed to the ring, and with a much smaller amplitude, is applied between the end-cap electrodes. By scanning the amplitude of the main r.f. voltage in the normal way (for a mass-selective instability scan), ions of different m/z are sequentially brought into resonance with the supplementary r.f. signal. These ions pick up translational energy and are then ejected from the trap in a process known as 'resonant ejection'.

A very useful feature of axial modulation is that ions leave the trap in m/z sequence at a much lower r.f. voltage than would ordinarily be required. This effect has been used, in conjunction with a reduced r.f. frequency, to increase the normal upper mass limit for the trap of 650 Da to a figure in excess of 70 kDa [33]. An analogous technique of resonant excitation provides the basis for MS–MS operation with the ion trap which has become one of the most significant recent areas of development (section 7.3.6). The most recently developed capability demonstrated with the ion trap is a greatly enhanced mass resolution, achieved by slow scanning in the mass-selective instability mode [34–36].

The long storage times used with the ion trap facilitate ion–molecule reactions of the type employed in chemical ionization at much lower reagent gas pressures than those used in conventional high pressure sources (section 3.2.4). The significant trapping times can also lead to ion–molecule reactions,involving analyte molecules, which result in abnormal $(M + 1)/M$ ratios under electron impact ionization conditions. This problem can only be overcome by a reduction in the concentration of sample entering the trap to an appropriate level [37]. The relatively poor accuracy of mass assignment achieved with the ion trap [38] is still a problem that needs to be addressed.

1.3.6 Fourier transform–ion cyclotron resonance [39]

In the simplest instrumentation for Fourier transform–ion cyclotron resonance mass spectrometry (FTMS), ions are formed, detected, and analysed within a single cell located within the solenoid of a superconducting magnet (Fig. 1.13). Once an ion is generated in the cell, it is constrained to move in a circular path perpendicular to the magnetic field axis with a frequency ω, which depends on the magnetic field strength B, and the mass-to-charge ratio:

$$\omega = 1.537 \times 10^7 \left(\frac{zB}{M}\right) \qquad (1.12)$$

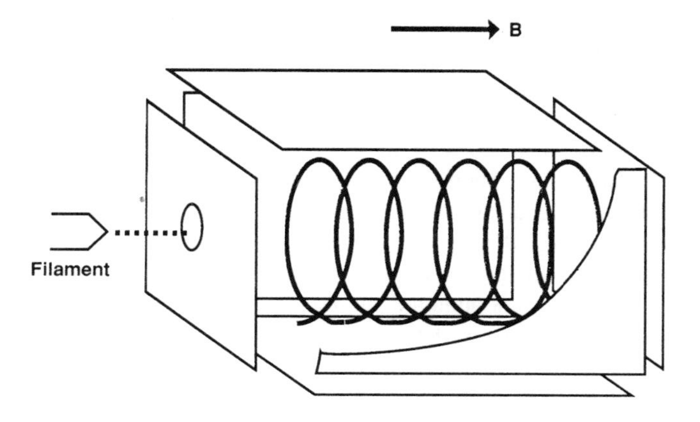

Fig. 1.13 Schematic diagram of Fourier transform–ion cyclotron resonance instrument. The opposing trap plates (left and right) are normal to the magnetic field while the opposing transmitter (front and back) and receiver (top and bottom) plates are parallel to the field. (Adapted from ref. 64)

where z is the number of charges on the ion, B is the magnetic field strength in tesla, M is the atomic mass, and ω is expressed in hertz. Ions are prevented from escaping axially by the application of a small d.c. field across the trapping plates. If an r.f. voltage with the same frequency, ω, is now applied to the transmitter plates, the appropriate ions absorb energy with a consequent increase in orbital radius and velocity but with no change in frequency. After a very few rotations all the ions are moving together coherently and it is this coherent motion that is then detected.

In practice, ion cyclotron motion is excited over the mass-to-charge range of interest by a very fast frequency sweep of the voltage applied to the transmitter plates. This cyclotron motion induces image currents in the receiver plates such that the frequency of each component of the induced current is the same as the ion cyclotron frequency, ω, and can therefore be related to mass. The induced current is detected as a 'time-domain' signal which is amplified, digitized, and finally converted to a frequency-domain spectrum by the application of a fast Fourier transform (FFT). The detection process in Fourier transform mass spectrometry is relatively inefficient so that 10–100 ions are required to produce a detectable signal while the electron multiplier used in conventional mass spectrometers can detect single ions. McLafferty and co-workers [40] have recently demonstrated that signal-to-noise for high mass ions may be improved by making multiple measurements on the same ions.

In FTMS, the longer the image current can be sampled, the higher the mass resolution. Although many factors can affect the image current decay rate, the most important factor is collisional damping which destroys the coherent motion of the ions of the ions in the cell. Thus, with a high pressure in the cell ($> 10^{-5}$ torr), the signal decays within milliseconds to produce a low resolution spectrum. At low pressures ($< 10^{-8}$ torr), on the other hand, the transient may last for tens of seconds to produce ultra high resolution. A unique feature of FTMS is the coincident increase of resolution and signal-to-noise as pressure is reduced [41].

Fourier transform mass spectrometry is, in principle, a high mass, high resolution technique. For example, a resolution of 53 000 has been reported [42] for an abundant ion at m/z 9746 and ions with a mass-to-charge ratio of 31 830 have been recorded [43]. Unfortunately, resolution is degraded with the smaller ion populations that are usually encountered at high mass and a mechanism to account for this signal loss has recently been proposed [44]. High resolution spectra have, however, been successfully recorded for high molecular weight compounds following the analysis of multiply charged ions generated by electrospray ionization [45].

The basic cyclotron equation (equation 1.12) is only sufficiently accurate for the calculation of unit mass due to space-charge effects and non-uniform electric fields. Low ppm accuracy is, however, generally available with the use of simple calibration procedures. The simplicity of these calibration procedures makes FTMS an attractive method for the accurate mass assignment of ions produced by less routine ionization methods.

A common application of FTMS is to pulsed laser desorption/ionization of involatile samples, since laser desorption mass spectrometry (LDMS) is compatible with the low pressure requirements of FTMS. The direct coupling of a Fourier transform mass spectrometer with ionization techniques which operate at a higher pressure is difficult because the gas load increases the pressure in the cell and causes a loss of resolution and sensitivity. These conflicts have been reconciled by the development of alternative source configurations. In the dual cell configuration [46], the two cells are connected by a small orifice in the common trapping plate so that a pressure differential can be maintained between them. Ions can be formed in the high pressure cell and subsequently transferred to the low pressure cell for detection. In the external cell configuration [47,48] a differentially pumped ion source, located outside the magnetic field, is interfaced to the analytical cell. Unfortunately, a further problem in FTMS is the perturbation of ion motion in the cell by space-charge effects [49]. Without a completely practical solution to this problem, Fourier transform mass spectrometers are not well-suited to combined chromatographic techniques such as GC–MS operation because of the resultant low dynamic range.

The ion trapping characteristics of a Fourier transform mass spectrometer provide a very suitable basis for studies of ion-molecule chemistry [50]. Complex MS–MS experiments are easily implemented and modified without hardware changes (section 7.3.7). As with the ion trap, ion–molecule reactions occur at relatively low gas pressures and may also occur between sample neutrals and sample ions.

1.4 Ion Detection Systems [51]

1.4.1 The electron multiplier

Ions produced in the source are separated according to their mass-to-charge ratio in the analyser section. As a result of this separation, ion currents of widely different intensities, ranging from approximately 10^{-9} A to a minimum of perhaps 10^{-18} A in selected ion monitoring (section 8.1.2), may reach the detector. The most widely used form of detection for these currents is the electron multiplier (Figs. 1.14 and 4.3).

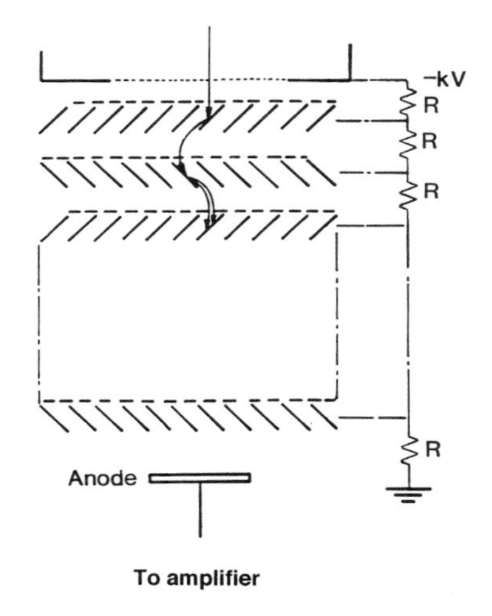

Fig. 1.14 Schematic diagram of Venetian blind electron multiplier (reproduced with permission from ref. 51)

The electron multiplier may be of either the discrete dynode or the continuous dynode type. A discrete dynode multiplier comprises some 12–20 dynodes made from beryllium–copper, which has good secondary emission properties, electrically connected through a resistive network [52]. The initial impact of energetic ions on the first conversion dynode results in the emission of electrons. These electrons are then accelerated by the applied voltage into further collisions at each stage of the dynode chain, eventually resulting in a greatly amplified electron emission. The modification of electron multiplier systems for negative ion detection is discussed in section 4.3.

The continuous channel multiplier [53] is made from a lead doped glass which has good secondary emission properties and is electrically resistive. A voltage applied between the ends of the tube will therefore create a uniform field along the length of the tube. An ion incident on the inner surface at one end will release secondary electrons which are then accelerated by the field so as to cause further collisions with the inner wall. Channeltrons are, in fact, curved to prevent spurious pulses due to the movement of positive ions formed within the detector. Channel electron multipliers are very compact and can also be made in the form of microchannel plates which are used as the basis for array detectors (section 1.4.4).

The gain of the electron multiplier, which may have a maximum value in excess of 10^6, depends on the applied voltage and is produced with virtually no noise and with a very small time constant. A collector plate, placed after the final dynode, is connected to a pre-amplifier which converts the output current into a voltage suitable for recording, e.g. by digitization. The output voltage (V) from this amplifier is given by

$$V = GIR_f \tag{1.13}$$

where G is the multiplier gain, R_f is the amplifier feedback resistance (typically 10^8 ohms), and I is the input current to the multiplier.

1.4.2 The definition and measurement of multiplier gain

An absolute determination of instrumental sensitivity (section 1.6) requires a knowledge of the multiplier gain used in the measurement. Multiplier gain is measured by a comparison of the apparent charge represented by a single ion peak as recorded after amplification by the multiplier with the known charge on a single ion (1.6×10^{-19} coulomb):

$$\text{Gain} = \frac{3.1 \times 10^{18} Vt}{R_f} \tag{1.14}$$

Method

Choose a relatively high multiplier gain setting and a relatively high bandwidth (approximately 3 kHz) so that single ion peaks are of measurable height. Too low a bandwidth will result in broad, weak, single ion peaks. Tune just off a major peak from a calibration compound in the m/z 200 region so that single ions are just detected. Make sure that the instrument is tuned far enough from the peak so as to register the arrival of a separated stream of single ions rather than multiple ions. These single ions may be recorded with an oscilloscope or an ultraviolet chart recorder run at high paper speed. A typical single ion peak will have a width of approximately 0.25 ms with a bandwidth of 3 kHz.

Measure the mean height and width over some 10–20 successive single ion peaks and convert these values to volts (V) and time (t) respectively. Do not ignore small single ions—a wide range of single ion peak sizes is expected due to the statistics of single ion amplification.

Multiplier gain is calculated using equation (1.14) with the value of the feedback resistor (R_f) used under the conditions chosen for recording. A typical value for R_f is 10^8 ohms. The gain at lower multiplier settings is measured by comparing the height of a calibration compound peak at these lower gain settings with the peak height at the gain setting used in the above determination.

A good multiplier should certainly be capable of a gain of 10^6. In addition to using the gain value in absolute sensitivity determinations, the operator should check the gain periodically during the life of the multiplier, and particularly after any vacuum accidents or other suspected contamination of the multiplier.

1.4.3 Post-acceleration detectors

If we consider the detection of high mass ions, the velocity of an ion on impact with the multiplier surface is given by

$$v = 1.39 \times 10^4 \left[(V_a - V_t)\left(\frac{z}{M}\right)\right]^{0.5} \qquad \text{m s}^{-1} \tag{1.15}$$

where V_a is the source accelerating voltage, V_t is the voltage on the target surface, and M/z is the mass-to-charge ratio of the ion. Thus, an ion of mass 1000 Da, which reaches the detector with an energy of 10 keV, will have a velocity of 4.4×10^4 m s^{-1}. Under these conditions detection by an electron multiplier is efficient. With an ion of higher mass-to-charge ratio, e.g. insulin (M = 5734, z = 1), under the same conditions the impact velocity becomes 1.8×10^4 m s^{-1} and detection efficiency falls dramatically. One solution to this problem is to increase the velocity of the ion to obtain efficient detection.

In a post-acceleration detector (PAD) (Fig. 1.15), the ion beam is accelerated, after mass analysis, to a separate metal target, which is usually placed off-axis. Electrons and negative ions [54] from the target are then accelerated to an electron multiplier. The target is operated at a potential of between −8 and −30 kV. Post-acceleration detectors are now widely used in organic mass spectrometry to facilitate the detection of high mass ions. Negative ion detection also often uses a form of the PAD in which the post-acceleration dynode is held at a high positive voltage. Under these conditions, negative ion impact causes the ejection of positive ions which are then accelerated to the multiplier. This technique is discussed more fully in section 4.3.

In fact, it has recently been reported that there is no universal ion velocity threshold at 1.8×10^4 m s^{-1} as previously suggested [55]. Instead, it has been shown that detection efficiencies measured for amino acids and peptides can be correlated into a single equation that contains both molecular weight and ion velocity as variables [56]. Further data on detection efficiencies for high mass ions has been

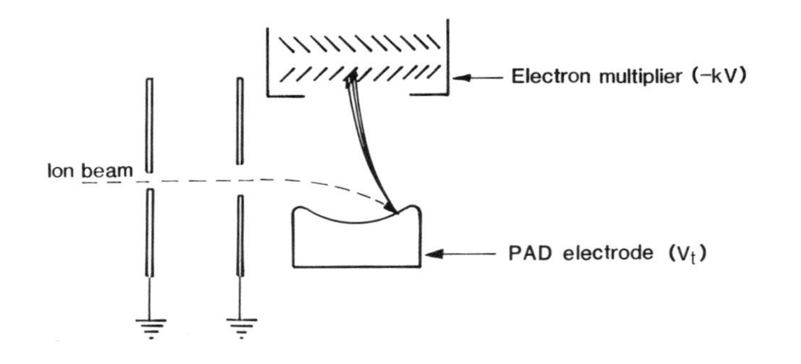

Fig. 1.15 Schematic diagram of post-acceleration detector. (Reproduced with permission from ref. 51)

reported by Hillenkamp *et al.* [57]. The detection of multiply charged ions, such as those produced by electrospray ionization (section 6.3.2.2), should be a more efficient process since these ions attain a higher velocity at the same accelerating voltage.

1.4.4 Array detectors [51]

An array detector for a magnetic sector instrument is designed to enhance detection limits by simultaneously recording ions over a number of adjacent mass values (Section 1.5). These detectors are based on microchannel plate multipliers installed so as to lie along the focal plane of the instrument. In the array described by Cottrell and Evans [58] (Fig. 1.16), electrons from the microchannel plates are accelerated to the phosphor-coated face of a fibre-optic coupling. Photons generated in the phosphor are transmitted via the fibre-optic to a photodiode array, mounted outside the vacuum system.

Another type of ion detector is the position and time resolved array [59]. In this device, the electrons produced by the microchannel plate are accelerated to a collector plate which consists of a number of conductive strips connected by capacitors. Charge-sensitive amplifiers located at each end of the collector plate register the current in the capacitor chain and, by comparing these two currents, the position of the electron cloud on the plate can be determined with good accuracy. With this array, the arrival time of an ion can also be measured and related to the simultaneous value of the magnetic field, so that the array may be used while the instrument is scanning.

1.5 Comparison of Analyser Operating Modes

The two most commonly used mass analysers, the magnetic sector and the quadrupole mass filter, are operated in one of two modes for most routine analyses.

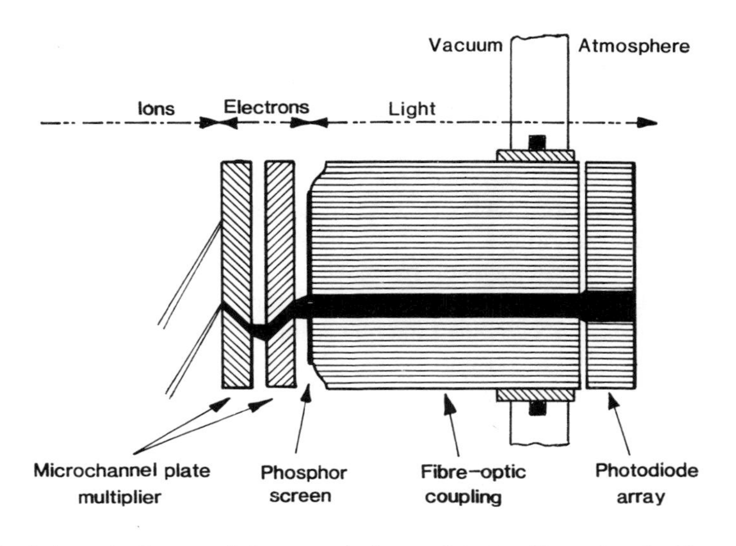

Fig. 1.16 Schematic diagram of electro-optical array detector. (Reproduced with permission from ref. 51)

The first of these is the scanning mode where the mass analyser is scanned over a wide mass range, usually repetitively, in order to record successive full spectra throughout an analysis. This type of analysis is a survey analysis where the spectra provide information on every sample component that enters the ion source. The number of recorded ions is, however, only a small fraction of the ions formed in the ion source since all ions with an m/z value outside the range reaching the detector at any instant have been rejected by the analyser.

The other operating mode is called selected ion monitoring (section 8.1). In this case, the instrument is set to repetitively monitor a limited number of m/z values chosen to be representative of the compounds sought. This type of analysis detects only targeted compounds, but generally does so with higher sensitivity because of the longer monitoring time devoted to chosen m/z values compared with the scanning mode. Such an improvement is, of course, only possible if scanning detection limits are limited by ion statistics and not by high background levels (cf. section 8.1.2).

With a magnetic sector instrument, sensitivity may also be increased by the use of an array detector placed in the focal plane. The use of an array detector ensures that a wider range of m/z values is detected at any one time compared with the use of a conventional point detector (electron multiplier). Again, the use of an array detector may not necessarily offer any improvement in detection limits where these are already determined by high background levels, e.g. from a fast atom bombardment matrix (section 5.4). With an ionization technique such as fast atom bombardment, an array detector has a more obvious advantage as the final detector in a four-sector instrument (section 7.3.3) where the use of a tandem mass spectrometer configuration effectively filters out chemical noise.

With a time-of-flight analyser the ratio of the number of ions recorded in a full spectrum to the number that enter the analyser is much larger than in quadrupole or magnetic sector analysers. The main reason for this is that the temporal separation principle used in time-of-flight, unlike spatial separation, does not direct any ions away from the detector. Magnetic sector or quadrupole analysers have compensated for their low recording efficiency by making use of ionization techniques, e.g. electron impact, chemical ionization, and fast atom bombardment which provide intense continuous ion currents from the sample. Again, ion trapping devices (sections 1.3.5 and 1.3.6) do not distinguish between full scan and selected ion monitoring modes since, in general, all trapped ions are available for analysis. Thus, the detection limit for such instruments is principally a function of the instrument duty cycle, i.e. the percentage of the sample introduction time devoted to ion generation rather than to ion analysis.

1.6 The Definition and Measurement of Sensitivity

Sensitivity is the most important parameter used in the evaluation of mass spectrometer performance. A sensitivity test will be used either to routinely assess the performance of an instrument installed in a laboratory or to compare the performance of different instruments, e.g. in purchase. The most basic test is a determination of absolute sensitivity, in which the sample is introduced on a direct insertion probe with the assumption that this method transfers the sample quantitatively to be ionized in the source. Alternatively such a test may be carried out using a directly coupled GC column where no sample loss is expected (section 2.2.1), although it should be appreciated that helium carrier gas introduced with the sample may reduce the source sensitivity.

The compound chosen for a probe sensitivity test should meet the following criteria:

(a) It should be relatively stable, to permit reproducible determinations without too much worry about inlet system performance.
(b) It should not be so volatile that evaporation begins before the solids insertion probe can be introduced into the ion source.
(c) It should be readily available, readily soluble and stable in solution.
(d) It should have an ion of reasonable intensity in the mass range where sensitivity is of most interest.

Methyl stearate (MW 298) meets most of these criteria and is now widely used as an industry standard for EI sensitivity, although it is somewhat volatile. This relative volatility does confer one advantage, i.e. methyl stearate can readily be analysed by gas chromatography so that a direct comparison of sensitivity using the probe and GC inlets becomes possible (section 2.2.5.1). Alternatively, a higher molecular weight ester, e.g. lauryl laurate[60], or cholesterol may be used.

1.6.1 Measurement of absolute sensitivity

Set up the instrument under standard low resolution operating conditions (section 1.6.3). Reduce the ion source temperature to 150°C. Load 100 ng methyl stearate (or other standard sample) dissolved in 1 μL of a solvent such as n-hexane or acetone on to the solids probe. Allow the solvent to evaporate and introduce the probe into the instrument as rapidly as possible. Bring the sample well into the ion source (section 2.3.3), making sure that the probe body is butted up to and seals against the source block. A suitable recording device should be available and the mass spectrometer set to monitor m/z 298 for methyl stearate. A narrow mass sweep may be superimposed to ensure that the appropriate mass value is monitored.

Slowly evaporate the sample by increasing the temperature of the probe or, if necessary, the ion source temperature. Note that the relative intensity of the molecular ion, even for a stable compound such as methyl stearate, is dependent on ion source temperature. For example, lowering the source temperature from 250 to 150°C can double the sensitivity recorded at m/z 298. Standard temperature conditions should therefore be used. It is also important to maintain a steady rate of evaporation during the sensitivity determination. Too high a rate of evaporation or the use of a larger sample size can lead to erroneous data through source saturation effects.

The absolute sensitivity S may then be calculated using

$$S = \frac{A}{R_f G n} \qquad \text{coulomb } \mu g^{-1} \tag{1.16}$$

where G is the multiplier gain used, n is the amount of sample consumed (in μg), A is the area under the evaporation profile, measured as the product of signal intensity (volts) and time, and R_f is the value of the feedback resistor.

1.6.2 Sensitivity measured as a signal-to-noise ratio

Sensitivity may also be measured as a signal-to-noise figure. A convenient noise level to measure against is single ion noise recorded under fast scanning, low resolution conditions. With this definition of sensitivity the sample flow rate into the source at the time of measurement must be established. The recording bandwidth must also be specified. For example, a decrease in bandwidth will markedly affect the height of the very narrow single ion peaks, while its effect on the sample peaks recorded for sensitivity measurement should be much less. The multiplier gain figure, on the other hand, is now not relevant since this is a relative rather than an absolute measurement.

Because of the need for a known sample flow, the signal-to-noise determination is perhaps more suited to a sample introduced through a gas chromatograph under conditions giving a known GC peak width. Again, in order to be able to measure a reasonable signal-to-noise figure of, say 50:1, it is necessary on most modern instruments to limit the sample size to approximately 100–500 pg. Such small quantities are more readily handled by means of a GC inlet than by a solids probe. As before, the mass spectrometer is set to monitor the mass chosen, with or without a mass sweep. Single ion peaks are also recorded using a bandwidth suitable for fast scanning (approximately 3 kHz) and their average height measured (section 1.4.2). The signal-to-noise figure for a sample flow rate of $F \mu g\ s^{-1}$ is then given by

$$S/N = \frac{h}{h_s} \tag{1.17}$$

where h is the signal intensity recorded at the given flow rate and h_s is the single ion peak intensity.

A signal-to-noise value determined in this way may be related to a sensitivity value (in coulombs per microgram). The general relationship is given by

$$S/N = \frac{FSt_s}{3.2 \times 10^{-19}} \tag{1.18}$$

where S is the sensitivity in coulombs per microgram, F is the sample flow rate into the ion source in micrograms per second, and t_s is the average base width of the single ion peaks. The quantity F is the sample flow rate into the ion source so that allowance must be made for the sample transfer efficiency of any GC interfacing device used.

The determination of an S/N value is a more practically relevant although less rigorously defined test than an absolute sensitivity determination. It should be noted, however, that detection limits in mass spectrometry are more often determined by chemical background than by ion statistics. Thus, theoretical detection limits may not be achieved, even in scanning experiments, unless special precautions such as the use of thermostable capillary columns are taken to ensure a low background level (section 8.1.2).

1.6.3 Mass discrimination and high mass sensitivity; standard operating conditions

Mass spectrometers may be tuned so as to give different sensitivities at different points on the mass scale (mass discrimination). It is therefore particularly important that an instrument can be set up to provide spectra that are comparable with those to be found in the major library search data bases. The EPA (Environmental Protection

Agency) spectrum validation test [61] is a standard test used to assess spectrum quality over a mass range of 40–450 daltons and is summarized below.

Method

The standard sample is decafluorotriphenyl phosphine (DFTPP). Select GC conditions that permit at least four scans during elution of the DFTPP. This will allow the selection of a spectrum that is reasonably free from intensity distortions due to rapidly changing sample pressure. Prepare for low resolution data acquisition by repetitive scanning with an ionizing voltage of 70 eV and inject 10–50 ng of the sample dissolved in acetone.

Select a spectrum on the front side of the GC peak as near to the apex as possible. Select a background spectrum immediately preceding the peak and compute the background subtracted spectrum. Spectrum averaging to minimize variations in ion intensity due to rapidly changing sample pressure may also be carried out if available. The subtracted spectrum must show ion intensities within the limits given for key ions in Table 1.1. Table 1.1 also lists some more up-to-date recommendations for these intensity values which are better substantiated and which do not introduce the same undesirable bias against sensitivity at higher mass as the old values.

Table 1.1 New and original intensity criteria for key ions

Mass	Intensity criteria	
	Original recommendations	New recommendations
51	30–80% of m/z 198	15–75% of m/z 198
68	$<2\%$ of m/z 69	$\leq 2\%$ of m/z 69
70	$<2\%$ of m/z 69	$\leq 2\%$ of m/z 69
127	30–70% of m/z 198	15–60% of m/z 198
197	$<1\%$ of m/z 69	$\leq 1\%$ of m/z 69
198	Base peak (100%)	Base or second most intense peak
199	5–9% of m/z 198	4.5–9% of m/z 198
275	0–30% of m/z 198	10–60% of m/z 198
365	$\geq 1\%$ of m/z 198	$\geq 0.5\%$ of m/z 198
441	Present but $< m/z$ 443	Delete criterion
442	$>40\%$ of m/z 198	$\geq 40\%$ of m/z 198 if 198 is base peak; alternatively may be base peak itself
443	17–23% of m/z 442	15–24% of m/z 442

The EPA system stability test [61] requires that a DFTPP spectrum meeting these same criteria should also be recorded on repeating the spectrum validation test after 20–28 hours without any intervening adjustments or recalibration of the system. An acceptable result from a spectrum validation test indicates the achievement of standard operating conditions under which sensitivity and other GC–MS tests (sections 1.6.1, 1.6.2, and 2.2.5) as well as routine analyses may then be carried out.

1.7 The Definitions and Measurement of Resolution

The resolution of a mass spectrometer may be defined in terms of its capacity to separate ions of adjacent mass number. A standard definition of separation has not, however, been adopted. The resolution necessary to separate two ions of mass m and $(m + \Delta m)$ respectively is given by

$$R = \frac{m}{\Delta m} \tag{1.19}$$

Resolution is sometimes expressed in terms of 'parts per million':

$$R(ppm) = 10^6 \frac{\Delta m}{m} \tag{1.20}$$

Resolution may be measured by reference to one or two peaks. For example, consider the 10% valley definition of resolution in which two peaks of equal intensity are considered to be resolved when they are separated by a valley which is just 10% of the height of either peak and which is made up from a 5% contribution from each component (Fig. 1.17). From the figure it can be seen that Δm can be measured from the separation of the two peaks or from the width of a single peak at 5% height. If a single peak is used, then the operator relies on a previous calibration of the peak display to measure resolution. If, on the other hand, a doublet of known separation is used, it is relatively easy to confirm that resolution has been achieved by measurement of the valley height between the peaks.

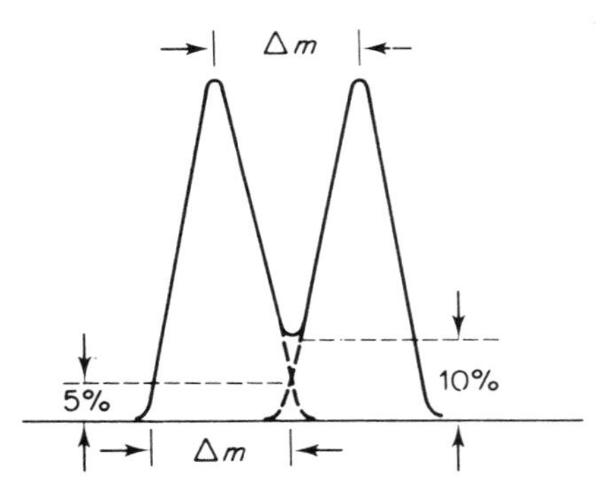

Fig. 1.17 Definitions of resolution

A convenient doublet for assessing the performance of medium resolution instruments is the $^{13}C—CH$ doublet which requires a resolution of approximately $200M$ at mass M. Such a doublet is readily produced from a mixture of two homologous aromatic compounds having the compositions C_xH_y and $C_{x+1}H_{y+2}$ by recording the molecular ion of the former and the ^{13}C isotope of the $(M—CH_3)^+$ ion of the latter. For example, a 10:1 mixture of xylene and toluene provides a doublet at m/z 92, requiring a resolution of 20 600 for separation. Lower resolution may be conveniently assessed at m/z 28. For example, the N_2/C_2H_4 doublet requires a resolution of 1110 whereas the N_2/CO doublet requires a resolution of 2490 (Table 1.2).

Table 1.2 Useful mass doublets

Composition		Exact masses (^{12}C = 12.000 00)	Resolution needed for separation
Ar	Both present in instrument	39.962 39	580
C_3H_4	background	40.031 30	
N_2		28.006 15	1110
C_2H_4		28.031 30	
CO		27.994 91	2490
N_2		28.006 15	
C_5H_5N	Pyridine	79.042 20	9750
$^{13}CC_5H_6$	Benzene	79.050 30	
$^{13}CC_6H_7$	Xylene	92.058 13	20600
C_7H_8	Toluene	92.062 60	

As mentioned previously, alternative definitions of separation may be used. Thus, while magnetic sector instruments use the 10% valley definition, the quadrupole definition is based on a peak width measured at 50% peak height, giving a resolution figure that is approximately double that calculated using the 10% valley definition. Quadrupole instruments are operated so that Δm is constant throughout the mass range. Typically the resolution of a quadrupole instrument will be 1000 at m/z 1000 and 300 at m/z 300. Magnetic sector instruments, on the other hand, operate with a constant value of $m/\Delta m$ so that the resolution is, for example, 1000 or 10 000 at both m/z 1000 and m/z 300.

References

1. Aston, F.W. (1919), *Phil. Mag.*, **38**, 707.
2. Dempster, A.J. (1918), *Phys. Rev.*, **11**, 316.
3. Thomson, J.J. (1913), *Rays of Positive Electricity*, Longmans, Green and Co., London.
4. Dempster, A.J. (1921), *Phys. Rev.*, **18**, 415.

5. Nier, A.O. (1947), *Rev. Sci. Instrum.*, **18**, 398.
6. Duckworth, H.E., Barber, R.C., Venkatasubramanian, V.S. (1988), *Mass Spectroscopy*, (Second. Edition), Cambridge University Press, Cambridge.
7. Cooks, R.G., Beynon, J.H., Caprioli, R.M., and Lester, G.R. (1973), *Metastable Ions*, Elsevier, Amsterdam.
8. Johnson, E.G., and Nier, A.O. (1953), *Phys. Rev.*, **91**, 10.
9. Morgan, R.P., Beynon, J.H., Bateman, R.H., and Green, B.N. (1978), *Int. J. Mass Spec. Ion Phys.*, **28**, 171.
10. Dawson P.H. (1976) *Quadrupole Mass Spectrometry and Its Applications*, Elsevier, New York.
11. Campana, J.E. (1980), *Int. J. Mass Spec. Ion Phys.*, **33**, 101.
12. Paul, W., Reinhard, H.P., and von Zahn, U. (1958), *Z. Phys.*, **152**, 143.
13. Brubaker, N.M. (1968), *Adv. Mass Spectrom.*, **4**, 293.
14. Cotter, R.J. (1989), *Biomed. Environ. Mass Spectrom.*, **18**, 513.
15. Cotter, R.J. (1992), *Anal. Chem.*, **64**, 1027A.
16. Chan, T.-W.D., Colburn, A.W., and Derrick, P.J. (1992), *Org. Mass Spectrom.*, **27**, 53.
17. Erickson, D.J., Enke, C.E., Holland, J.F., and Watson, J.T. (1990), *Anal. Chem.*, **62**, 1079.
18. Wiley, W.C., and McLaren, I.H. (1955), *Rev. Sci. Instrum.*, **26**, 1150.
19. Mamyrin, B.A., Karataev, V.I., Schmikk D.V., and Zagulin, V.A. (1973), *Sov. Phys. JETP*, **37**, 45.
20. Grix, R., Kutscher, R., Li, G., Grüner, U., and Wollnik, H. (1988), *Rapid Commun. Mass Spectrom.*, **2**, 83.
21. Chen, K.-H., and Cotter R.J. (1988), *Rapid Commun. Mass Spectrom.*, **2**, 237.
22. Opsal, R.B., Owens, K.G., and Reilly, J.P. (1985), *Anal. Chem.*, **57**, 1884.
23. Grotemeyer, J., and Schlag, E.W.(1987), *Org. Mass Spectrom.*, **22**, 758.
24. Grix, R., Grüner, U., Li, G., Stroh, H., and Wollnik, H. (1989), *Int. J. Mass Spectrom. Ion Processes*, **93**, 323.
25. Boyle, J.G., Whitehouse, C.M., and Fenn, J.B. (1991), *Rapid Commun. Mass Spectrom.*, **5**, 400.
26. Dawson, J.H.J., and Guilhaus, M. (1989), *Rapid Commun. Mass Spectrom.*, **3**, 155.
27. Boyle, J.G., and Whitehouse, C.M. (1992), *Anal. Chem.*, **64**, 2084.
28. March, R.E., and Hughes, R.J. (1989), *Quadrupole Storage Mass Spectrometry*, Wiley, New York.
29. Todd, J.F.J. (1991), *Mass Spectrom. Rev.*, **10**, 3.
30. Todd, J.F.J. (1991), *Mass Spectrom. Rev.*, **10**, p.29.
31. Todd, J.F.J. (1991), *Mass Spectrom. Rev.*, **10**, p.45.
32. Yost, R.A., McClennen, W., and Snyder, A.P. (1987), in *Proceedings of the 35th ASMS Conference on Mass Spectrometry and Allied Topics*, Denver, Colorado, p. 789.
33. Kaiser, R.E., Jr, Cooks R.G., Stafford, G.C., Jr, Syka, J.E.P., and Hemberger, P.H. (1991), *Int. J. Mass Spectrom. Ion Processes*, **106**, 79.
34. Schwartz, J.C., and Jardine, I. (1992), *Rapid Commun. Mass Spectrom,*. **6**, 313.
35. Williams, J.D., Cox, K.A., Cooks, R.G., Kaiser R.E., Jr, and Schwartz J.C.(1991), *Rapid Commun. Mass Spectrom.*, **5**, 327.
36. Schwartz, J.C., Syka, J.E.P., and Jardine, I. (1991), *J. Am. Soc. Mass Spectrom.*, **2**, 198.
37. Pannell, L.K., Pu, Q.-L., Fales, H.M., Mason, R.T., Stephenson, J.L. (1989), *Anal. Chem.*, **61**, 2500.
38. Williams, J.D., and Cooks, R.G. (1992), *Rapid Commun. Mass Spectrom.*, **6**, 524.
39. Asamoto B. (Ed.) (1991), *FT–ICR MS: Analytical Applications of Fourier Transform Ion Cyclotron Resonance Mass Spectrometry*, VCH, Weinheim.

40. Williams, E.R., Henry, K.D., and McLafferty, F.W. (1990), *J. Am. Chem. Soc.*, **112**, 6162.
41. White, R.L., Ledford, E.B., Jr, Ghaderi, S., Wilkins, C.L., and Gross, M.L. (1980), *Anal. Chem.*, **52**, 1525.
42. Lebrilla, C.B., Amster, I.M., and McIver, R.T., Jr (1989), *Int. J. Mass Spectrom. Ion Processes*, **87**, R7.
43. Lebrilla, C.B., Wang, D.T.-S., Hunter, R.L., and McIver, R.T., Jr (1990), *Anal. Chem.*, **62**, 878.
44. Holliman, C.L., Rempel, D.L., and Gross, M.L. (1992), *J. Am. Soc. Mass Spectrom.*, **3**, 460.
45. Henry, K.D., Quinn, J.P., and McLafferty, F.W. (1991), *J. Am. Chem. Soc.*, **113**, 5447.
46. Cody, R.B., Kinsinger, J.A., Ghaderi, S., Amster, I.J., McLafferty, F.W., and Brown, C.E. (1985), *Anal. Chim. Acta*, **178**, 43.
47. Hunt, D.F., Shabanowitz, J., McIver, R.T., Hunter, R.L., and Syka, J.E.P. (1985), *Anal. Chem.*, **57**, 765.
48. Kofel, P., Alleman, M., Kellerhals, H., and Wanczek, K.P. (1985), *Int. J. Mass Spectrom. Ion Processes*, **65**, 97.
49. Hogan, J.D. and Laude, D.A., Jr. (1990), *Anal. Chem.*, **62**, 530.
50. Freiser, B.S. (1988), in *Techniques for the Study of Ion Molecule Reactions*, Vol. 20, (Eds.J.M.Farrar and W.H. Saunders, Jr), Ch. 2, p.61, Wiley, New York.
51. Evans, S., (1990) in *Methods in Enzymology*, Vol. 193: Mass Spectrometry (Ed. J.A. McCloskey), Ch.3, Academic Press, San Diego.
52. Allen, J.S. (1947), *Rev.Sci.Instrum.*, **18**, 739.
53. Kurz E.A. (1979), *American Laboratory*, March.
54. Wang, G.H., Aberth, W., and Falick, A.M. (1986), Int. J.Mass Spectrom. Ion Processes, **69**, 233.
55. Hedin, H., Håkansson, P., and Sundqvist, B.U.R. (1987), *Int. J. Mass Spectrom. Ion Processes*, **75**, 275
56. Geno, P.W., and Macfarlane, R.D. (1989), *Int. J. Mass Spectrom. Ion Processes*, **92**, 195.
57. Hillenkamp, F., Karas, M., Bahr, U., and Ingendoh, A. (1989), *Proceedings of Vth International Conference Ion Formation from Organic Solids* (Ed. B.U.R. Sundquist), p. 1.
58. Cottrell, J.S., and Evans, S. (1987), *Anal. Chem.*, **59**, 1990.
59. Pesch, R., Jung G., Rost K., and Tietje K.-H. (1989), in *Proceedings of the 37th ASMS Conference on Mass Spectrometry and Allied Topics*, p.1079.
60. Stout, S.J., Cardaciotto, S.J., and Millen, W.G. (1983), *Biomed. Mass Spectrom.*, **10**, 103.
61. Budde, W.L., and Eichelberger, J.W. (1980), *Performance Tests for the Evaluation of Computerized Gas Chromatography/ Mass Spectrometry Equipment and Laboratories*, Environmental Protection Agency, EPA-600/4-80-025.
62. Cooks, R.G., Glish, G.L., McLuckey, S.A., and Kaiser, R.E. (1991), *Chem. and Engng.* News, 25 March, p.26.
63. Nourse, B.D., and Cooks, R.G. (1990), *Anal. Chim. Acta*, **228**, 1.
64. Marshall, A.G. (1989), *Adv. Mass Spectrom.*, **11A**, 651.
65. McFadden, W.H. (1973), *Techniques of Combined Gas Chromatography/Mass Spectrometry: Applications in Organic Analysis*, Wiley—Interscience, New York.
66. Howe, I., Williams, D.H., and Bowen, R.D. (1981), *Mass Spectrometry, Principles and Applications* (Second Edition) McGraw-Hill, New York.

Chapter 2

Sample Introduction

2.1 Sample Flow and Pressure in the Mass Spectrometer

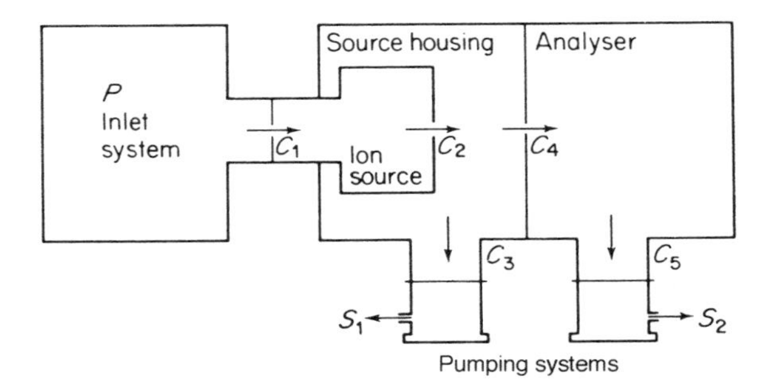

Fig. 2.1 Schematic representation of a mass spectrometer vacuum system

In the representation of a mass spectrometer vacuum system presented in Fig. 2.1 the total conductance or pumping speed (C_{tot}) of the system with the exception of the analyser is given by

$$\frac{1}{C_{tot}} = \frac{1}{C_1} + \frac{1}{C_2} + \frac{1}{C_3} + \frac{1}{S_1} \tag{2.1}$$

With most inlet systems from which flow is restricted the approximation given by

$$\frac{1}{C_{tot}} = \frac{1}{C_1} \qquad \text{since } \frac{1}{C_1} \gg \frac{1}{C_2} \gg \frac{1}{C_3} + \frac{1}{S_1} \tag{2.2}$$

also applies. Therefore, the flow rate of sample vapour from a reservoir inlet system or of carrier gas from a chromatographic inlet into the source region is determined only by the conductance of the inlet system (C_1 litre s^{-1}) and the pressure in the inlet system (P torr):

$$\text{Flow} = C_1 P \qquad \text{litre torr s}^{-1} \tag{2.3}$$

For a given flow rate of sample vapour or carrier gas the pressure in the source housing is then determined by the speed of the source pump S_1 and the conductance C_3. The pressure in the ion source is determined by the conductance C_2. Ideally the pump is situated as close as possible to the source housing and connected to it by wide-bore pipework so that the decrease in conductance caused by connections between the source housing and pump is as small as possible. In the example, illustrated in Fig. 2.1, the pumping speed is reduced from its original value S_1 to a new value S'_1, where S'_1 can be calculated from

$$\frac{1}{S'_1} = \frac{1}{S_1} + \frac{1}{C_3} \tag{2.4}$$

The flow into the analyser section is determined by the conductance of the connecting orifice (C_4), compared with the conductance of the source pumping system. The size of this orifice usually ensures that less than 1% of the total vapour flow through the source housing enters the analyser section, since low pressures are required for adequate analyser performance.

In gas chromatography–mass spectrometry (GC–MS), we are considering the flow of relatively large volumes of helium carrier gas into the mass spectrometer. Helium is particularly chosen for GC–MS work because its low molecular weight makes it suitable for removal by various enrichment devices and its low ionization cross-section means that relatively few ions are formed from helium compared with most other materials.

The flow rate of helium that can be accommodated in the source region is limited. First, the pumping speed of diffusion and turbomolecular pumps drops rapidly above an operating pressure of 10^{-3} torr. Thus, a more tolerable source housing pressure of 2.5×10^{-4} torr (note that the observed pressure with helium on an ionization gauge would be almost ten times less than this figure because of the low ionization cross-section of helium) corresponds to a flow rate of $7\,\text{cm}^3\,\text{min}^{-1}$ with a pumping speed of 350 litre s^{-1} (Equation 2.5) or $5\,\text{cm}^3\,\text{min}^{-1}$ with a pumping speed of 250 litre s^{-1}.

$$\text{Flow rate} = 350 \times 2.5 \times 10^{-4} = 0.088 \text{ litre torr s}^{-1}$$

and thus

$$\frac{0.088 \times 1000 \times 60}{760} = 7 \text{ cm}^3 \text{ min}^{-1} \tag{2.5}$$

Second, while the source housing is pumped at a high speed, an ion source is relatively tightly sealed ($1/C_2 \gg 1/S_1 + 1/C_3$) (sections 1.2 and 3.2). Thus, the introduction of a flow of 7 $cm^3\ min^{-1}$ helium into a commonly used combination electron impact/chemical ionization source with a conductance of 0.2 litre s^{-1} (section 3.2) results in an ion source pressure of nearly 0.5 torr. At these pressures sensitivity is markedly reduced. It should be pointed out that although the pressures in the ionization chamber are now at least as great as those commonly employed in chemical ionization work (Chapter 3), there is not too much effect on the quality of an electron impact spectrum. This is because helium acts as a charge exchange reagent gas (section 3.1.1.2), transferring sufficient energy in this process to ensure electron impact-like fragmentation as a result.

2.2 Gas Chromatography–Mass Spectrometry (GC–MS)

The flow from a conventional narrow bore capillary GC column of i.d. 0.25–0.32 mm is of the order of 1–2 $cm^3\ min^{-1}$ and is generally introduced directly into the mass spectrometer ion source. No attempt is made to enrich the sample with respect to the helium carrier gas. The interfacing methods used for these capillary columns are described in sections 2.2.1 and 2.2.2. For capillary columns with flow rates of 5 $cm^3\ min^{-1}$ and above, and particularly for packed columns, problems can arise, as mentioned in the previous paragraphs. In this case, the use of an enrichment device, such as the jet separator (section 2.2.3), is preferred.

2.2.1 Direct coupling

With the lower flow rates used by capillary columns, the simplest method of interfacing is to couple the column exit directly to the mass spectrometer vacuum system. A form of capillary restriction can be interposed between the column and ion source to provide an approximately normal exit pressure for the column and to ensure that the source and source housing are not let up to atmospheric pressure when the column is removed. Direct coupling is limited to capillary columns which provide an appropriate flow rate. Some loss of sensitivity may be noticed if the flow rate into the ion source rises much above 2 $cm^3\ min^{-1}$ [1].

The advent of flexible fused silica capillary columns has provided columns which can readily be extended beyond the confines of the GC oven to terminate much closer to the ion source. In fact, with these columns the restriction may be removed totally and the column extended to terminate in the ion source a few millimetres back from the electron beam. Under these circumstances a relatively inert path is provided by the column almost up to the site of ionization so that analysis of labile materials is facilitated [2]. In addition, this simple method of interfacing is entirely without any form of narrow capillary restriction which may be liable to blockage. Direct coupling with a flexible column introduced into the ion source is simple and offers 100% transfer of sample in a relatively inert environment without any effects due to dead volumes. This form of direct coupling is by far the most common method of GC–MS interfacing in use today.

Flexible silica is also a practical material for the manufacture of capillary restrictions where these are used [3]. Care must be taken to ensure that the flow from the column is presented directly to the sampling capillary if chromatographic resolution is not to be lost. A simple and effective butt coupling may be made from a length of shrink polytetrafluouroethylene (PTFE) tubing [4]. Excellent presentation may also be achieved if the inside diameter of the column is sufficiently large for the sampling capillary to be introduced a short distance into the end of the column.

Direct coupling without a restriction, so that the column exit is at the same low pressure as the ion source, reduces the maximum column efficiency by approximately 10% [5,6]. In addition, optimum resolution is obtained at a higher gas velocity than under normal GC conditions with a column outlet at atmospheric pressure [5]. Thus, while some resolution is lost, analysis times are reduced by coupling directly into vacuum. In fact, the use of a longer column to recover the original resolution still allows a more rapid analysis than with the outlet at atmospheric pressure [5].

The use of shorter, wider bore capillary columns without an intervening restriction is more complex in practice since the inlet pressure required to maintain column flow can be sub-atmospheric. Yost [7] has promoted the use of shorter capillary columns for rapid, high sensitivity GC–MS analyses and has described a means of regulating a sub-atmospheric inlet pressure for these columns [8].

Operation at higher temperature can lead to thermal degradation of the polyimide outer coating of fused silica columns and in this case, aluminium clad columns are a viable alternative. The coupling of aluminium clad columns to high voltage magnetic sector instruments has been implemented recently [9,10]. Blum *et al.* [4] have also discussed the problems of high temperature capillary GC–MS operation including chemical ionization operation (Chapter 3).

2.2.2 Open-split coupling [3,11,12]

A schematic diagram of the open-split coupling is shown in Fig. 2.2. The input restriction to the mass spectrometer samples a constant flow of helium (say 1–2 $cm^3 \, min^{-1}$) from the end of the column which is at atmospheric pressure in an

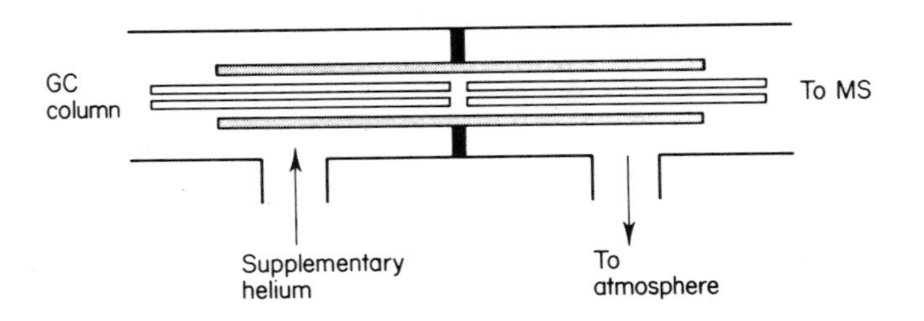

Fig. 2.2 Schematic diagram of an open-slit coupling

unsealed system. The input restriction may be a fused silica capillary similar to that used in direct coupling.

Column flows that are smaller than the sampling rate are supplemented by a helium flow to exclude the possibility of sampling air. This supplementary flow is controlled by a valve. Excess flow from column flows that are greater than the sampling rate is allowed to escape through the vent to atmosphere. Solvent peaks may be accommodated by the temporary use of the supplementary helium flow to dilute the solvent with helium so that less solvent is sampled.

The open-split coupling has the major advantage that the end of the column is at atmospheric pressure so that column conditions are completely unchanged from normal GC practice. It has the disadvantage that only a fixed flow is sampled so that if the column flow is in excess of this figure, the sample transfer efficiency falls below 100%.

2.2.3 The jet separator

The use of the jet separator (Figs. 2.3 and 2.4) as a GC–MS interface was pioneered by Ryhage [13,14]. It has been the most consistently successful and popular basis

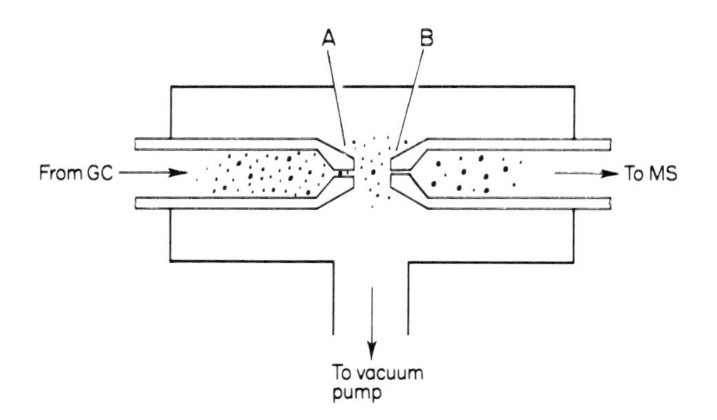

Fig. 2.3 Schematic diagram of jet separator

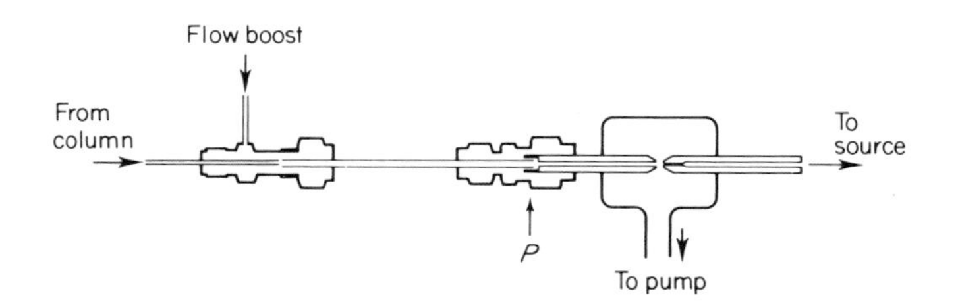

Fig. 2.4 Jet separator installation

for an enrichment interface. The mixture of helium and sample vapour from the GC acquires a substantial forward velocity emerging from the first narrow orifice (A). For the lighter helium molecules, this forward velocity is superimposed on an already large thermal velocity so that little directional motion is given to the helium molecules and these are largely (approximately 95%) removed by pumping. The heavier sample molecules acquire a relatively large forward motion and many of these (approximately 50%) are able to pass through the counter-orifice (B).

The performance of the separator, or of any enrichment device, is defined in terms of its yield (Y) and enrichment (E):

$$Y = 100\left(\frac{M_2}{M_1}\right) \quad \% \tag{2.6}$$

$$E = \frac{M_2 V_1}{V_2 M_1} \tag{2.7}$$

where M_1 is the amount of sample leaving the GC, M_2 is the amount of sample entering the MS, V_1 is the helium flow rate leaving the GC, and V_2 is the helium flow rate entering the MS.

Typical values of yield and enrichment are, for methyl stearate as sample, $Y = 50\%$ and $E = 10$. The original design of Ryhage provided two stages of enrichment, but this is no longer necessary with modern, fast-pumped source housings. Construction is generally all of glass. An operating temperature of 250°C is satisfactory for most GC–MS work.

A minimum helium flow rate of 15–20 $cm^3\,min^{-1}$ is usually necessary for satisfactory operation and in the case of capillary columns a supplementary helium flow is provided, as shown in Fig. 2.4. This method has the added advantage of providing a large flow of gas to rapidly sweep out any dead volumes between the end of the capillary column and the separator. By this means, identical to one often used in the coupling of capillary columns to GC detectors, the high resolution of the column is faithfully maintained. The column should be installed with its exit well past the flow boost inlet to ensure that this is so.

A dump valve is often introduced between the end of the column and the separator to divert solvent or sample overloads away from the source region. The conductance of the first jet is usually such that flows of 20–30 $cm^3\,min^{-1}$ are achieved when the pressure at the column exit is 0.5–1 atmosphere. GC–MS conditions are therefore as nearly as possible the same as normal GC conditions, i.e. a column exit pressure of 1 atmosphere.

The advantage of the jet separator is its versatility in allowing the use of the full range of columns from high flow rate packed columns to low flow rate capillary columns without any change of hardware. One disadvantage of the jet separator is a reduced transfer efficiency with the most volatile samples. Another disadvantage is a tendency for the narrow first jet to block in use.

2.2.4 General constructional details

2.2.4.1 Materials and methods of interface construction

Materials that are suitable for the construction of sample lines in interfaces may be listed in terms of increasing surface activity: deactivated silica or glass < silica < glass < glass-lined stainless steel. Glass has more active sites than silica. Glass-lined tubing (GLT) can become damaged internally on working and therefore may have much greater surface activity towards samples than a piece of glass tube after working.

Even flexible silica lines show some surface activity. A silica or glass line will, however, fairly rapidly become coated with column bleed and therefore deactivated in use, particularly if packed columns are used (this is one of the few advantages of packed columns over capillary columns in mass spectrometry!). Otherwise the line may be permanently deactivated by deliberately lightly coating with a thermostable stationary phase material [4], or temporarily deactivated by the use of silanization reagents commercially available for column deactivation. In general, the internal surface area with which the sample comes into contact should be a minimum so that tubulation with the smallest internal diameter consistent with the expected flow rates should be used. The advantages of a capillary column introduced directly into the ion source are obvious from the preceding discussion.

One further constructional material of importance is that used for ferrules in interface couplings. Graphite is an excellent material for temperature stability and sealing but has a tendency to flake off and eventually block the narrowest interface lines. Vespel is also thermostable but is perhaps too hard a material. Graphitized vespel is probably the best material available commercially and is recommended in all cases.

Unswept volumes in sample lines should be avoided; this is most important in the region surrounding the column exit. In any case the internal volume of sample lines on the high pressure side of the interface should be kept to a minimum, especially where low flow rates are encountered, if chromatographic resolution is not to be lost.

2.2.4.2 Interface heating

Interface components such as a capillary restrictor or a jet separator may be maintained at a temperature that is somewhat lower, perhaps as much as 50°C, than that of the column, since retention of the analyte by the relatively clean interface surfaces should be small compared with the effect of the chromatographic liquid phase. If the column is introduced directly into the ion source, then an interface temperature which is more or less equal to the maximum column temperature is preferred. If an excessive interface temperature is required to maintain column performance, this is either because of cold spots compensated for by overheating or because of contaminating material absorbing the sample. A build-up of contaminating material may itself be initiated by a cold spot.

Interface temperatures should be monitored by a thermocouple in each section of the interface. Because of the importance of constant temperatures throughout the interface, smaller components, such as lengths of narrow capillary, are preferably surrounded, within the heated enclosure, by a body having a much greater thermal mass, e.g. thicker walled tubing or a block through which the capillary can pass. It is also particularly important to ensure that any junction between two separate ovens, e.g. between the GC and interface, is evenly heated. Supplementary gas for a jet separator or open-split interface should always be pre-heated, e.g. by passing the supply lines through the interface oven.

The ion source itself must also be adequately and evenly heated. Although the pressure in this region is low, nevertheless the ion source temperature must be high enough to promote a rapid re-evaporation of sample molecules as they collide with the source walls. On the other hand, overheating the ion source block will cause a noticeable degradation in mass spectrum quality (cf. section 2.3.4).

2.2.5 Standard tests of GC–MS performance

2.2.5.1 Sensitivity

The determination of instrument sensitivity, expressed as a signal-to-noise figure, towards a sample introduced via the GC has already been dealt with in section 1.6.2. An alternative sequence of sensitivity experiments comprises a sensitivity test of the basic instrument using a sample introduced on the solids probe followed by a determination of relative sensitivity using the same sample introduced via the GC. Information is then available regarding basic instrumental sensitivity and the efficiency of the GC interfacing.

Method

Set up the instrument under standard low resolution operating conditions (section 1.6.3). Ensure that the carrier gas flow is entering the source for both probe and GC sensitivity determinations as the pressure of gas may affect the source sensitivity. The carrier flow rate through capillary columns may conveniently be checked by introducing an unretained sample such as methane on to the column and measuring its elution time by monitoring the peak at m/z 16.

Complete a probe sensitivity determination as described in section 1.6.1 using approximately 10 ng methyl stearate. When the sample is completely evaporated from the probe, introduce the same quantity of methyl stearate via the GC without removing the probe so that a determination of GC sensitivity is made under exactly the same conditions as used for the probe sample. A comparison of the area under the GC record with that from the probe sample gives a figure for the transfer efficiency of the GC–MS interface. The effect of carrier gas on source sensitivity may be judged by comparison with a probe sensitivity determined in the absence of a carrier gas flow.

The use of a chromatographically easy compound such as methyl stearate means that a working system should routinely give a good transfer efficiency and a symmetrical peak shape without any preparation by conditioning. Loss of transfer efficiency can generally be ascribed to obstruction of the column or other capillary restriction, leaks in couplings at the inlet or outlet of the GC column, or misalignment of the sampling capillary and capillary column. The same comparative sensitivity test may alternatively be carried out with a more difficult, polar analyte where, for example, optimum performance with that particular analyte is required.

2.2.5.2 Chromatographic resolution and interface activity

Most capillary GC–MS systems will be used for the trace analysis of a range of compounds, many of which are considerably more polar and subject to problems of absorption on active sites than, say, methyl stearate. A more practical test for this kind of work is provided by the following adaptation of a standardized test for capillary columns described by Grob *et al.* [15].

Method

Set up the instrument under standard low resolution operating conditions (section 1.6.3). Prepare a solution containing approximately 200 ng μL^{-1} each of n-undecane, n-dodecane, octan-l-ol, 2,6-dimethylphenol, 2,6-dimethylaniline, decanoic acid methyl ester, and undecanoic acid methyl ester in n-hexane. Inject 1.5 μL of the solution, either further diluted 1:50 with solvent or with a split ratio of l:50, to introduce approximately 6 ng of each component on to the column. Acquire fast repetitive scans throughout the elution of the mixture. Typical GC conditions are given in Fig. 2.5.

The following measurements may be made from the total ion current trace obtained from the analysis:

(a) The relative heights of other peaks compared with those due to the non-polar hydrocarbons and esters, indicating possible losses of more polar materials. A satisfactory test with a primary alcohol such as octan-1-ol indicates that the column should be suitable for other polar compounds. The mixture also contains dimethylaniline and dimethylphenol to check for acidic and basic active sites respectively. Other 'more difficult' compounds used by Grob, e.g. dicyclohexylamine, 2-ethylhexanoic acid, and 2,3-butanediol, may be added to the test mixture when the ability to analyse similar compounds is particularly important.
(b) Column resolution measured from the total ion current trace, e.g. as a *TZ* value [15] for the methyl ester peaks:

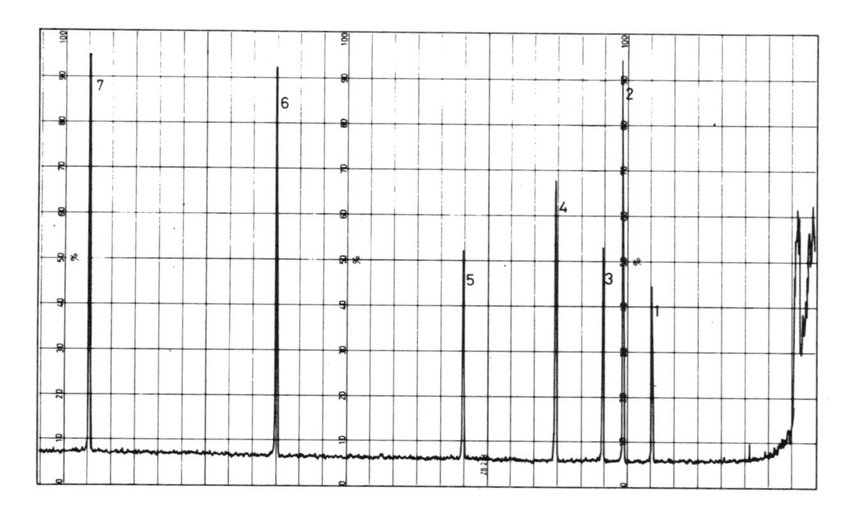

Fig. 2.5 Total ion current trace from injection of Grob-type mixture (20 ng each component) on a 50 m × 0.25 mm non-polar column installed in-source. Temperature programme 80–150°C at 3° min^{-1}. (1) Octan-1-ol; (2) 2,6-dimethylphenol; (3) n-undecane; (4) 2,6-dimethylaniline; (5) n-dodecane; (6) C_{10} methyl ester; (7) C_{11} methyl ester. *TZ* value of peaks (6) and (7) is approximately 50. (Courtesy Kratos Analytical Instruments)

$$TZ = \frac{t_{(E11)} - t_{(E10)}}{\omega_{(E11)} + \omega_{(E10)}} - 1 \qquad (2.8)$$

where t is the retention time and ω is the peak width at half-height. For this measurement, the scan cycle time should be rapid enough compared with the peak width so as not to degrade the recorded column resolution.

(c) A GC sensitivity using the molecular ion of either ester perhaps following an initial correlation with methyl stearate.

Other, more specific, tests of GC–MS performance may sometimes be considered appropriate. For example, the peak shape and relative intensity over a series of n-alkanes up to $C_{40}H_{82}$ or $C_{44}H_{90}$ gives a good indication of the ability of the system to handle less volatile materials [16]. Again, 2-hexyl-5-pentyl-pyrrolidine has been described as a sensitive test material for dehydrogenation occurring in the interface [17].

2.2.5.3 EPA tests

The Environmental Protection Agency (EPA) has published a series of ten general-purpose tests which can be used to evaluate the performance of computerized gas chromatograph–mass spectrometry systems [18]. Two of these tests (spectrum validation and system stability) have already been described in section 1.6.3. Two of the other tests, viz. detection limit and saturation recovery, provide useful supplementary GC–MS tests.

Detection limit. A determination of GC–MS sensitivity in coulombs per microgram (section 2.2.5.1) can be used to calculate a theoretical detection limit (section 8.1.2). In practice, however, the detection limit is usually determined by the background level, e.g. from column bleed, chemical ionization reagent gas, or the vacuum system itself at the masses of interest. Exceptionally, with low bleed, high resolution capillary columns, the theoretical detection limit may be achieved in scanning experiments.

The EPA test aims to give an assessment of detection limits based on the summed intensity of selected peaks in the spectrum of the test compound compared with the same number of the most intense background peaks. Thus, it is a test of the complete system. Background is defined as extraneous peaks that have not been removed by standard data processing methods. Background equivalent to 10% or more of the total ion intensity of the spectrum of the test material is considered too high to allow routine spectrum interpretation.

Method

Make several injections of the standard compound (DFTPP—section 1.6.3) on to the column in the concentration range 100 ng to 100 pg. After the spectra have been acquired for these samples, they should be processed (background corrected) and examined to see if they meet the criteria of Table 1.1. All spectra not meeting these criteria should be discarded.

Sum the relative intensities of the ions at m/z 127, 255, 275, 441, 442, and 443 in a suitable spectrum of DFTPP for each injection made. Sum the relative intensities for the same number of the most intense ions not due to the sample (within the mass range 40–450) in each case. Calculate a ratio R:

$$R = \frac{\text{summed sample ions}}{\text{summed background ions}}$$

and plot this ratio against the amount of sample injected. The detection limit is then defined as the amount injected that gives an R value of 5. This definition is consistent with the requirement that background ions should be equivalent to less than 10% of the total ion intensity in a usable spectrum.

Saturation recovery test. This test measures the ability of the system to acquire the mass spectrum of a low concentration test compound following the elution of a large amount of another component of comparable molecular weight, a situation that occurs frequently with real samples.

Method

Set up the instrument under standard low resolution operating conditions (section 1.6.3). Recovery from saturation is measured using a binary mixture where the relative quantities are in the ratio 250:1. The GC conditions are chosen so that the minor component elutes within 2 min following the major component. EPA recommends a mixture of *p*-bromobiphenyl (major component) and DFTPP (minor component) analysed on a 6 × 2 mm 1% SP 2250 column (helium flow 30 cm^3 min^{-1}, temperature programming from 120 to 230°C at 10°C min^{-1}).

The three most abundant ions from the spectrum of the major component (*m/z* 152, 232, and 234 in the case of *p*-bromobiphenyl) should each be less than 5% relative intensity in the background corrected spectrum of the minor component. Analysis conditions that are more appropriate to higher resolution capillary columns may, of course, be substituted.

2.2.5.4 Summary of GC–MS tests

Which of the tests described should be employed by the GC–MS user? The use of the Grob mixture (section 2.2.5.2) provides a valuable test of the condition of the column and interface that is relevant to most analytical situations. If this mixture is also used to assess system sensitivity as described, then it provides the basis for a comprehensive test which can be carried out regularly.

Routine checking of sensitivity is important since remedial action, e.g. source cleaning, should be taken when the sensitivity falls to one-third of the normal figure or less. An exceptionally low GC–MS sensitivity might be further investigated by a check on basic instrument sensitivity via the probe (section 1.6.1) in case the fault lies in the GC or the interface. The multiplier gain should be checked (section 1.4.2) if source cleaning does not recover basic sensitivity or following a vacuum accident in the analyser region.

Spectrum validation (section 1.6.3) is particularly valuable for quadrupole instruments. This test ensures that the instrument can generate spectra that are compatible with the major databases which contain a large number of magnetic sector instrument spectra. Other tests described in Chapters 1 and 2 should be used as appropriate.

2.3 Probe Inlets [19]

2.3.1 Introduction

Volatile samples may be introduced into the mass spectrometer via a GC or, if they are sufficiently volatile and thermostable and do not require separation, via a heated reservoir inlet system. In the latter, a sample is vaporized and the vapour expanded into a reservoir whose capacity may vary from 50 to 2000 cm^3. The flow from this reservoir into the ion source is controlled by a fixed restriction which may be fabricated as a capillary, pin-hole or porous leak.

Many samples submitted for mass spectrometric examination, however, have such a low vapour pressure or thermal stability that they must be introduced directly into the ion source to permit analysis. With a sample of low vapour pressure, an adequate sample flow rate can then be ensured merely by controlling the temperature of evaporation rather than by interposing a restriction. For example, vapour pressure data for saturated hydrocarbons [20] indicates an order of magnitude increase in vapour pressure and hence evaporation rate in vacuum for an increase of 50°C in sample temperature. An order of magnitude increase in vapour pressure is also found for each extension of the carbon chain by five methylene units, so that fractional distillation of mixed samples is quite practicable with adequate temperature control [21].

A typical direct insertion probe is shown in Fig. 2.6. The sample is loaded on the probe tip and the probe introduced through the vacuum lock, also shown in the figure. In addition to being more suitable for involatile and labile samples, the direct insertion probe requires far less sample than a reservoir inlet. It should be noted that the probe body seals against the ion source both to prevent loss of sample and to ensure that the source remains pressurized in chemical ionization (CI) operation (section 3.2). Thus, the probe or some equivalent seal must be in position during CI operation whatever inlet system is used to introduce the sample.

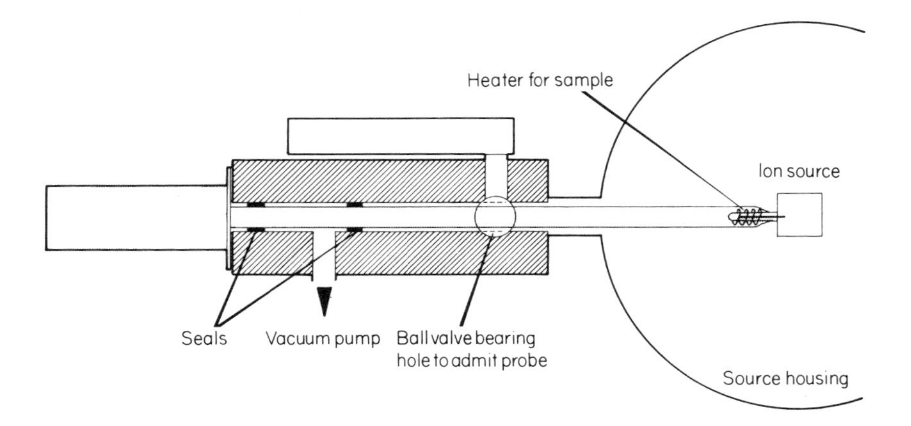

Fig. 2.6 Direct insertion probe and vacuum lock

2.3.2 Sample heating

Sample heating is usually provided by a heater built into the probe. Ideally, this heater is situated close to, and is in good thermal contact with, the tip carrying the sample, since rapid heating favours volatilization at the expense of decomposition (section 5.1.1). If the heater is distant from the sample and the heating period unduly prolonged, decomposition of labile samples becomes more probable and the sample flow rate is likely to be inadequate for a successful analysis. An alternative form of sample heating provided by the use of a laser is described in section 5.3.4.

Heat from the ion source itself can play an important part in the evaporation process [22]. For example, if the source is cooler and the sample positioned so that repeated condensation and re-evaporation cannot be avoided prior to ionization (Fig. 2.7) then more decomposition will occur. Making the ion source very much hotter will result in increased fragmentation in the spectra, so that a compromise temperature of approximately 200°C is usually employed. The filament alone will provide an equilibrium temperature of approximately 150°C in an EI source and a somewhat lower temperature in a CI source because of the cooling effect of the reagent gas flow. Probe temperatures under chemical ionization conditions may also be somewhat lower than anticipated because of this same cooling effect.

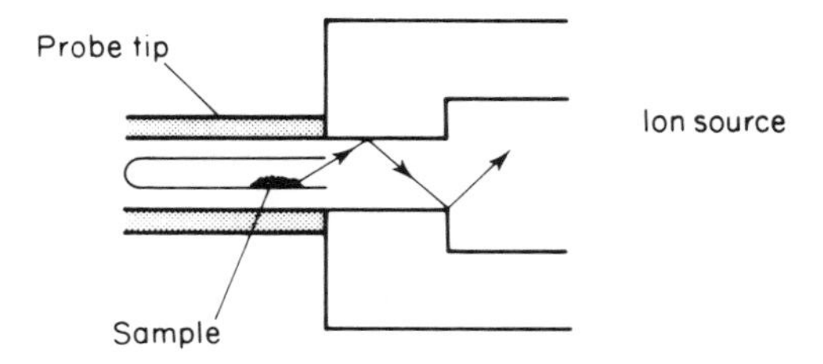

Fig. 2.7 Decomposition resulting from repeated condensation and re-evaporation of a sample positioned outside the ion source

More volatile samples will evaporate quite readily in vacuum even at room temperature and under these circumstances cooling of the probe is a valuable additional facility. A practical arrangement uses a nitrogen flow which is freed from condensible vapours and is then passed through a cooling spiral immersed in a dry-ice/acetone mixture in a Dewar flask and finally through the probe body. The small thermal mass of the probe tip ensures that such a system has a very rapid response.

Most direct insertion probes can be used with programmable heating supplies which allow the accurate control of heating rates and temperature limits. This facility is particularly useful for the repetitive analysis of a large number of samples under predetermined conditions, e.g. in quantitative analysis or for the fractionation of mixtures.

A programmable heater supply is also useful for the direct pyrolysis into the mass spectrometer source of samples such as polymers that are not themselves volatile enough for conventional analysis. For example, synthetic polymers [23] and cellulose and its derivatives [24] have been characterized by analysis of the products found during temperature programming. The pyrolysis occurs in high vacuum and close to the electron beam so that the products formed are rapidly removed from the pyrolysis site and ionized with the minimum of further interaction.

2.3.3 The probe tip and sample loading

Horning and co-workers [25] list three characteristics of samples that cannot easily be volatilized without decomposition, viz.

(a) ionic functional groups, e.g. a quaternary ammonium group;
(b) non-ionic functional groups that participate in hydrogen bonding including surface bonding and subsequent decomposition reactions;
(c) a large molecular size providing sufficient degrees of freedom to accommodate a large amount of internal energy that promotes cleavage rather than volatilization on heating [26].

It is evident that the nature of the surface on which the sample is deposited and the manner in which it is deposited affect the quality of the spectra recorded. A strategy that involves a change in the probe tip surface material to provide better spectra is discussed in the present section since it is readily implemented with a standard direct insertion probe. The extension of the probe tip so that the sites of evaporation and ionization are brought much closer is another useful modification. This approach is also considered here and again in more detail in section 5.2.1 together with rapid heating as a method of promoting volatilization with minimum decomposition.

Beuhler *et al.* [27] have shown that the activation energy required to desorb the small peptide, thyrotropin releasing hormone, deposited as a thin film from solution on various probe tips, decreases as the surface is changed from glass to copper to Teflon (PTFE). This reflects a decreased ability of the surface to hydrogen bond with the peptide, with the immediate practical consequence of a higher desorption rate at the same surface temperature as the activation energy is decreased. Similarly, the use of silanization techniques to modify an existing glass surface [28,29] or the deposition of a siloxane layer on a probe tip [25] have been shown to provide surfaces free from absorption effects, resulting in better sensitivity and improved quality of spectra for polar compounds. Another example of the effect of the surface is found in desorption chemical ionization where the effective desorption of samples from a rapidly heated wire is prevented if the wire is not cleaned of decomposition products from the previous sample (section 5.3.1.1).

The choice of a probe tip material has to be reconciled with the need for an effective transfer of heat to the sample [22]. Therefore the tip should have an inert surface but an acceptable thermal conductivity overall. Less decomposition is observed if the sample can be deposited on a surface of the tip from which it may evaporate directly into the ionization region rather than reaching this region by repeated condensation and re-evaporation. Glass, deactivated glass, quartz, or gold tips are commonly used for this purpose [22,30].

The rate of evaporation is further enhanced when the sample is loaded from solution as a thin layer rather than as a thick deposit on the probe tip. Overloading the probe is a common mistake and one which quickly causes source contamination and a reduced source sensitivity. A few micrograms is sufficient to give adequate spectra in most cases. If the sample is easily visible on the probe tip then there is probably too much present. A high rate of evaporation into the ion source can result in spectra in which the sample vapour acts as a chemical ionization reagent gas (section 3.1), giving $(M+H)^+$ ions of increased relative intensity.

Spectra of improved quality, i.e. more intense molecular ions and reduced fragmentation, are observed when the probe tip is brought closer to the ionizing electron beam. Under electron impact (EI) conditions, if the probe tip is extended too close to the electron beam, then an insulating tip will become negatively charged and will begin to repel the electron beam, causing an attenuation of the ion beam. At a distance of 1–2 mm, however, excellent spectra can be recorded [31]. The optimum tip position may be different for a probe that is heated solely by the source block [22].

This technique of placing the sample close to the electron beam in EI conditions was first described by Reed [32] and later rediscovered as 'in beam' mass spectrometry by Dell and other workers [33,34]. A closely related chemical ionization (CI) technique was used by Baldwin and McLafferty[35], who also reported that introduction of the probe tip into the CI ion plasma lowered the apparent temperature needed for sample volatilization by as much as 150°C. This CI technique was subsequently combined with rapid heating and named desorption chemical ionization (DCI). Both EI and CI in beam techniques and DCI are discussed more fully in Chapter 5.

2.3.4 Test of probe performance

The use of the direct insertion probe to measure instrument sensitivity has already been discussed in section 1.6.1 when it was assumed that the probe provides a means of quantitatively introducing the test compound into the ion source with little or no thermal degradation. If, however, the test compound is less volatile or more labile then the intensity of the molecular ion may be considerably reduced.

Cholesterol has been used as a probe sample to evaluate thermal decomposition. For example, an EPA test (section 2.2.5.3) requires that cholesterol (250 ng) be loaded on to the probe and spectra acquired by repetitive scanning. The m/z 386/368 ratio for cholesterol should be three or greater if thermal decomposition is not excessive. A more exacting test of the direct insertion probe performance and of ion source heating is provided by the use of bilirubin as a test sample.

Method

Prepare a chloroform solution containing 100 ng μL^{-1} of bilirubin. Load 1 μL of this solution onto the direct insertion probe. Begin recording repetitive EI mode scans and increase the probe temperature at approximately 20°C min^{-1} to a maximum of 300–350°C. The bilirubin should evaporate, giving a reasonably symmetrical ion current profile over approximately 3–4 min. If the ion source contains cold spots or the sample itself is insufficiently heated the evaporation period will be greatly prolonged and may not be adequately completed. Sensitivity may also be low due to loss of sample.

A source temperature of 200°C should be quite adequate for the clean evaporation of bilirubin and under these conditions the molecular ion at m/z 584 should be at least 50% of the base peak at m/z 286 (Fig. 2.8). If, however, all or part of the ion source is too hot then the relative size of the molecular ion will be reduced. For example, at a source temperature of 260°C the relative intensity of the m/z 584 ion becomes approximately 15–20% of that at m/z 286.

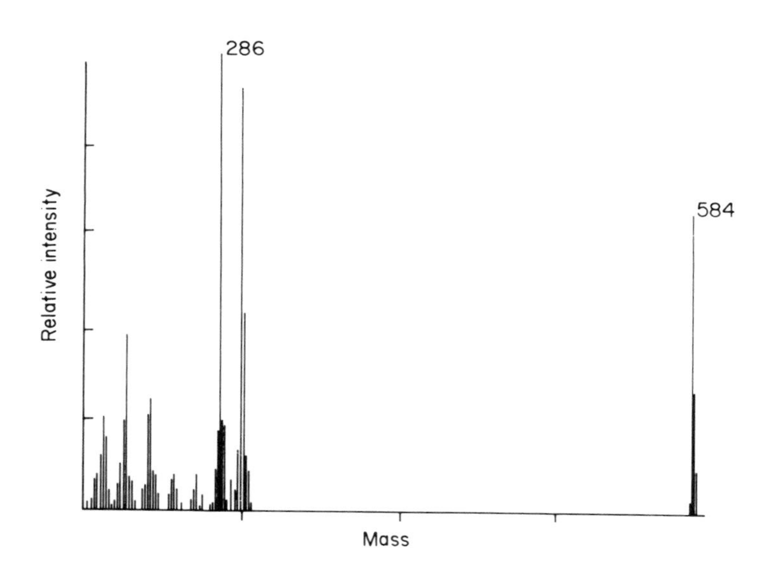

Fig. 2.8 Mass spectrum of bilirubin (source temperature 200°C)

2.4 Liquid Chromatography–Mass Spectrometry (LC–MS)

The practical problems associated with the development of an LC–MS interface are much more severe than those associated with GC–MS interfacing. There are three particular difficulties: (a) high flow rates; (b) involatile and labile samples; (c) use of involatile buffers.

The effective flow rates are much higher in LC than in GC and a considerable energy input is required to separate the flow of solvent from the solute as vapour (Table 2.1). A conventional source housing pumping system gives a pressure of 2.5×10^{-4} torr when the flow rate of vapour into the system is of the order of 5–7 $cm^3\ min^{-1}$; thus, this figure represents a typical upper limit to the vapour flow rate that can be allowed to enter the source housing.

Table 2.1 Flow considerations in GCMS and LCMS

GC	Packed column	Capillary column
	30 cm^3 helium min^{-1}	1 cm^3 helium min^{-1}
LC	Conventional column	Microbore column
Methanol	1 cm^3 liquid min^{-1} requiring 16 W to provide 550 cm^3 vapour min^{-1}	60 μL liquid min^{-1} requiring 1 W to provide 34 cm^3 vapour min^{-1}
Water	1 cm^3 liquid min^{-1} requiring 43 W to provide 1240 cm^3 vapour min^{-1}	60 μL liquid min^{-1} requiring 2.6 W to provide 74 cm^3 vapour min^{-1}

With the exception of the specially designed thermospray ion source (Section 2.4.3), all of the flow that enters the ion source to be ionized is then directed into the source housing to be pumped away (Figure 2.1). Thus, the figure of 5–7 $cm^3\ min^{-1}$ also represents the flow of vapour that can be admitted into the mass spectrometer ion source under normal circumstances. A flow of 5 $\mu L\ min^{-1}$ water when volatilized is equivalent to a vapour flow of 6.25 $cm^3\ min^{-1}$. A similar vapour flow results from the complete volatilization of 15 $\mu L\ min^{-1}$ of acetonitrile.

There are a number of possible solutions to the problem of restricted solvent flow rate:

(a) Splitting the flow from a conventional LC column and introducing only a fraction into the mass spectrometer. This course of action wastes a high proportion of the sample and is not acceptable if the sample quantity is limited.
(b) Increasing the source housing pumping speed, usually by the addition of a cryopump. Under these circumstances, the maximum permissible flow rate may be increased to perhaps 50 $\mu L\ min^{-1}$.
(c) Using low flow narrow-bore packed capillary or open tubular columns that eliminate the need for flow splitting.
(d) Removal of the solvent prior to introduction of the sample into the mass spectrometer. This approach is used in the transport (section 2.4.1) and particle beam (section 2.4.5) interfaces.

(e) Attaching additional pumping to the ion source itself which can then accept much higher flow rates. This approach is used in the thermospray interface (section 2.4.3).
(f) Ionization at atmospheric pressure. This method is discussed in section 2.4.6.

In many cases the materials being analysed are relatively involatile and/or labile since they are generally compounds that are not amenable to GC–MS analysis. Conventional ionization techniques such as electron impact and chemical ionization are gas-phase ionization techniques which require thermal volatilization of the sample. Although this limitation is acceptable for some compounds that can be analysed by LC, it is unacceptable for a larger number. For this reason, considerable use is made of liquid-phase ionization techniques such as thermospray ionization (section 2.4.3 and Chapter 6), continuous-flow fast atom bombardment (section 2.4.4), and electrospray ionization (section 2.4.6 and Chapter 6).

Most LC–MS interfaces and/or ion sources do not accept the use of involatile buffer systems. Unless a transport interface (section 2.4.1) is being used, volatile buffers based on materials such as ammonium acetate, ammonium formate, ammonium hydroxide, and acetic or trifluoroacetic acid should be used.

A number of different approaches have been used in interfacing LC and MS. The earliest attempts focused on methods of overcoming the incompatibility of the liquid flow rate and maintenance of the mass spectrometer high vacuum. More recently, more attention has been focused on the practical use of ionization techniques that do not require sample volatilization. Perhaps the most appropriate comment at this stage is to stress that no one interfacing technique is suitable for all LC–MS applications.

2.4.1 Transport interfaces [36]

In a transport interface, the flow from the LC is carried by mechanical means from the end of the LC column to the ion source with the solvent being completely removed before introduction of the sample into the ion source. This principle was the basis of the first commercial LC–MS interface. In the most recent design of transport interface (Fig. 2.9), the flow is deposited on an endless polyimide belt. The belt first passes under an optional infrared heat source to evaporate the solvent, then through two vacuum locks and into the ion source of the mass spectrometer. In the ion source the sample is either flash evaporated or desorbed, e.g. by fast atom bombardment, from the belt. On the return path the belt passes over a clean-up heater to remove residual solvent and sample and finally through a wash bath to remove non-volatile materials.

Problems associated with sample deposition on the transport medium can be minimized by spraying the flow on to the belt. The use of a pneumatic nebulizer with a pre-heated gas provides a fine spray which is uniformly coated on to the belt; at the same time a large fraction of the solvent is evaporated during

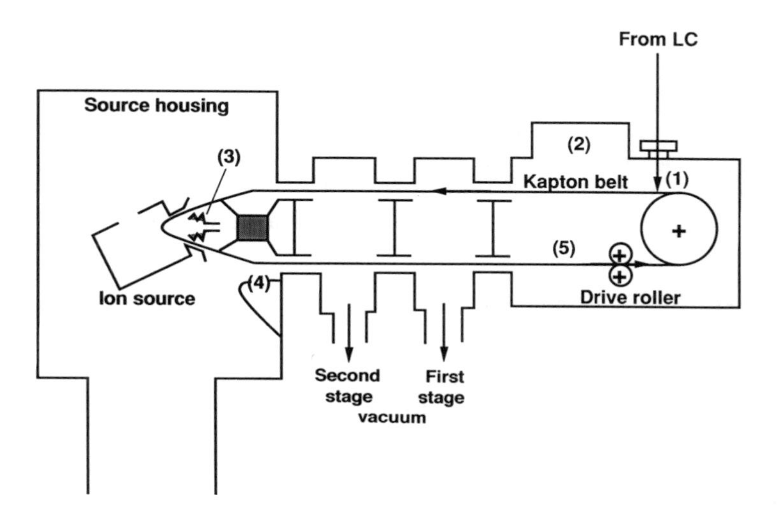

Fig. 2.9 Schematic representation of a moving belt interface. (1) Contact depositor or jet sprayer (gas nebulized, thermospray); (2) solvent evaporation. Infrared heater in earlier versions; (3) sample heater; (4) clean-up heater for EI/CI modes. Not used with FAB; (5) belt cleaner (mechanical scrubber, wash bath)

spraying so that the infrared heater is no longer needed. More recently, thermospray nebulization (section 2.4.3) has proved to be an effective replacement for pneumatic nebulization, especially when solvents with high water content and/or high solvent flow rates are used. The use of microcolumns with the moving belt has also been found to be advantageous since depositing and evaporating a reduced solvent flow rate is obviously easier.

The moving-belt interface has two particular advantages. The principal advantage, which it shares with the particle beam interface (section 2.4.5), is that since the mobile phase has been evaporated by the time the sample reaches the ion source, the belt interface is compatible with a variety of ionization modes including electron impact and chemical ionization. The second advantage is that use of the moving-belt system places few restrictions on the LC system. In particular, since buffers are washed from the belt after it has left the ion source, non-volatile buffers can be used. This is not the case with most other LC–MS interfaces.

There are a number of practical problems associated with the belt interface in addition to its relative expense and complexity. One problem is a gradual build-up of background from the belt during use, since it is particularly difficult to ensure efficient removal of involatile residues from solvent and sample before recycling. The second problem is more fundamental and is that involatile or thermally labile samples cannot be analysed, at least not by evaporation from the belt. Under these conditions, such compounds may give spectra only at very high (microgram) sample levels, but most often give no useful data at all. As a consequence, the transport interface has declined in popularity as an LC–MS interface.

The combination of the moving belt with fast atom bombardment (FAB) ionization was proposed as a promising way of extending the application of the transport interface to the analysis of more polar and thermally labile compounds. However, the method has never really reached the stage of routine applicability and has, to a great extent, been superseded by continuous-flow FAB (section 2.4.4).

2.4.2 Direct liquid introduction [37]

In direct liquid introduction (DLI) LC–MS, the LC column flow is introduced directly, without any prior enrichment of sample with respect to solvent, into the mass spectrometer through some form of restriction. Typically, a laser-drilled orifice, 2–5 μm in diameter (Fig. 2.10), is used as a restrictor. So long as there is an adequate liquid flow through this orifice, a liquid jet forms and then immediately breaks into small droplets. The tip of the interface is water-cooled prior to the orifice to prevent premature evaporation of the solvent. This jet then passes into a heated desolvation chamber where the droplets are vaporized, and the solvent vapour and sample enter the ion source of the mass spectrometer.

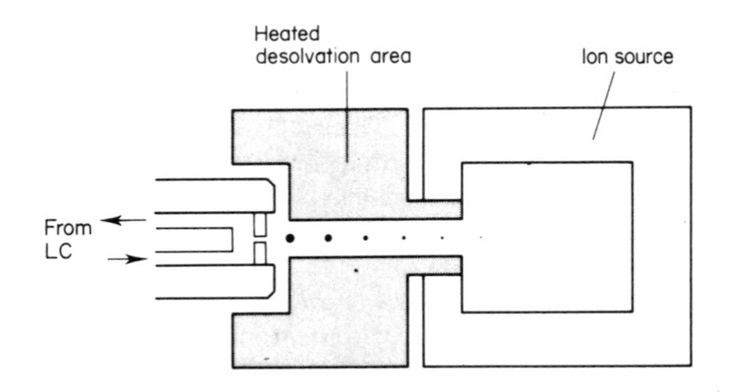

Fig. 2.10 Schematic diagram of direct liquid introduction LC–MS interface

As indicated in section 2.4, the mass spectrometer vacuum system can only tolerate a limited solvent flow if suitable operating pressures are to be maintained in the source housing and in the analyser. DLI interfaces designed for use with conventional LC columns (approximately 1 cm^3 min^{-1} solvent flow rate) incorporate a flow splitter, usually a needle valve located downstream from the orifice. This allows 5–15 μL min^{-1} of the solvent flow to enter the ion source and provides the back-pressure necessary for jet formation. If cryopumping or larger pumps are employed, flow rates up to 20–50 μL min^{-1} can then be accommodated.

The desolvation chamber is interfaced to a conventional chemical ionization ion source. The high source pressures that result yield CI spectra, both positive and negative ion, with the solvent vapour acting as the reagent gas. The character of the chemical ionization spectra may be changed by the use of additives, introduced directly into the ion source, in both the positive and negative ion modes.

The DLI interface can be used with most common LC solvents and with volatile buffer materials such as ammonium acetate, ammonium formate, ammonium hydroxide, and acetic and trifluoroacetic acids. The continued use of highly aqueous phases may, however, lead to a reduced lifetime for the ion source filament. A particular practical difficulty with the DLI interface is the very small size of the orifice which makes it very susceptible to plugging by particulate matter. Location of the orifice in a replaceable diaphragm is a considerable advantage in this case.

In general, the use of the DLI interface has been restricted to the analysis of more volatile samples. With careful attention to interface design and experimental conditions, however, some research groups have demonstrated that DLI can accommodate less volatile samples. For example, vitamin B_{12} (MW 1354) introduced via a DLI interface shows a negative ion spectrum which compares very favourably with those recorded using established soft ionization methods such as FAB [38]. Just as with the transport interface, the direct liquid introduction interface presents certain practical problems in operation and has declined in popularity compared to other LC–MS systems.

An alternative method of introducing a liquid flow directly into a chemical ionization source uses a porous frit, as in some forms of continuous-flow fast atom bombardment (section 2.4.4). The frit is maintained at a high temperature by heat from the adjacent ion source so that sample brought to the surface of the frit by the liquid flow is continually evaporated into the ion source [39]. Ionization is again effected by the use of solvent vapour as a chemical ionization reagent gas.

There is another variant of direct liquid introduction that has not been offered in any commercial form. In this system, a fused silica microcolumn, which may be open tubular or packed, and which is drawn to a fine tip at the exit, is then introduced directly into the ion source. De Wit and co-workers [40, 41] have used coated open-tubular columns mounted within a heated probe assembly (Fig. 2.11) as the separating columns. When the column tip is heated as well as tapered, vaporization occurs at the tip rather than inside the column, resulting in much better control of both solute and solvent evaporation. The solvent flow through these open tubular columns is so low (30 nL min^{-1}) that electron impact or chemical ionization (with the addition of reagent gas) spectra may be obtained.

Stenhagen and Alborn [42] have used packed capillary columns (0.22 mm i.d.) with a flow rate of 1–2 μL min^{-1}. The tapered column is led up to, but not introduced into, a conventional EI ion source on a magnetic sector mass spectrometer. By the simple expedient of creating an electrostatic field (cf. Chapter 6) at the column tip it is possible to obtain stable nebulization of the flow directly

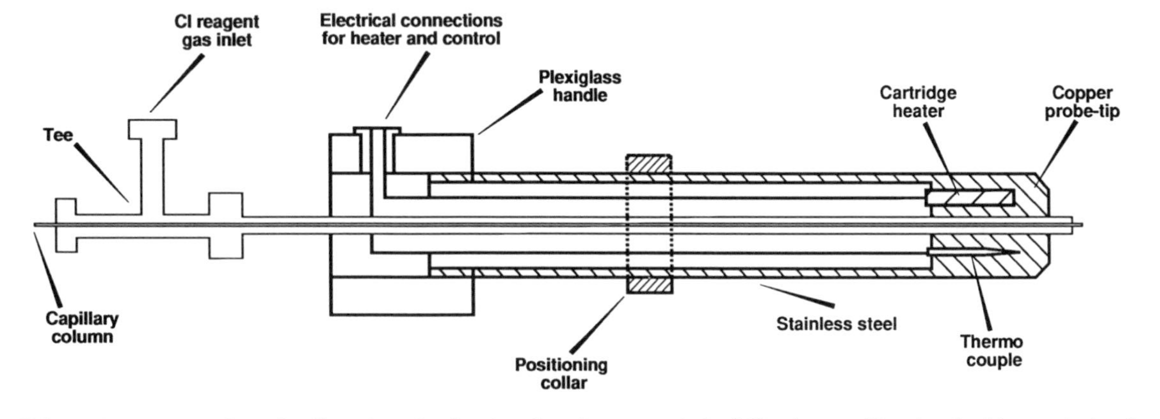

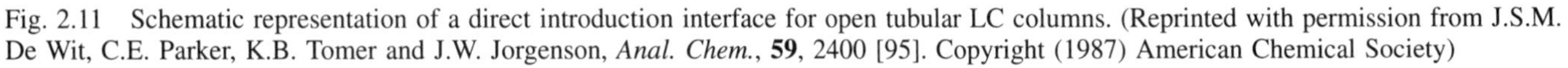

Fig. 2.11 Schematic representation of a direct introduction interface for open tubular LC columns. (Reprinted with permission from J.S.M. De Wit, C.E. Parker, K.B. Tomer and J.W. Jorgenson, *Anal. Chem.*, **59**, 2400 [95]. Copyright (1987) American Chemical Society)

into the electron beam, even at low ion source temperatures. By this means relatively labile, biologically active compounds could be analysed on a routine basis.

2.4.3 Thermospray and discharge ionization [43,44]

The term thermospray has been applied both to a method of introducing high flow rates of liquid into a mass spectrometer vacuum system and to a method of ionization (section 6.2), adventitiously discovered by Vestal during his experiments with the liquid introduction system [45]. Figure 2.12 shows a complete thermospray interface. In this interface, a flow of 1–1.5 cm^3 min^{-1} solvent is rapidly vaporized in a resistively heated metal capillary tube with an inside diameter of approximately 100 μm.

The power supplied to the capillary is controlled so as to maintain a constant temperature, T_1, measured by a thermocouple attached to the capillary near the inlet end where no vaporization occurs. It can be shown [46] that T_1 is linearly related to the fraction vaporized so that maintaining T_1 constant keeps a constant fraction volatilized for a given solvent composition as required despite flow rate changes. The temperature T_1 may also be programmed to achieve a constant fraction volatilized during a gradient elution LC experiment. Normally, a high percentage ($\geq$90%) of the liquid flow is volatilized and under these conditions the vapour which expands from the end of the capillary has sufficient energy to transform the remainder of the liquid flow into a mist of fine droplets.

The other important feature of the inlet system is the rotary pump attached to the ion source. This pump removes the great majority of the solvent vapour that enters the ion source (and certainly some of the sample), while only a small fraction

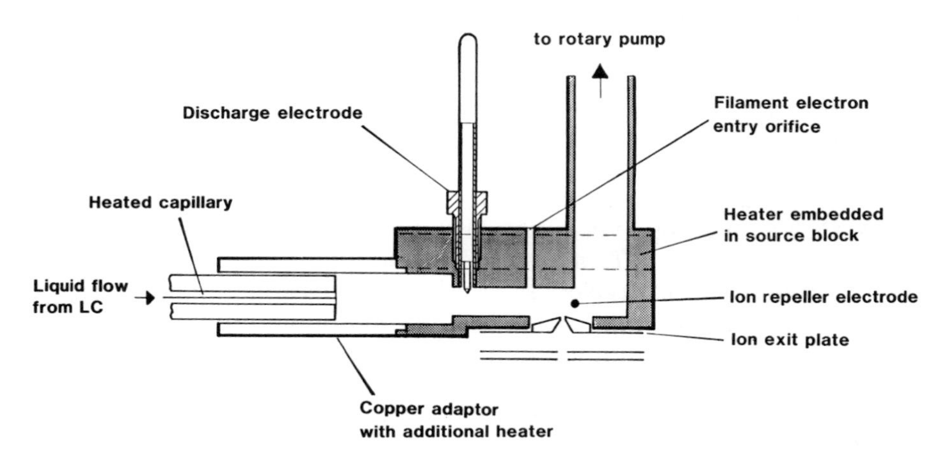

Fig. 2.12 Schematic representation of a thermospray ion source with discharge ionization facilities. (Courtesy Kratos Analytical Instruments)

enters the source housing via the orifice in the ion exit plate. It is the use of this pump, which may be supplemented by a cold trap, that allows the introduction of such high liquid flow rates directly into the ion source region.

The jet that carries the droplets expands into the ion source where additional heat is added to complete the vaporization process. Independent heaters surrounding the capillary tip and the main ion source block allow the optimization of desolvation conditions and prevent sample condensation. The process by which sample ions then originate from these droplets in the presence of a volatile electrolyte, such as ammonium acetate—thermospray ionization—is described in section 6.2.

The original thermospray interface offered no form of ionization other than the use of a volatile electrolyte and, as a consequence, was restricted to the analysis of samples in solutions with a relatively high water content. Most sources now include a filament or discharge electrode to assist in ion production (Fig. 2.12). Electrons from the filament or the discharge ionize the solvent vapour and hence the sample by a chemical ionization process. These changes have greatly increased the versatility of the system so that samples in mainly organic solvents and solutions that do not contain a volatile electrolyte may also be analysed. Another advantage of discharge ionization is that, since its performance is much less dependent on the nature of the solvent, it is generally more suitable than thermospray ionization for LC–MS experiments where a gradient that covers a wide range of solvent compositions is used. Most thermospray ion sources also include an ion repeller electrode (section 6.3.3.2) (Fig. 2.12) which can be used to improve sensitivity, particularly for higher mass compounds, as well as to induce controlled fragmentation of quasi-molecular ions (section 3.1.1.1) in some cases.

Although thermospray comprises a special inlet system and a special ionization technique it is, nevertheless, the most popular LC-MS technique in use today. This is because it is directly compatible with conventional LC flow rates (1–1.5 cm^3 min^{-1}) and can deal, especially in conjunction with discharge ionization, with most solvents. Many compounds, including those that are thermally labile, with a molecular weight (MW) between 200 and 1000 can be readily studied and this mass range can often be extended, although not usually with good sensitivity [47,48]. Thus, thermospray will deal with a wider range of analytes under conventional LC conditions than any other interfacing technique currently available.

Nevertheless, thermospray does have its disadvantages. For example, involatile buffers cannot be used with thermospray. Again, thermospray ionization is a relatively mild ionization technique so that many compounds fail to give fragment ion information, although the use of MS–MS techniques in conjunction with thermospray LC–MS has proved quite useful here. There is a variation in sensitivity from compound to compound and the operating temperatures for optimum sensitivity depend quite critically on the sample to be analysed and on the solvent system used. In this respect, it is found that the discharge ionization mode requires less scrupulous setting of operating temperatures.

Figure 2.13 shows an example of an LC–MS analysis of a substituted

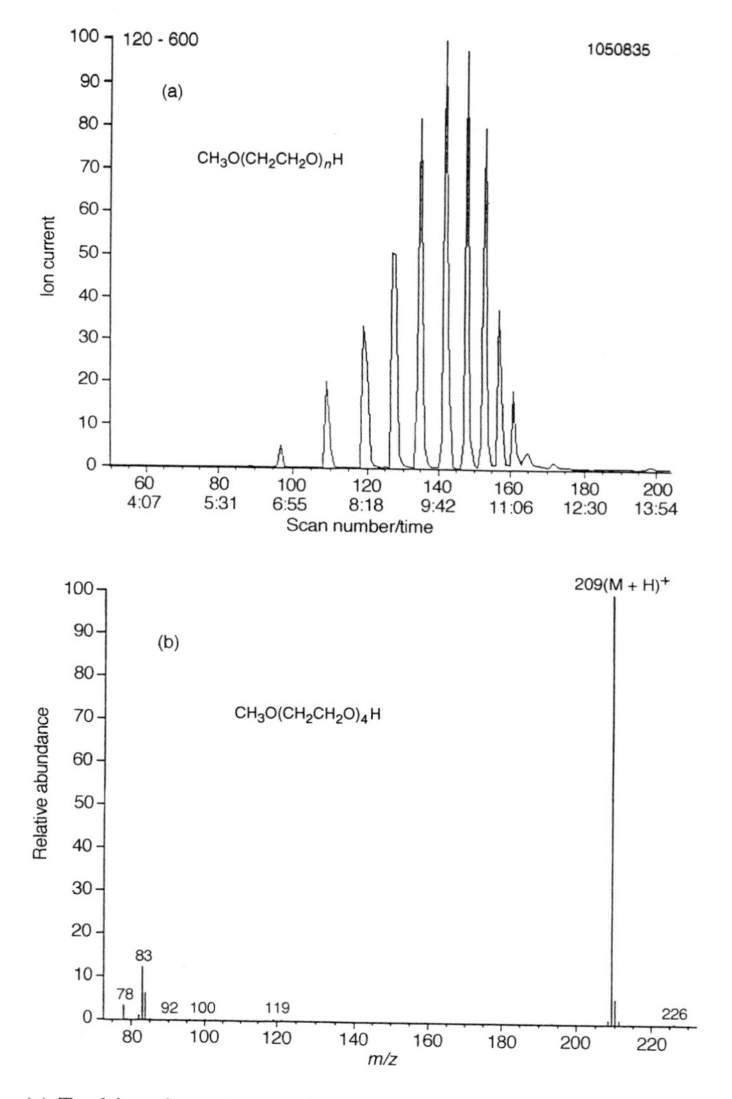

Fig. 2.13 (a) Total ion chromatogram from reverse phase gradient separation of methoxy polyethylene glycol oligomer using plasma discharge ionization with a thermospray interface. (b) Spectrum of the component eluting at 8 min 14 s (MW 208) in (a). (Courtesy Kratos Analytical Instruments)

polyethylene glycol using a thermospray interface with discharge ionization. The symmetrical distribution of peaks in this total ion chromatogram suggests that each component of the glycol mixture is detected with similar sensitivity under discharge ionization conditions, despite the change in solvent composition during the analysis.

2.4.4 Continuous-flow fast atom bombardment [49]

Continuous-flow fast atom bombardment (CF-FAB) is an important interfacing technique which is also sometimes known as dynamic FAB. It is, in fact, another form of direct liquid introduction, but one that has been developed almost totally for use with FAB ion sources (cf. sections 5.2.2.1 and 5.3.2.4). A typical CF-FAB interface is shown in Fig. 2.14.

The technical principle is very straightforward. A liquid flow, typically 5–10 μL min^{-1} with standard pumping systems, enters a fast atom bombardment ion source through a narrow fused silica capillary (50–75 μm i.d.). This capillary is generally contained within a probe which can be brought up to or removed from the ion source by means of a standard vacuum lock. Again, as in any direct liquid introduction system, flow splitting is required unless some form of microcolumn is used. A very convenient design of flow splitter has been published for this application [50]. Involatile buffers are not recommended with CF-FAB.

A small percentage of a FAB matrix material, such as glycerol, is added to the liquid flow either at the column exit or, in some cases, directly to the aqueous component of the chromatographic mobile phase. The liquid emerging from the capillary forms a target on the metal end of the probe which is bombarded by the atom or ion beam as in a conventional FAB source. Some heat is applied

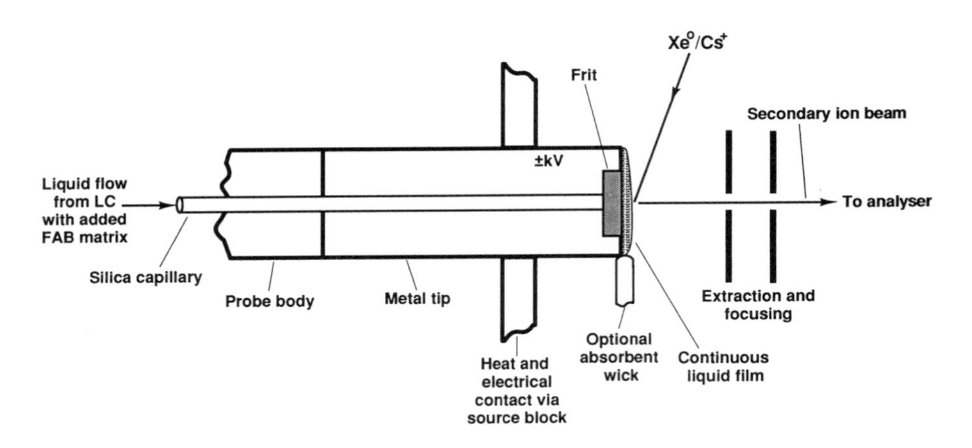

Fig. 2.14 Schematic representation of an interface probe for continuous-flow fast atom bombardment

to the probe tip solely to prevent freezing as the liquid evaporates in the mass spectrometer vacuum. In the case of open-tubular microcolumns where the flow is, in fact, too low for CF-FAB, the FAB matrix can be added in a make-up flow in another capillary which is coaxial with and surrounds the column itself [51].

For the optimum use of CF-FAB as an LC–MS interface, two conditions should be met: (a) the liquid flow should form a stable, continuous film on the probe tip in which the previously eluted sample is removed from the area where fast atom bombardment takes place (Fig. 2.14) and (b) the liquid flow rate should be equal to the liquid evaporation rate. Probably the most satisfactory method of accomplishing this is to interpose a small metallic frit between the end of the capillary and the mass spectrometer vacuum [39]. Not only is the metal surface of the frit easily wetted but also the improved thermal conductivity of the metal surface and the narrow orifices in the frit encourage a very stable liquid evaporation process. A further advantage of the frit construction is that the glycerol concentration required can be reduced from approximately 5 to 0.5%. Another technique used to promote a stable liquid film is to place a layer of absorbent material against the edge of the probe to remove liquid from the tip (Fig. 2.14).

Continuous-flow FAB is becoming more frequently used since the range of compounds that may be analysed is the same as for conventional FAB and includes thermolabile and ionic materials as well as non-volatile biopolymers such as peptides (e.g. Fig. 2.15), oligosaccharides and oligonucleotides. In addition, CF-FAB has some advantages compared with conventional FAB which are discussed in section 5.3.2.4. CF-FAB is most compatible with the use of lower flow packed microbore columns and has also been proposed as an interfacing technique for high resolution separations by capillary electrophoresis. An example of a separation on a packed microbore column monitored by CF-FAB is shown in Fig. 2.15.

2.4.5 Particle beam interface

The first particle beam LC–MS interface was introduced by Willoughby and Browner [52] under the acronym MAGIC (monodisperse aerosol generating interface for chromatography). In the original device the liquid flow from the LC was broken into a stream of small droplets in an atmospheric pressure enclosure by a flow of gas introduced at right angles to the liquid stream. Most devices now use either a pneumatic nebulizer with pre-heated gas or a thermospray nebulizer to achieve the same effect. Subsequently, solvent evaporates from the droplets as they drift through the desolvation chamber (Fig. 2.16) which is warmed just sufficiently to compensate for the latent heat of vaporization of the solvent.

Most solvent is lost in the desolvation chamber but the remaining sample particles which are entrained in the gas stream are directed towards the inlet jet of a two-stage momentum separator. This device, which is very similar to the jet separator used for packed column GC-MS, acts to enrich the heavier sample particles with respect to the surrounding gas. As a result of this process a well-defined sample

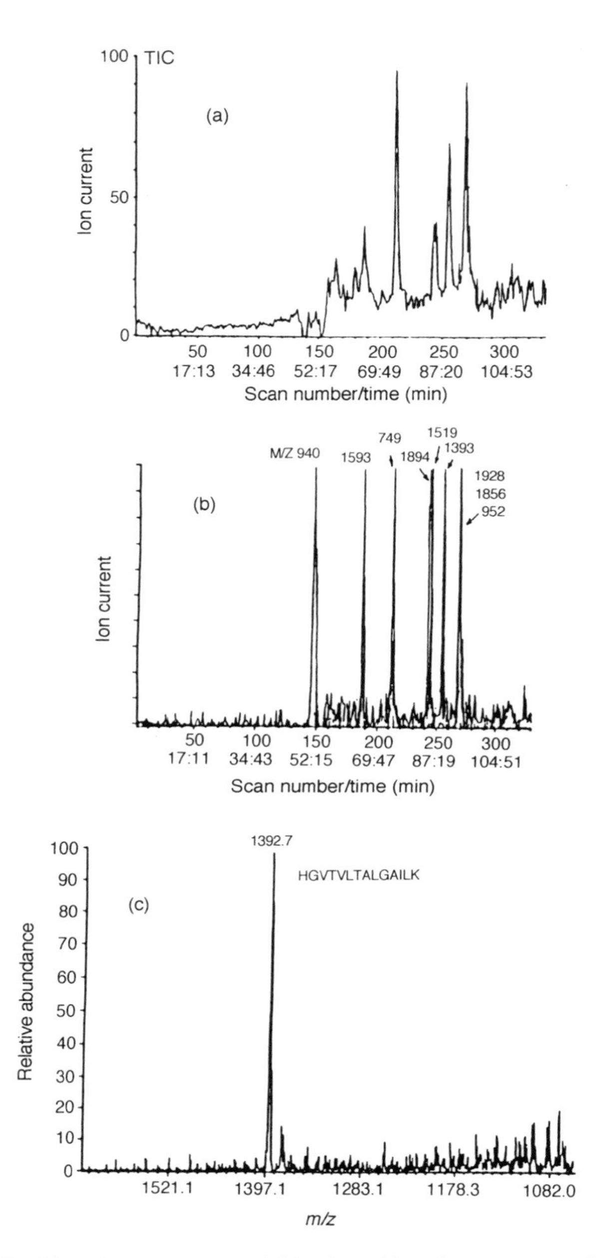

Fig. 2.15 (a) Total ion chromatogram and (b) selected ion chromatograms from continuous-flow FAB analysis of 100 pmol sperm whale myoglobin tryptic digest. (c) High mass region of the spectrum of the component eluting at scan 256 in (b). (Reproduced with permission from ref. 94)

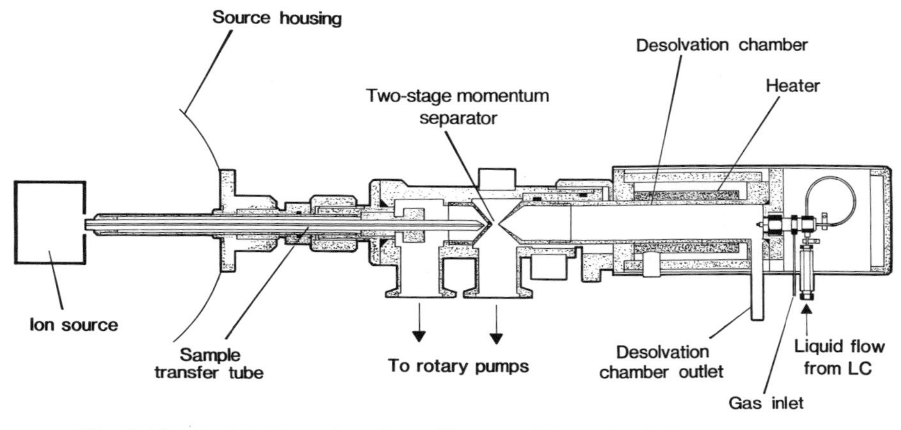

Fig. 2.16 Particle beam interface. (Courtesy Kratos Analytical Instruments)

particle beam, which is virtually devoid of solvent, enters the ion source of the mass spectrometer. A recently published variant of the particle beam interface interposes a membrane separator, which takes the form of a length of Teflon tubing, between the spray chamber and momentum separator [53]. In this design, the majority of the solvent is lost through the membrane, which is permeable to solvent but much less so to the sample.

Most published applications of particle beam LC–MS have used electron impact (EI) or chemical ionization (CI), with added reagent gas, based on a standard EI/CI source. Under these circumstances, the ion source is usually heated sufficiently to vaporize material in the beam prior to ionization (e.g. ref. 54) although a separate heated target for flash vaporization may be placed in the ion source opposite the particle beam (e.g. ref. 55). Preliminary efforts have been made to interface particle beam with a FAB source, using the FAB sample probe itself as a target [56], but more development is needed in this area.

The most important advantage of the particle beam interface is that complete removal of solvent gives the user a free choice between standard EI and CI operation. In this mode, the particle beam interface is generally preferred to the moving-belt interface, although sensitivity, particularly with more aqueous solvents, can be disappointing. Normal LC flow rates (approximately $1\,cm^3\,min^{-1}$) of any solvent can be accommodated in a well-designed particle beam interface, so that flow splitting is not required. The use of involatile buffers is, however, not really practicable.

2.4.6 Atmospheric pressure ionization

In atmospheric pressure ionization (API) LC–MS, ions are formed from a liquid flow introduced into a source region maintained at atmospheric pressure. Thus, the ion source contains sample ions together with solvent vapour and the ambient gas (usually nitrogen). The coupling of a liquid flow inlet to an API source offers some advantages over other approaches. For example, it avoids problems associated with the introduction of a liquid flow directly into high vacuum. Again, it readily allows liquid chromatography systems, especially low flow systems such as capillary electrophoresis columns, to be operated under 'normal' conditions, i.e. with the column exit at atmospheric pressure.

There are, in fact, three forms of ionization used in API sources for LC–MS: (a) heated nebulizer with a corona discharge [57], (b) electrospray [58], and (c) ion

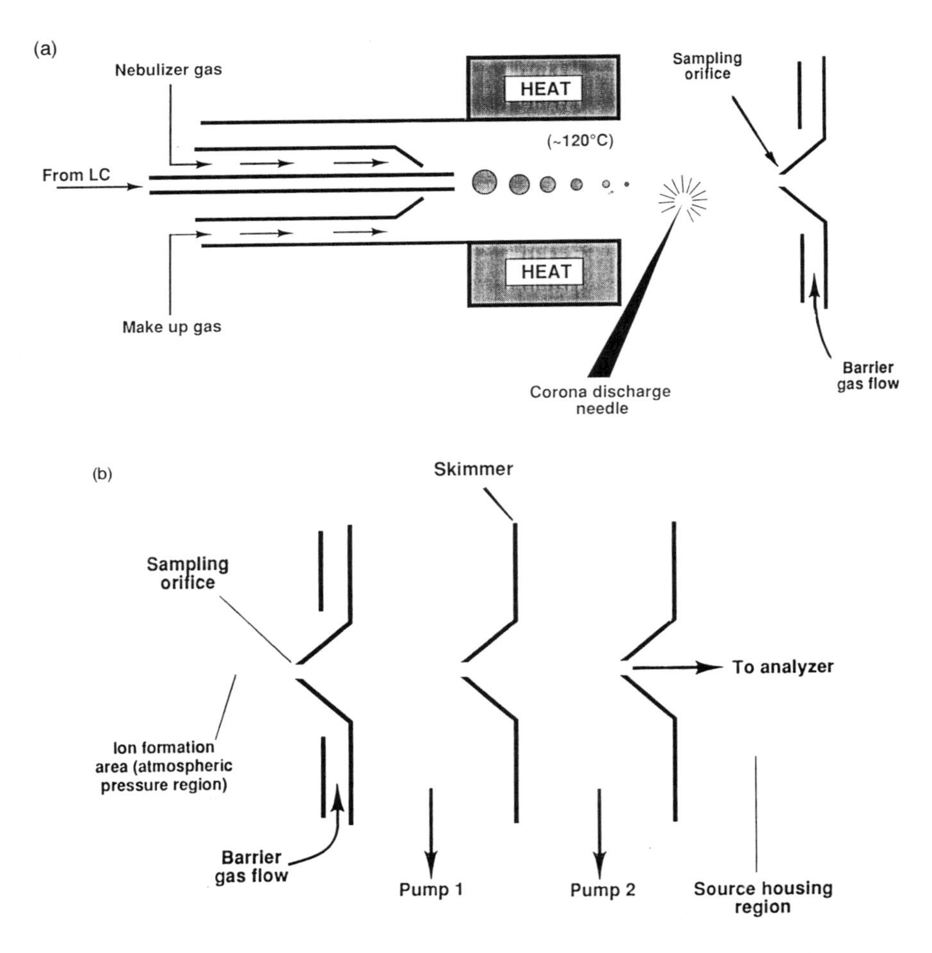

Fig. 2.17 (a) Schematic representation of a heated nebulizer for atmospheric pressure ionization. (adapted from ref. 57) (b) Sampling from an atmospheric pressure ionization source using a differentially pumped system

spray [59]. The electrospray and ion-spray systems are discussed in detail in section 6.3.2. The heated nebulizer interface (Fig. 2.17a) uses pneumatic nebulization to convert the liquid flow into droplets which are then swept by means of a sheath gas through a quartz tube heater to vaporize the solvent and analyte. The mixture of vaporized solvent and analyte flows towards the ion formation region where a corona discharge initiates chemical ionization at atmospheric pressure (APCI) using the vaporized solvent as the reagent gas. These CI processes take place within a wall-less reaction region which is defined by the gas flow in the source volume. Flow rates from 0.5 to $2\,cm^3\ min^{-1}$ of any solvent and the use of volatile or involatile buffers are permitted. Memory effects are minimized by the large-scale, open construction of the source volume.

Figure 2.17a also illustrates one method used in transferring ions from any form of API source into a mass analyser. Ions are driven towards a sampling orifice (approximately $125\,\mu m$ diameter), which leads directly from atmosphere into the mass spectrometer vacuum system, by a combination of gas flow and electric fields. The flow of an inert barrier gas across the sampling orifice can be used to minimize the number of neutral solvent molecules that enter the vacuum system. The same barrier gas will also quench any ion–molecule reactions initiated by the corona discharge. With this sampling method, a very high pumping speed, provided by cryopumping and cryotrapping, is used to maintain a suitable operating pressure [60].

An alternative interfacing method for API sources (Fig. 2.17b) uses a larger sampling orifice (approximately $200\text{–}300\,\mu m$ diameter) which leads into an intermediate vacuum stage. Successive vacuum stages may be used to improve the ion-to-neutral ratio in the stream which eventually enters the mass analyser by preferentially pumping away neutral molecules while ions are directed towards the next orifice by means of electric fields. As in all LC–MS interfaces, the neutral flow has to be reduced to a level at which the mass spectrometer pumping system can maintain a good working pressure.

Sakairi and Kambara have described a very similar sequential nebulization and vaporization interface which also uses a corona discharge [61]. In addition, they have described the simple conversion of this APCI interface, by removal of the vaporization stage, to effect thermospray nebulization at atmospheric pressure, which they term 'APSI' [62]. Interestingly, under APSI conditions, sample ionization does not necessarily require the presence of a high level of an electrolyte, such as ammonium acetate, that is conventionally used for thermospray ionization. In a more recent communication, Sakairi and Yergey [63] have demonstrated the coupling of this nebulizer system to a conventional EI source via a differentially pumped interface system. The use of thermospray nebulization and ionization at atmospheric pressure has also been reported by Henion and co-workers [64,65].

A comparison of the three atmospheric pressure techniques reveals differences in the type of compounds that can be analysed and in the liquid flow rates (see this section and section 6.3.2 for flow rate data) that can be accommodated. Both the electrospray and ion-spray techniques are based on the desorption of ions from small droplets with virtually no heat input and are therefore very suitable for highly polar molecules. The heated nebulizer, on the other hand, depends on a gas-phase ionization process and is therefore only suitable for moderately involatile or labile molecules. Interestingly, however, in a recent comparison of LC–MS methods for the analysis of carbamate pesticides [66] the heated nebulizer APCI technique provided considerably better detection limits than either ion-spray or thermospray, which both showed very similar sensitivities.

The most exciting development of interest to LC–MS coupling is the emergence of electrospray and ion-spray as viable techniques for very high molecular weight and highly polar samples. Electrospray ionization has also been proposed as an interfacing technique for high resolution separations by capillary electrophoresis [67]. An example of an API technique, in this case ion-spray, used in LC–MS is provided in Fig. 2.18. The ion-spray technique is a very suitable method of providing a spectrum with an intact quasi-molecular ion (Fig. 2.18b) from ionic samples; the acidic nature of the sulphate group suggests the use of negative ions in this case. An MS–MS scan (Chapter 7) of collision induced fragments from the quasi-molecular ion (Fig. 2.18c) offers more structural information, with ions at m/z 97 and 80 being typical of aliphatic sulphate conjugates.

2.5 Supercritical Fluid Chromatography–Mass Spectrometry (SFC–MS) [68]

Supercritical fluid chromatography can be effected with fused silica capillary columns containing bonded, highly cross-linked stationary phases or with high performance liquid chromatography (HPLC) columns of varying internal diameters using conventional packing materials. Typical mobile phase flow rates used for capillary column SFC can be accommodated directly by mass spectrometers that are suitable for CI operation. Packed column SFC requires some alternative means, e.g. a transport interface (section 2.4.1) or thermospray source (section 2.4.3), to deal with the higher flow rates.

In a capillary supercritical fluid chromatograph–mass spectrometer interface, a restrictor is used to separate the mass spectrometer vacuum from the high pressure of the chromatographic column. The expansion of mobile phase at the restrictor cools the surrounding area; therefore, heat must be added at this point to assist in vaporization of the flow. Of several types of restrictors that have been reported, three are most commonly used. These are the multipath frit [69], the tapered restrictor [70], and the integral restrictor [71].

Abrupt restriction is desirable in order to minimize analyte precipitation and the necessity for extensive heating and therefore a laser drilled disk restrictor would appear to be an ideal choice. Difficulties in fabrication have, however, so far precluded the use of an interface with this type of restrictor [72]. Sheeley and Reinhold [72] have made extensive use of the integral restrictor—a 1 μm hole in

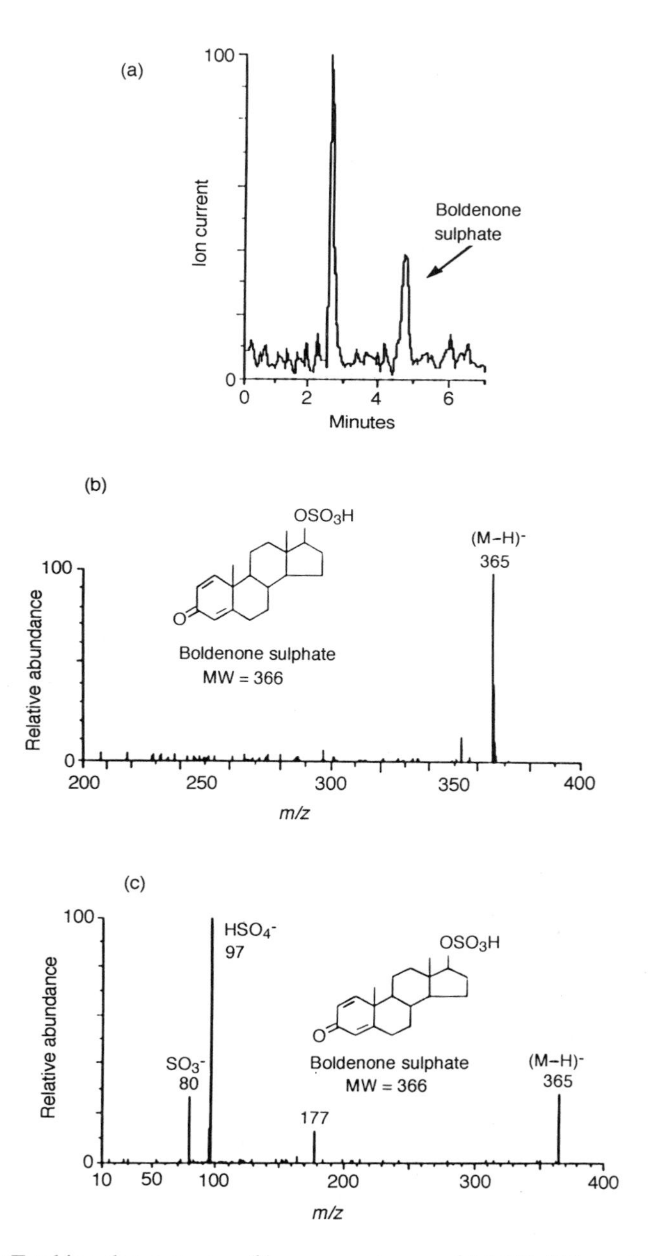

Fig. 2.18 (a) Total ion chromatogram (b) mass spectrum and (c) MS–MS spectrum from the quasi-molecular ion of boldenone sulphate. Analysis using an ionspray interface. (Courtesy Sciex Inc)

the otherwise sealed exit of the capillary column itself—as the best approximation to a laser-drilled orifice for the analysis of high molecular weight analytes by SFC–MS. In their experiments, a platinum heating wire of the same type as used in a DCI probe (section 5.3.1) is wrapped directly around the last millimetre of column to supply heat over the length of the integral restrictor. Low heater currents resulted in tip plugging and the operating current was set slightly above these values. Over a rather broad range above this temperature, no change in the recorded spectra was detected. The interface proved to be effective for the analysis of derivatized oligosaccharides with a molecular weight in excess of 5 kDa.

Alternatively, the restrictor, of various design types, may be formed in a separate capillary which is butt-coupled to the SFC column. For example, the coupling of glass capillary columns via a micrometer-adjustable interface which includes an integral pressure restrictor formed in a separate glass capillary has been described [73]. A number of authors [74,75] have used a multipath frit restrictor, housed in a separate capillary, where the high-temperature region (350°C) of the interface was limited to the frit portion of the restrictor. Satisfactory transfer of high molecular weight and labile materials was achieved and, in a study of the analysis of labile tricothecenes by SFC–MS [74], it was again found that spectra did not change over the restrictor temperature range 225–300°C.

A robot-pulled, tapered restrictor, heated to approximately 300°C over a length of about 4 cm where restriction occurs, also gave good transmission of a range of thermally labile and non-volatile materials [76]. Recently, another probe-mounted interface, which incorporates a heated zone for the SFC flow restrictor, has been described [77]. The three types of SFC flow restrictor were used with comparable ease, although a performance comparison was not given.

The most effective method of interfacing packed columns used for SFC with a mass spectrometer has proved to be the thermospray interface [78,79]. With this technique, the thermospray vaporizer capillary itself acts as the restrictor which separates the mass spectrometer vacuum from the chromatographic column. Ionization of the analyte is effected by operating the thermospray system in the discharge ionization mode (section 2.4.3).

2.5.1 Characteristics of SFC–MS spectra

Capillary SFC–MS enables conventional chemical ionization or electron impact mass spectra to be obtained. Most often, the spectra recorded are a result of chemical ionization reactions of the analyte with the carbon dioxide, widely used as an SFC mobile phase, and any other reagent gas that may be added.

If no additional reagent gas is added to the carbon dioxide, electron impact or charge exchange ionization contributions to the spectra predominate, depending on the gas-tightness of the ion source. Charge exchange from carbon dioxide parallels electron impact ionization in the fragmentation it produces. Thus, derivatized inositol triphosphate [76] and labile pharmaceuticals, including benzyl penicillin

[77], have been analysed under these conditions to give spectra with extensive fragment ion information. Other investigators have added another inert gas such as helium [80] or argon [77] to the carbon dioxide to produce spectra with abundant fragment ions.

Work reported in ref. 72 with glycosphingolipids has indicated that while ammonia chemical ionization provides molecular ions, with little fragmentation, use of a mixture of carbon dioxide and methanol as chemical ionization reagent can induce very informative fragmentations in this particular class of compound. Chester and Innis[81] and Pinkston *et al.* [82] have extended high molecular-weight studies to carbohydrates using trimethyl-silylation to block the polar hydroxyl groups. Chemical ionization spectra were obtained using ammonia as the reagent

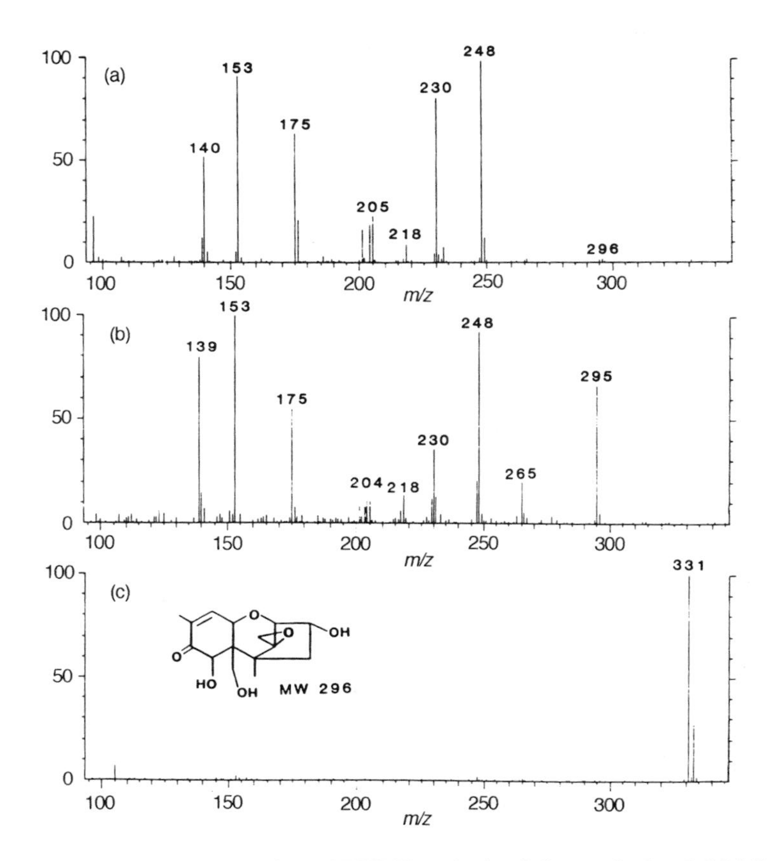

Fig. 2.19 Negative ion spectra from SFC-MS analysis of deoxy-nivalenol. Mobile phase carbon dioxide. Additional reagent gases gave (a) resonance electron capture; (b) hydroxide ion proton abstraction; (c) methane chloride attachment conditions. See text for further details. (Reproduced with permission from ref. 74)

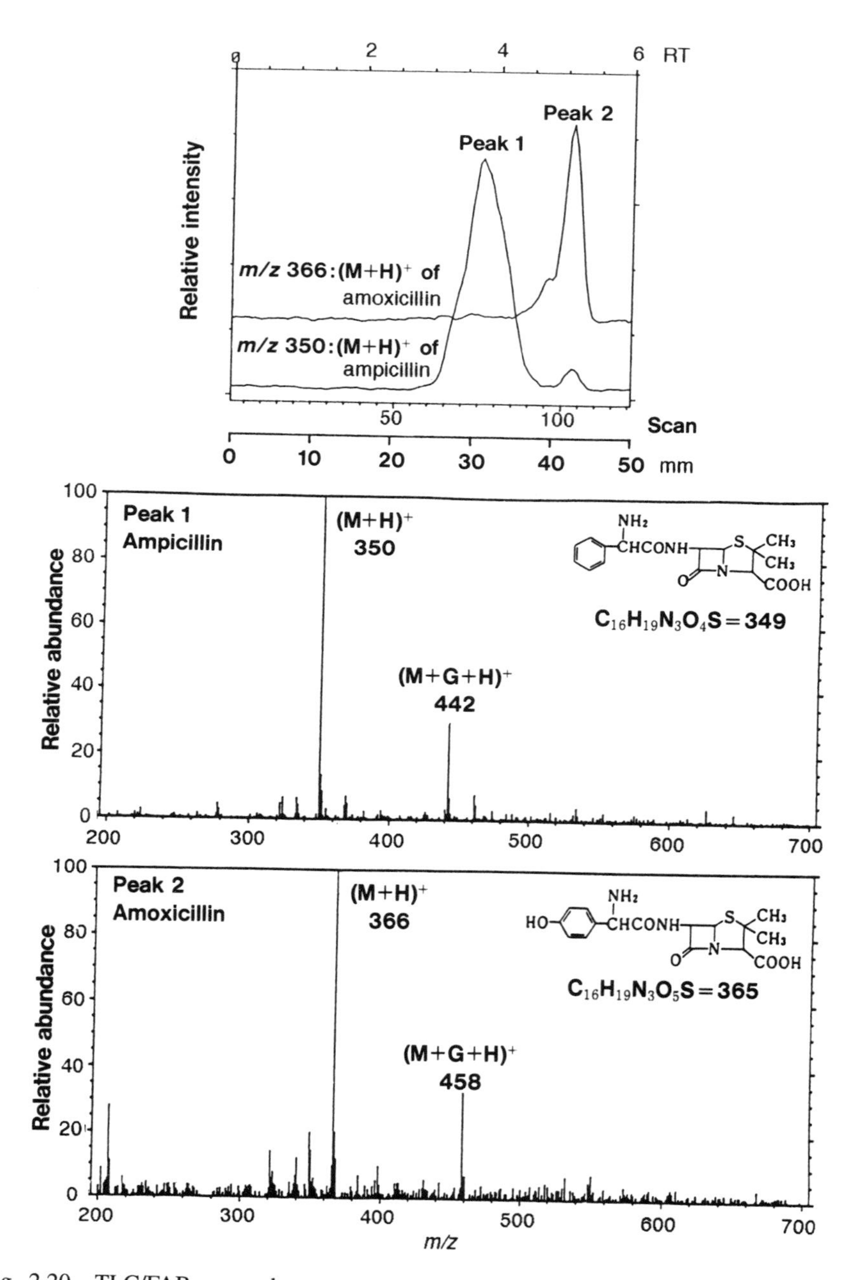

Fig. 2.20 TLC/FAB mass chromatogram and mass spectra of cephalosporin antibiotics. (Courtesy JEOL Ltd)

gas [82]. The $[M+NH_4]^+$ ions were detected up to m/z 2830 which was the expected ion for $(glucose)_7$. The direct injection of poly-dimethylsiloxanes dissolved in a supercritical fluid under ammonia chemical ionization conditions is a reasonable tuning and calibration tool for SFC–MS and has given quasi-molecular ions up to m/z 4000 [83].

In an SFC–MS study of tricothecenes under negative ion chemical ionization conditions [74], it was found that the carbon dioxide mobile phase did not significantly affect spectrum quality when other chemical ionization reagent gases were present. Thus, spectra recorded under methane electron capture or methane–nitrous oxide proton abstraction conditions showed extensive fragmentation while chloride attachment from methane–difluorodichloromethane showed virtually no fragmentation (Fig. 2.19). Huang *et al.* [75] have presented negative ion chemical ionization data for the cyclic peptide valinomycin (MW 1111) obtained using methane as reagent gas.

The use of other supercritical fluids such as ammonia for porphyrin analysis [84], carbon dioxide modified with 5% isopropanol [83,84], and propane modified with 10% isopropanol [84] has also been reported. Nitrous oxide can also be used as a combined SFC mobile phase and reagent gas [74].

2.6 Thin Layer Chromatography–Mass Spectrometry (TLC–MS)

Early attempts to interface TLC with MS involved the introduction of separated components on a direct insertion probe after extracting or scraping the appropriate area of the TLC plate. Subsequent MS analysis mainly used EI or CI although fast atom bombardment (Chapter 5) has also been used successfully [85]. Direct interfacing of intact TLC plates with mass spectrometry, without extensive sample preparation and with the capability of volatilizing thermally labile samples, has mainly come with the use of energetic particle desorption methods, especially fast atom bombardment (FAB) or liquid secondary ion mass spectrometry (LSIMS) and laser desorption mass spectrometry (Chapter 5).

A commercial device has been described [86] which can carry out direct measurements by either automatic or manual scanning of the TLC plate which is simply attached to the tip of a direct insertion FAB probe. With this device, the developed and dried TLC plate is coated evenly with a matrix which is suitable for the analytes before being introduced into the ion source. Care must be taken to ensure that the matrix does not completely evaporate before the end of the analysis. This system has been used for the analysis of a wide range of samples that are normally amenable to ionization by FAB (Fig. 2.20). Lawson and co-workers [87] have recently described an optimized procedure whereby individual bands are cut from a TLC plate together with their aluminium backing and then treated with an appropriate mixture of solvent and matrix prior to LSIMS analysis.

Masuda *et al.* [88] have reported a method for reconcentrating developed spots on TLC plates for FAB analysis after visualization by appropriate detection methods.

Thioglycerol, which has a rapid diffusion rate, is applied around the developed spot on the TLC plate. After a few minutes the thioglycerol concentrates the analyte in the centre of the spot and an appropriate FAB matrix liquid may be dropped on the condensed sample zone. This method has been applied to the analysis of food dyes and can provide satisfactory spectra from 0.1 to 0.5 μg per spot compared with a loading in excess of 5 μg per spot before reconcentration. Further improvements to this method have been published recently [89].

DiDonato and Busch [90] have developed a method for analysis of TLC plates that can limit diffusion of the analyte in the plane of the plate. Movement of the analyte to the surface of the plate without excessive diffusion is accomplished by the addition of a matrix, such as threitol, that can be controllably cycled through the solid/liquid-phase transition. In the solid phase, no diffusion of analyte through the matrix occurs. With liquefaction, analyte is extracted to provide a reservoir of material for ionization. In some experiments, the threitol is maintained in the liquid state by heat transfer from the heated sample support. At higher primary ion current densities, heating by the primary ion beam is sufficient to melt the threitol at the point of analysis, providing a spatially discrete extraction.

Laser desorption mass spectrometry is also well suited to coupling with TLC. Hercules and co-workers [91] used an Nd:YAG laser to ablate and ionize separated compounds directly from polyamide TLC plates. Direct analysis from polyamide plates is practical because these plates only produce low mass background ions($<m/z$ 150). Because laser desorption is a microprobe technique, it has the ability to perform mapping of the TLC plate to obtain spatial information about compound distribution. Typically, the limit of detection for TLC–laser desorption mass spectrometry is at the picogram level or below.

Similar laser microprobe instrumentation has been used by Dewey and Finney [92] who found that the TLC stationary phase has a large effect on spectrum quality. Dewey and Finney conclude that the relatively weak interaction between the sample molecules and a polyamide stationary phase allows the sample to be desorbed using a low laser power so that molecular and structural information are obtained. Overcoming the stronger adsorption on silica requires a higher laser power which causes fragmentation and loss of the molecular ion.

A laser has also been used to desorb neutral analyte molecules from a TLC plate [93]. These neutral molecules are then entrained in a supersonic jet expansion (section 5.3.4) and carried into a time-of-flight mass spectrometer, where resonant two-photon ionization (R2PI) (section 5.2.4.1) is performed with a second laser.

References

1. Alexander, L.R., Maggio, V.L., Gill, J.B., Green V.E., Turner, W.E., Patterson D.G., Jr, Green, B.N., Gray, B.W., Guyan, S.A., Krolik, S.T., and Nicolaysen, L.C. (1989), *Chemosphere*, **19**, 241.

2. Hurst, R.E., Settine, R.L., Fish, F., Roberts, E.C. (1981), *Anal. Chem.*, **53**, 2175.
3. Koller, W.D., and Tressl, G. (1980), *J. High Resoln Chromatogr., Chromatogr. Commun.*, **3**, 359.
4. Blum, W., Ramstein, P., and Eglinton, G. (1990), *J. High Resoln Chromatogr., Chromatogr. Commun.*, **13**, 85.
5. Cramers C.A., and Leclerq, P.A. (1988), *CRC Critical Reviews in Analytical Chemistry*, **20**, 117.
6. Hail M.E., and Yost R.A. (1989), *Anal. Chem.*, **61**, 2402.
7. Trehy, M.L., Yost, R.A., and Dorsey, J.G. (1986), *Anal. Chem.*, **58**, 14.
8. Hail, M.E., and Yost, R.A. (1989), *Anal. Chem.*, **61**, 2410.
9. Bowen D.V., and Pullen, F.S. (1989), *Rapid Commun. Mass Spectrom.*, **3**, 67.
10. Evershed R.P., and Prescott, M.C. (1989), *Biomed. Environ. Mass Spectrom.*, **18**, 503.
11. Henneberg, D., Henrichs, U., and Schomburg, G. (1975), *J. Chromatogr.*, **112**, 343.
12. Andresen, B.D., Ng, K.J., Wu, J., and Bianchine, J.R. (1981), *Biomed. Mass Spectrom.*, **8**, 237.
13. Ryhage, R. (1964), *Anal. Chem.*, **36**, 759.
14. Ryhage, R. (1966), *Arkiv för Kemi*, **26**, 305.
15. Grob, K., Jr, Grob, G., and Grob, K. (1978), *J.Chromatogr.*, **156**, 1.
16. Rose, K. (1983), *J. Chromatogr.*, **259**, 445.
17. Fales, H.M., Comstock, W., and Jones, T.H. (1980), *Anal. Chem.*, **52**, 980.
18. Budde, W.L., and Eichelberger, J. (1980), *Performance Tests for the Evaluation of Computerized Gas Chromatography/Mass Spectrometry Equipment and Laboratories*, Environmental Protection Agency, EPA- 600/4-80-025.
19. Yergey, A.L. (1979), *Biomed. Mass Spectrom.*, **6**, 465.
20. Handbook of Chemistry and Physics (63rd Edition), Weast, R.C. and Astle, M.J. (Eds.) (1982) p. D203, Chemical Rubber Co., Cleveland.
21. Schronk, L.R., Grigsby, R.D., Scheppele, S.E. (1982), *Anal. Chem.*, **54**, 748.
22. Traldi, P., Vettori, U., and Dragoni, F. (1982), *Org. Mass Spectrom.*, **17**, 587.
23. Garozzo, D., and Montaudo, G. (1985), *J. Anal. Appl. Pyrolysis*, **9**, 1.
24. Franklin, W.E. (1979), *Anal. Chem.*, **51**, 992.
25. Carroll, D.I., Dzidic, I., Horning, M.G., Montgomery, F.E., Nowlin, J.G., Stillwell, R.N., Thenot, J-P., and Horning, E.C. (1979), *Anal. Chem.*, **51**, 1858.
26. Adam, N.K. (1956), *Physical Chemistry*, p.458, Oxford University Press, Oxford.
27. Beuhler, R.J., Flanigan, E., Greene, L.J., and Friedman, L. (1972), *Biochem. Biophys. Res.Commun.*, **46**, 1082.
28. Raaymakers, J.G.A.M., and Engel, D.J.C. (1974), *Anal. Chem.*, **46**, 1357.
29. Webb, K.S., Wood, B.J., and Davis, R. (1980), in *Advances in Mass Spectrometry*, Vol 8B (Ed. A. Quayle), p.1921, Heyden & Son, London.
30. Constantin, E., and Hueber, R. (1982), *Org. Mass Spectrom.*, **17**, 460.
31. Ohashi, M., Yamada, S., Kudo, H., and Nakayama, N. (1978), *Biomed. Mass Spectrom.*, **5**, 578.
32. Reed, R.I. (1958), *J. Chem. Soc.*, **1958**, 3432.
33. Dell, A., Williams, D.H., Morris, H.R., Smith, G.A., Feeney, J., and Roberts, G.C.K. (1975), *J. Am. Chem. Soc.*, **97**, 2497.
34. Ohashi, M., Tsujimoto, K., and Yasuda, A. (1976), *Chem.Lett.*, 439.
35. Baldwin, M.A., and McLafferty, F.W. (1973), *Org. Mass Spectrom.*, **7**, 1353.
36. Arpino, P. (1989), *Mass Spectrom. Rev.*, **8**, 35.
37. Arpino, P.J., Krien, P., Vajta S., and Devant, G. (1981), *J.Chromatogr.*, **203**, 117.
38. Dedieu, M., Juin, C., Arpino, P.J., and Guiochon, G. (1982), *Anal. Chem.*, **54**, 2372.
39. Ito, Y., Takeuchi, T., Ishi, D., and Goto, M. (1985), *J. Chromatogr.*, **346**, 161.

40. De Wit, J.S., Tomer, K.B., and Jorgenson, J.W. (1989), *J.Chromatogr.*, **462**, 365.
41. Escoffier, B.H., Parker, C.E., Mester, T.C., De Wit, J.S.M., Corbin, F.T., Jorgensen, J.W., and Tomer, K.B. (1989), *J. Chromatogr.*, **474**, 301.
42. Stenhagen, G., and Alborn, H. (1989), *J. Chromatogr.*, **474**, 285.
43. Arpino, P. (1990), *Mass Spectrom. Rev.*, **9**, 631.
44. Arpino, P. (1992), *Mass Spectrom. Rev.*, **11**, 3.
45. Blakley, C.R., and Vestal, M.L. (1983), *Anal. Chem.*, **55**, 750.
46. Vestal, M.L., and Fergusson, G.J. (1985), *Anal. Chem.*, **57**, 2373.
47. Jones, D.S., and Krolik, S. (1987), *Rapid Commun. Mass Spectrom.*, **1**, 67.
48. Straub, K., and Chan, K. (1990), *Rapid Commun. Mass Spectrom.* **4**, 267.
49. Caprioli, R.M. (Ed.) (1990) *Continuous-flow Fast Atom Bombardment Mass Spectrometry*, John Wiley, Chichester.
50. Mizuno, T., Matsuura, K., Kobayoshi, T., Otsuka, K., and Ishhi, D. (1988), *Analytical Sciences*, **4**, 569.
51. De Wit, J.S.M., Deterding, L.J., Moseley, M.A., Tomer, K.B., and Jorgenson, J.W. (1988), *Rapid Commun. Mass Spectrom.*, **2**, 100.
52. Willoughby, R.C., and Browner, R.F. (1984), *Anal. Chem.*, **56**, 2625.
53. Vestal, M.L. (1990), in *Methods in Enzymology*, Vol. 193: Mass Spectrometry (Ed. J.A. McCloskey), Ch.5, Academic Press, San Diego.
54. Jones, G.G., Pauls, R.E., and Willoughby, R.C. (1991), *Anal. Chem.*, **63**, 460.
55. Sheehan, E.W., Ketkar, S., and Willoughby, R.C. (1991) in *Proceedings of the 39th ASMS Conference on Mass Spectrometry and Allied Topics*, Nashville, Tennessee, p. 1306.
56. Kirk, J.D., and Browner, R.F. (1989), *Biomed. Environ. Mass Spectrom.*, **18**, 355.
57. *The API BOOK* (1991), Perkin-Elmer, Sciex.
58. Fenn, J.B., Mann, M., Meng, C.K., Wong, S.F., and Whitehouse, C.M. (1990), Mass Spectrom. Rev., **9**, 37.
59. Bruins, A.P., Covey, T.R., and Henion, J.D. (1987), *Anal. Chem.*, **59**, 2642.
60. Lane, D.A., Thomson, B.A., Lovett, A.M., and Reid, N.M. (1980), in *Advances in Mass Spectrometry*, Vol. 8B (Ed. A. Quayle), p. 1480, Heyden & Son, London.
61. Sakairi, M., and Kambara, H. (1988) *Anal. Chem.*, **60**, 774.
62. Sakairi, M., and Kambara, H. (1989) *Anal. Chem.*, **61**, 1159.
63. Sakairi, M., and Yergey, A.L. (1991), *Rapid Commun. Mass Spectrom.*, **5**, 354.
64. Covey, T.R., Bruins, A.P., and Henion, J.D. (1988), *Org. Mass Spectrom.*, **23**, 178.
65. Henion, J., and Lee, E. (1990), *Pract. Spectrosc. (Mass Spectrom. Biol. Mater.)*, **8**, 469.
66. Pleasance, S., Anacleto, J.F., Bailey, M.R., and North, D.H. (1992), *J. Am. Soc. Mass Spectrom.*, **3**, 378
67. Smith, R.D., Loo, J.A., Barinaga, C.J., Edmonds, C.G. and Udseth, H.R. (1989), *J. Chromatogr.*, **480**, 211.
68. Smith, R.D., Wright, B.W., and Kalinoski, H.T. (1989), *Prog. in HPLC*, **4**, 111.
69. Markides, K.D., Fields, S.M., and Lee, M.L. (1986), *J. Chromatogr.Sci.*, **24**, 254.
70. Chester, T.L., Innis, D.P., and Owens, G.D. (1985), *Anal. Chem.*, **57**, 2243.
71. Guthrie, E.J., and Schwartz, H.E. (1986), *J. Chromatogr. Sci.*, **24**, 236.
72. Sheeley, D.M., and Reinhold, V.N. (1989), *J. Chromatogr.*, **474**, 83.
73. Blum, W., Grolimund, K., Jordi, P.E., and Ramstein, P. (1988), *J. High Resoln Chromatogr., Chromatogr. Commun.*, **11**, 441.
74. Roach, J.A.G., Sphon, J.A., Easterling, J.A., and Calvey, E.M. (1989), *Biomed. Environ. Mass Spectrom.*, **18**, 64.
75. Huang, E.C., Jackson, B.J., Markides, K.E., and Lee, M.L. (1988), *Anal. Chem.*, **60**, 2715.

76. Pinkston, J.D., Bowling, D.J., and Delaney, T.E. (1989), *J. Chromatogr.*, **474**, 97.
77. Kalinoski, H.T., and Hargiss, L.O. (1989), *J. Chromatogr.*, **474**, 69.
78. Chapman, J.R. (1988), *Rapid Commun. Mass Spectrom.*, **2**, 6.
79. Niessen, W.M.A., Bergers, P.J.M., Tjaden, U.R., and van der Greef, J. (1988), *J. Chromatogr.*, **454**, 243.
80. Lee, E.D., Hsu, S-H., and Henion, J.D. (1988), *Anal. Chem.*, **60**, 1990.
81. Chester, T.L., and Innis, D.P. (1986), *J.High Resoln Chromatogr., Chromatogr. Commun.*, **9**, 178.
82. Pinkston, J.D., Owens, G.D., Burkes, L.J., Delaney, T.E., Millington, D.S., and Maltby, D.A. (1988), *Anal. Chem.*, **60**, 962.
83. Pinkston, J.D., Owens, G.D., and Petit, E.J. (1989), *Anal. Chem.*, **61**, 777.
84. Wright, B.W., and Smith, R.D. (1989), *Org. Geochem.*, **14**, 227.
85. Monaghan, J.J., Morden, W.E., Johnson, T., Wilson, I.D., and Martin, P. (1992), *Rapid Commun. Mass Spectrom.*, **6**, 608.
86. Tamura J., Sakamoto, S., and Kubota, E. (1988), *Analusis*, **16**, 64.
87. Chai, W., Cashmore, G.C., Carruthers, R.A., Stoll, M.S., and Lawson, A.M. (1991), *Biol. Mass Spectrom.*, **20**, 169.
88. Masuda, K., Harada, K.-I., Suzuki, M., Oka, H., Kawamura, N., and Yamada, M. (1989), *Org. Mass Spectrom.*, **24**, 74.
89. Oka, H., Ikai, Y., Kondo, F., Kawamura, N., Hayakawa, J., Masuda, K., Harada, K-I., and Suzuki, M. (1992), *Rapid Commun. Mass Spectrom.*, **6**, 89.
90. DiDonato, G.C., Busch, K.L. (1986), *Anal. Chem.*, **58**, 3231.
91. Kubis, A.J., Somayajula, K.V., Sharkey, A.G., and Hercules, D.M. (1989), *Anal. Chem.*, **61**, 2516.
92. Dewey, C.R., and Finney, R.W. (1990), *Anal. Proc.*, **27**, 125.
93. Li, L., and Lubman, D.M. (1989), *Anal. Chem.*, **61**, 1911.
94. Caprioli, R.M., DaGue, B., Fan, T., and Moore, W.T. (1987), *Biochem. Biophys. Res. Commun.*, **146**, 291.
95. De Wit, J.S.M., Parker, C.E., Tomer, K.B., and Jorgenson, J.W. (1987), *Anal. Chem.*, **59**, 2400.

Chapter 3

Chemical Ionization: Ion–Molecule Reactions

3.1 Introduction

In 1966, Munson and Field [1] described a new technique for generating spectra—chemical ionization (CI) mass spectrometry—in which ions characteristic of the sample are produced by ion–molecule reactions rather than by electron impact ionization. The technique requires a high pressure (classically approximately 1 torr, but cf. section 3.2.4) of a reagent gas in the ion source. Electron impact ionization of the reagent gas produces ions that are either non-reactive or only very slightly reactive with the reagent gas, but which readily react to ionize sample molecules. The sample itself, as in electron impact ionization, is introduced at a much lower pressure so that the partial pressure of sample is rarely greater than 0.01% of the reagent gas pressure and usually much less.

Chemical ionization mass spectrometry has, since its introduction, become a widely used technique. One of its principal attractions has been its ability, as a so-called 'soft' ionization technique, to provide molecular weight information in many cases where electron impact mass spectrometry fails to do so (section 3.3.1). One reason for this difference is the fact that whereas in electron impact ionization the energy transfer distribution may include a small fraction with energies more than 10 eV above the ionization potential, the energy transfer in chemical ionization processes other than charge exchange does not exceed 5 eV, even with the more energetic protonating reagents. Another reason is the greater stability of even-electron protonated ions (e.g. $(M+H)^+$ formed by chemical ionization) compared with radical molecular ions ($M^{+\cdot}$ formed by electron impact ionization).

Much of the additional power of chemical ionization mass spectrometry arises from the fact that the characteristics of a CI mass spectrum are highly dependent on the nature of the reagent gas used to ionize the sample. As a consequence, it is possible to control the structural information observed by varying the nature of the reagent gas used (section 3.3.2).

Ion–molecule reactions are an integral part of many ionization techniques other than chemical ionization itself. For example, atmospheric pressure chemical ionization (APCI), which is of increasing importance as an LC–MS technique (section 2.4.6), relies on ion–molecule reactions which generally involve ions derived from the solvent. Again, in the so-called ‘energy-sudden’ techniques, such as FAB/LSIMS (Chapter 5), which are used for the ionization of involatile analytes, the formation, from the initial impact, of a dense plasma that contains ions and analyte molecules encourages ion–molecule reactions. Gas-phase ion–molecule reactions are also an integral part of other high pressure ionization techniques such as thermospray ionization (Chapter 6).

3.1.1 Reactions in chemical ionization [2]

Reagent ions may be molecular ions, fragment ions or the product of ion–molecule reactions between these ions and reagent gas molecules. For example, the principal reagent ions formed by electron bombardment of methane at 1 torr are CH_5^+ and $C_2H_5^+$ [3]:

$$CH_4 \longrightarrow CH_4^{+\cdot},\ CH_3^+,\ CH_2^{+\cdot} \tag{3.1}$$

$$CH_4^{+\cdot} + CH_4 \longrightarrow CH_5^+ + CH_3^{\cdot} \tag{3.2}$$

$$CH_3^+ + CH_4 \longrightarrow C_2H_5^+ + H_2 \tag{3.3}$$

Other reactions that do not lead to useful ions also occur:

$$CH_2^{+\cdot} + CH_4 \longrightarrow C_2H_3^+ + H_2 + H^{\cdot} \tag{3.4}$$

$$C_2H_3^+ + CH_4 \longrightarrow C_3H_5^+ + H_2 \tag{3.5}$$

The subsequent reactions between positively charged ions and sample molecules can conveniently be grouped into four major categories, viz.:

(a) Proton transfer $M + BH^+ \rightarrow MH^+ + B$
(b) Charge exchange $M + X^{+\cdot} \rightarrow M^{+\cdot} + X$
(c) Electrophilic addition $M + X^+ \rightarrow MX^+$
(d) Anion abstraction $AB + X^+ \rightarrow B^+ + AX$

For example the CH_5^+ and $C_2H_5^+$ ions formed from methane react with and ionize sample molecules principally by proton transfer:

$$M + C_2H_5^+ \longrightarrow MH^+ + C_2H_4 \tag{3.6}$$

Each type of reaction is discussed in more detail in the following sections, with examples chosen to illustrate the use of the more common reagent gases. Published applications of these same reagent gases listed by compound type are documented later in Appendices 3.4 and 3.5.

Most reagent ions are capable of participating in more than one of the listed reactions. For example, NH_4^+ from ammonium may act as a proton donor to give $(M+H)^+$ ions from the sample or may enter into an addition reaction to give $(M+NH_4)^+$ ions. Similarly NO^+ from nitric oxide may act as a charge transfer reagent or anion abstractor or may enter into an addition reaction to give $(M+NO)^+$ ions.

Another important consideration is that the performance of reagent gases can be fundamentally affected by the presence of quite low levels of other components, whether these are present as impurities (section 3.2.3) or added deliberately. The chemical ionization reactions of a number of mixed reagent gas systems are also discussed in the following sections.

3.1.1.1 Proton transfer

The tendency for a reagent ion BH^+ to protonate a particular sample molecule may be assessed from a knowledge of proton affinity values. The proton affinity (PA) of compound B, [PA(B)], is the negative of the change in enthalpy or heat content accompanying reaction:

$$B + H^+ \rightarrow BH^+, \qquad PA(B) = -\Delta H \tag{3.7}$$

Thus, observation of the reaction

$$M + BH^+ \rightarrow MH^+ + B \tag{3.8}$$

implies that $PA(M) > PA(B)$, i.e. the reaction is exothermic.

Table 3.1 Selected proton affinities

Conjugate base (B)	Reagent ion	PA(B) kJ mole^{-1b}
H_2	H_3^+	423
CH_4	CH_5^+	551
C_2H_6	$C_2H_7^+$	601
H_2O	H_3O^+	697
CH_3OH	$CH_3OH_2^+$	761
CH_3CN	CH_3CNH^+	787
$(CH_3)_2C{=}CH_2$[a]	$(CH_3)_3C^{+}$[a]	824
NH_3	NH_4^+	854
CH_3NH_2	$CH_3NH_3^+$	896
$NH_2(CH_2)_2NH_2$	$NH_2(CH_2)_2NH_3^+$	945

[a] Reagent ion from isobutane.
[b] 100 kJ mole^{-1} ≡ 23.9 kcal mole^{-1} ≡ 1.04 eV.

Representative proton affinity values taken from the very extensive data compilations now available [4] are listed in Table 3.1. From the data given in the

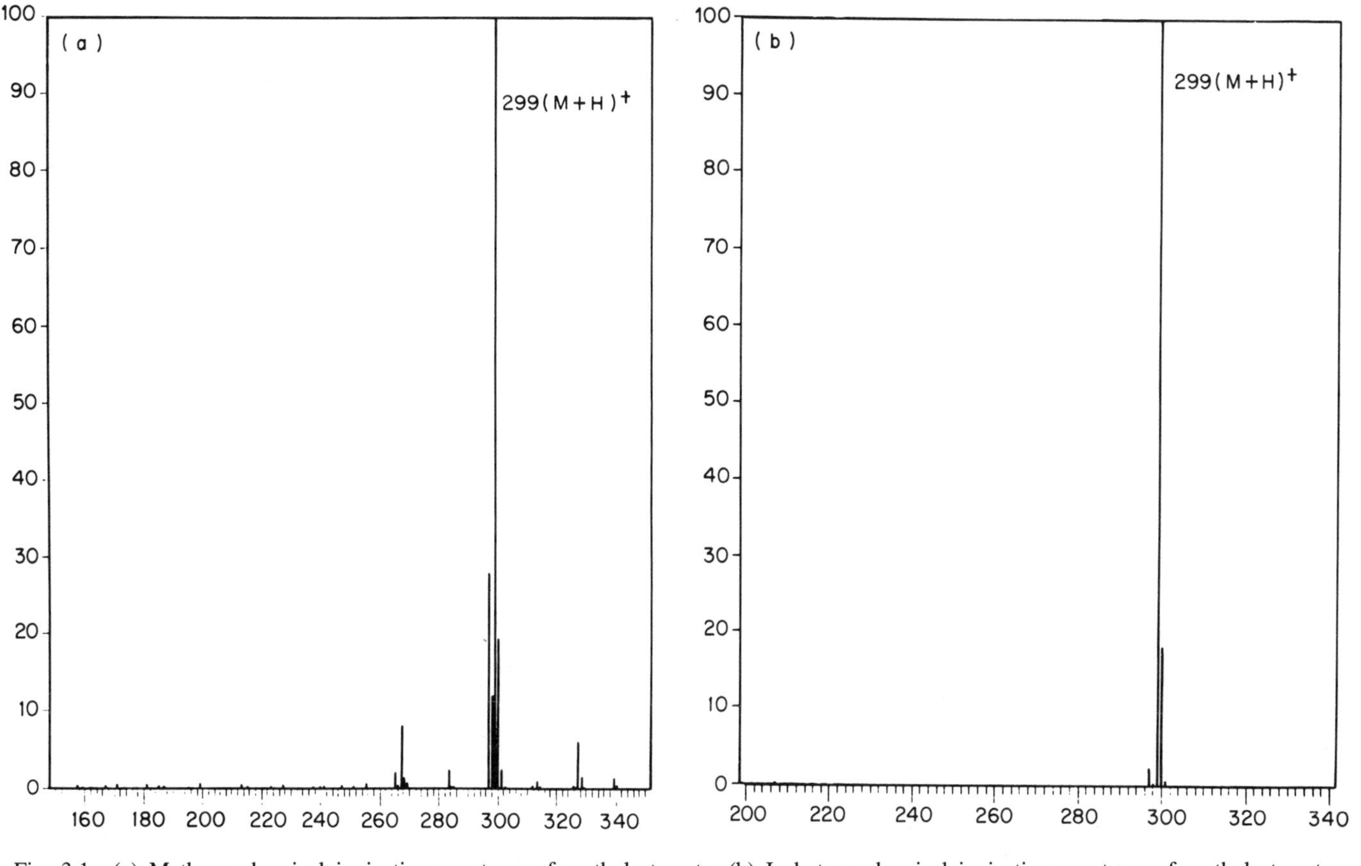

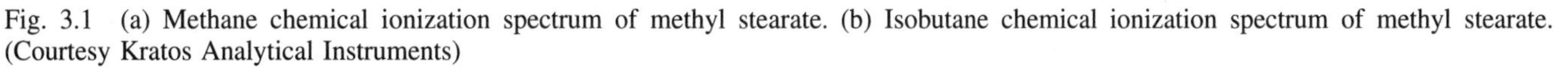

Fig. 3.1 (a) Methane chemical ionization spectrum of methyl stearate. (b) Isobutane chemical ionization spectrum of methyl stearate. (Courtesy Kratos Analytical Instruments)

table we see that while methane reagent gas can be used to effect the protonation of all the other compounds in the table, with the exception of hydrogen, isobutane reagent gas will not protonate water or ethane and is generally unreactive towards normal alkanes. Not only is isobutane more selective in the types of compound that it can protonate, but protonation of a particular compound by isobutane is considerably less exothermic than protonation by methane. Thus, fragmentation may occur with methane while with isobutane the spectrum often consists solely of an $(M+H)^+$ ion (quasi-molecular ion)(Fig. 3.1).

Solvents such as water, methanol and acetonitrile also form ions that are relatively mild protonating agents [5]. The flow from a liquid chromatograph vaporized directly into a conventional ion source (section 2.4.2) or ionized as the vapour with a corona discharge at atmospheric pressure (section 2.4.6) is a common source of these ions which are then able to ionize dissolved analytes by conventional CI processes. H_3O^+ clustered with a small number of water molecules is also the most abundant positive reagent ion formed by a corona discharge in ambient air.

Hydrogen is the most exothermic of all the protonating reagent gases, but the fragmentation that ensues often does so via hydrogenation reactions [6–8]. Although the spectra produced using hydrogen are therefore less predictable, the reactions observed have sometimes proved to be of practical value, e.g. in the analysis of tetrapyrroles [6]. At the other end of the scale, proton transfer from NH_4^+ is restricted to compounds with a proton affinity greater than that of ammonia, such as amines and amides [9]. The utility of ammonia is, however, greatly enhanced by its ability to add NH_4^+ to compounds containing a wide range of functional groups (section 3.1.1.3).

3.1.1.2 Charge exchange

With reagent gases that do not contain available hydrogen a major alternative to proton transfer is ionization to produce molecular ions by charge exchange (electron transfer):

$$M + X^{+\cdot} \rightarrow M^{+\cdot} + X \qquad (3.9)$$

The rate of decomposition of these molecular ions depends on their internal energy E_{int} calculated using

$$E_{int} = RE(X^{+\cdot}) - IP(M) \qquad (3.10)$$

where $RE(X^{+\cdot})$ is the recombination energy (RE) of the reagent ion and IP(M) is the ionization potential (IP) of the sample molecule.

Relatively intense molecular ions are expected if $RE(X^{+\cdot})$ is slightly greater than IP(M) and extensive fragmentation, similar to electron impact spectra, occurs when $RE(X^{+\cdot})$ is much greater than IP(M). Typical recombination energies of major ions are listed in Table 3.2 [10]. Ionization potentials for the majority of organic molecules are in the range 7–11 eV.

Table 3.2 Selected recombination energies

Gas	Ion ($X^{+\cdot}$)	RE($X^{+\cdot}$) (eV)
He	$He^{+\cdot}$	24.6
Ar	$Ar^{+\cdot}$	15.8
N_2	$N_2^{+\cdot}$	15.3
CO	$CO^{+\cdot}$	14.0
CO_2	$CO_2^{+\cdot}$	13.8
CS_2	$CS_2^{+\cdot}$	~10
NO	$NO^{+\cdot}$	9.3
C_6H_6	$C_6H_6^{+\cdot}$	9.2

It is apparent that charge exchange spectra will be produced simply by introducing sample diluted with helium carrier gas into a tightly sealed CI source from a GC inlet [11]. Charge exchange spectra produced by helium show extensive fragmentation although molecular ions in these spectra are relatively more intense than would be expected from a consideration of the data in Table 3.2 [12].

Other common gases, such as nitrogen and argon, also cause extensive fragmentation so that they are principally used as components of mixed reagent gas systems such as argon–water [13] and nitrogen–nitric oxide [14]. Electron bombardment of argon containing a small percentage of water affords abundant ions at m/z 40 (Ar^+) and m/z 19 (H_3O^+). Thus, while Ar^+ causes charge exchange followed by fragmentation, H_3O^+ functions as a proton transfer reagent giving abundant quasi-molecular $(M+H)^+$ ions that undergo little fragmentation (Fig. 3.2). For certain classes of compounds, N_2–NO charge exchange spectra also show enhanced molecular ions as well as retention of major fragment ions rather than the extensive fragmentation found in nitrogen charge exchange or electron impact spectra [14].

A mixture of nitrogen and carbon disulphide is a useful low energy charge exchange system giving CS_2^+ as the major reactant ion. Thus, a mixture of vinyl methyl ether, carbon disulphide and nitrogen (5:20:75) gives a high yield of vinyl methyl ether cations which is then a suitable reagent for the location of double bonds in olefins (section 3.1.1.3). CS_2^+ charge exchange spectra can be used to distinguish monoepoxide isomers derived from polyunsaturated fatty acids through characteristic ions that are very much weaker in conventional EI spectra [15].

Charge exchange reagents may also be used to selectively ionize particular classes of compounds. For example, $C_6H_6^+$ ions formed from benzene will ionize esters of unsaturated fatty acids but not esters of saturated fatty acids, since their ionization potentials are greater than 9.2 eV [16]. Similarly, chlorobenzene can

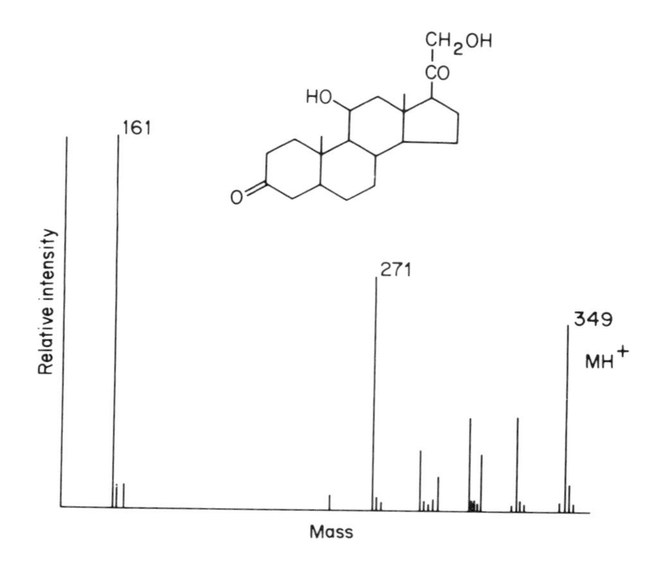

Fig. 3.2 Argon–water chemical ionization spectrum of 5α-dihydrocorticosterone. (Reprinted with permission from D.F. Hunt and J.F. Ryan III, *Anal. Chem.*, **44**, 1306 [13]. Copyright (1972) American Chemical Society)

be used to selectively ionize aromatic hydrocarbons in the presence of aliphatic hydrocarbons in petroleum products [17].

3.1.1.3 Electrophilic addition

The major reaction of sample molecules of higher proton affinity in reagent gases such as methane, isobutane, and ammonia is protonation (section 3.1.1.1). Cations produced by these gases can, however, also be involved in other reactions such as electrophilic addition with sample molecules. For example, methane produces $C_2H_5^+$ and $C_3H_5^+$ ions (equations 3.3 and 3.5) which frequently provide $(M+C_2H_5)^+$ and $(M+C_3H_5)^+$ ions of low intensity in the spectrum.

Probably the most frequently encountered example of electrophilic addition is the formation of intense $(M+NH_4)^+$ ions from a wide range of compounds. This reaction has been most extensively documented using ammonia [18] as a reagent gas, but the use of ammonium acetate as an additive in LC–MS techniques such as thermospray is an equally common source of the ammonium ion in positive ion operation.

If the sample molecule has a sufficient proton affinity then it is protonated by the ammonium ion:

$$M + NH_4 \rightarrow [M- - - -H- - - -NH_3]^{+*} \tag{3.11}$$

$$[M- - - -H- - - -NH_3]^{+*} \rightarrow [M + H]^+ + NH_3 \quad (3.12)$$

If, however, the molecule is somewhat less basic (proton affinity $< 854\,kJ\,mole^{-1}$) then a stable addition complex is formed [9] (see also equation 3.11):

$$[M- - - -H- - - -NH_3]^{+*} \rightarrow [M + NH_4]^+ \quad (3.13)$$

The presence of a nitrogen or oxygen atom, or in some cases merely a centre of unsaturation [19], can provide electrons to form this type of complex with NH_4^+. Rudewicz and Munson [20] have shown that the use of 1% NH_3 in CH_4 in these circumstances provides a greater sensitivity with some oxygenated analytes than the use of ammonia. Since these addition complexes are formed under very mild conditions, this technique permits the observation of quasi-molecular ions from relatively labile materials (e.g. see Fig. 3.8). The formation of stable complexes is, as may be expected, a relatively temperature-sensitive process (section 3.2.3).

Isobutane reagent gas will protonate a low level of ethylene diamine to produce reagent ions $NH_2CH_2CH_2NH_3^+$ which will then form addition ions $(M{+}NH_2CH_2CH_2NH_3)^+$ with a sample [21]. More basic reagent gases such as ethylene diamine or dimethylamine [22] are sometimes more suitable than ammonia for the formation of stable addition complexes. Another example of electrophilic addition is provided by the formation of $(M{+}NO)^+$ ions using nitric oxide [14,23].

The further reactions of the products of an addition reaction may sometimes lead to characteristic ions of analytical value. For example, the vinyl methyl ether cation, generated by charge exchange from CS_2^+, will add across a double bond to form an unstable ion radical complex which then decomposes giving fragment ions (a and b) from which the double bond position may be determined [24,25]:

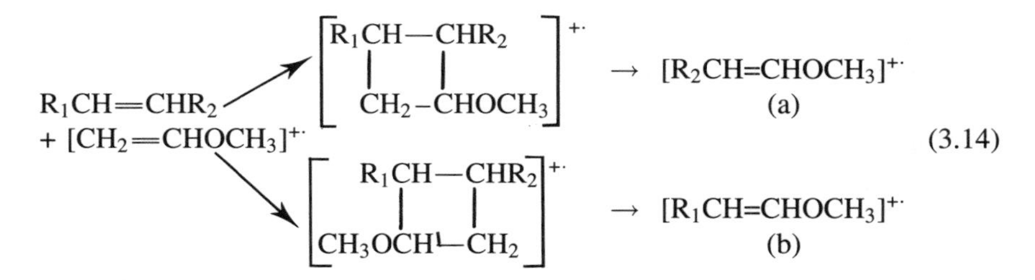

Double bond addition products formed using NO^+ from nitric oxide also fragment in a manner which gives information on the location of the double bond [26,27].

3.1.1.4 Anion abstraction

A further reaction available to proton transfer reagents with samples of low proton affinity or to charge exchange reagents with samples for which charge transfer is an endothermic process is anion abstraction [12,28]. For example, aliphatic hydrocarbons form $(M - H)^+$ ions by a process of hydride abstraction under methane chemical ionization conditions:

$$M + C_2H_5^+ \rightarrow (M - H)^+ + C_2H_6 \quad (3.15)$$

Nitric oxide reagent gas reacts with tertiary alcohols to form a single stable ion by hydroxyl abstraction [29]:

$$R_1R_2R_3COH + NO^+ \rightarrow R_1R_2R_3C^+ + HONO \quad (3.16)$$

The use of this reagent gas system, or a more practical mixture of 2–10% NO in H_2 [30], provides a method for the direct analysis of these difficult compounds (cf. section 4.4).

3.2 CI Instrumentation

Chemical ionization experiments with scanning instruments make use of an ion source which is capable of operation at pressures up to approximately 1 torr. Chemical ionization with ion trapping instruments is carried out under somewhat different conditions which are considered in section 3.2.4. Source components for scanning instruments are manufactured so that the construction is relatively gas-tight and sample entry ports are tightly sealed to the corresponding inlet system. Thus the conductance of a CI source, which is less than that of an EI source, is almost entirely determined by the dimensions of the ion exit and electron entry slits. The ion exit slit must be large enough to allow efficient ion extraction but small enough to maintain an adequate differential between the ion source and source housing pressures. A relatively high pumping speed is also required to maintain a suitable source housing pressure in CI work.

Many CI sources are actually designed as combined EI/CI sources; that is to say in the absence of a reagent gas they function as EI sources. The conductance of a combined source is low, like that of a dedicated CI source (approximately 0.2 litre s^{-1}) to allow it to function in the CI mode. Its EI sensitivity is lower, although perhaps only a factor of two, than that of a dedicated EI source operating at the same electron current. The combined source may be switched between EI and CI operation and vice versa in times of 1 second or less, limited only by the need to pressurize or depressurize the ion source. An alternative configuration, which does not permit such rapid switching, employs a mechanical linkage external to the vacuum system to change between EI and CI modes.

Electron energies considerably higher than the 70 eV normally employed in electron impact work are used to provide better penetration of the high pressure gas within the ionization chamber. The CI sensitivity optimizes at an ion repeller voltage which is close or equal to the ion source potential.

Prolonged operation in the CI mode with a continuous input of reagent gas can lead to the build-up of insulating layers in the source, evidenced by altered tuning conditions and a reduced sensitivity. The materials that form these layers can be substantially removed by regular baking of the source, but are otherwise polymerized into more intractable deposits by the action of the electron beam. In this case, cleaning of the source is indicated. The use of ammonia rather than methane or isobutane avoids the formation of these insulating layers.

CI reactions may sometimes be initiated under conditions that do not permit the use of a heated filament as the primary ionizing source. For example, the use of a discharge [27,31,32] in conventional chemical ionization is indicated when strongly oxidizing materials (e.g. O_2, NO), undiluted by an excess of inert gas, are to be used as reagent gases. Again, operation at atmospheric pressure (section 2.4.6) uses a corona discharge as the source of electrons. A glow discharge, established in a region of reduced pressure, can also be used to initiate ion–molecule reactions which permit the analysis of trace organics in ambient air [33]. A similar low pressure discharge has been used very extensively as an important accessory for thermospray ion sources. It is also possible to establish a glow discharge at atmospheric pressure when helium is used as the support gas [34].

3.2.1 Introduction of reagent gas into the ion source

Most routine chemical ionization experiments with scanning instruments use source pressures of from 0.1 to 0.5 torr. With a CI source conductance of 0.2 litre s^{-1}, flow rates of 0.02–0.1 litre torr s^{-1} (1.6–8 cm^3 min^{-1}) are required to maintain these pressures. Some CI sources are of more open construction and will require higher flow rates to maintain the same pressures. While higher flow rates may be controlled simply by interposing a fixed restriction between the reagent gas cylinder and the ion source, such a system is not generally applicable.

Figure 3.3 shows a more generally useful reagent gas introduction system with a low pressure gas regulator between the cylinder and restriction. The use of solenoid valves in such a system allows automatic switching between EI and CI operation. Special considerations apply to the design of reagent gas and sample inlet systems for CI work on sector instruments in order to prevent electrical breakdown. These

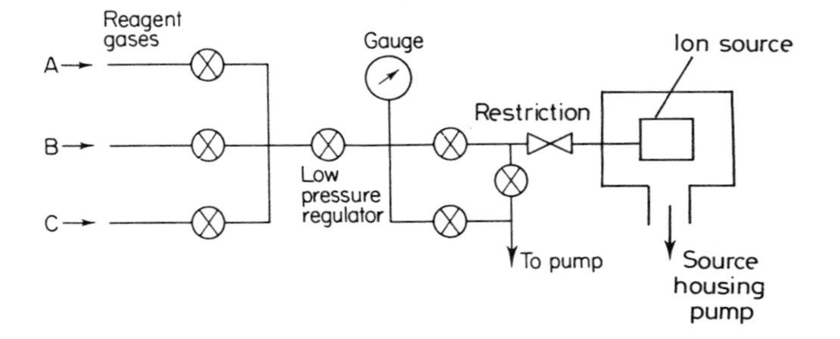

Fig. 3.3 Chemical ionization reagent gas introduction system

points are discussed in section 3.2.2. Details of a reagent gas introduction system with automatic feedback control to maintain a pre-set pressure have been published [35].

Other inlet systems may be used to introduce reagent gases. For example, methane may be used both as GC carrier gas and reagent gas. A smaller flow of methane or other reagent gas may be introduced by mixing it coaxially with the flow from a GC column [36]. Direct insertion probes have been designed to also admit a flow of reagent gas to sweep the sample evaporating from the probe into the ion source. Sample should, however, be transferred quantitatively to a CI source at high pressure, even without this gas flow through the probe, provided that the probe maintains a gas-tight seal to the source.

The use of less volatile materials as components of reagent gas systems presents special problems. For example, the introduction of water as part of an argon–water mixture may be accomplished by allowing the water to evaporate at room temperature from an ampoule via a fine metering valve into the argon stream at reduced pressure [13]. Ethylene diamine, which is considerably less volatile, may be introduced by measuring a known quantity into a heated reservoir system connected to a source by a fixed restriction. Formic acid has been used as a reagent gas by allowing the helium GC carrier gas to be saturated with this material [37].

The admixture of solid samples with an ammonium salt can be used to promote localized chemical ionization reactions on heating, even in a conventional EI source (cf. sections 4.3.1 and 5.3.1.2). Thus, both the potassium salt of penicillin G and lithocholic acid sodium salt surprisingly provide molecular ions when evaporated together with ammonium chloride from a direct insertion probe under EI conditions [38].

3.2.2 Electrical breakdown in CI operation

At low pressures, which include the usual range of CI pressures, a self-sustaining discharge can be produced by a potential difference maintained between appropriately spaced electrodes in the gas. The actual voltage required for the initiation of this discharge is a function of the product of the gas pressure and the electrode spacing, as can be seen from Fig. 3.4. While of no consequence with quadrupole instruments, whose ion sources operate at a very low voltage, the electrical conductivity of reagent gases is a problem with magnetic sector instruments. This problem is apparent in the design of inlet systems suitable for CI work since inlet systems are generally at earth potential while the ion source is at a high voltage.

Some of the problem areas with magnetic sector instruments may be illustrated with reference to Fig.3.3:

(a) The gas in the line from the ion source to the restriction is at a pressure of 0.1–1 torr and is therefore an excellent electrical conductor. The construction of this line should therefore everywhere offer a high resistance to earth.

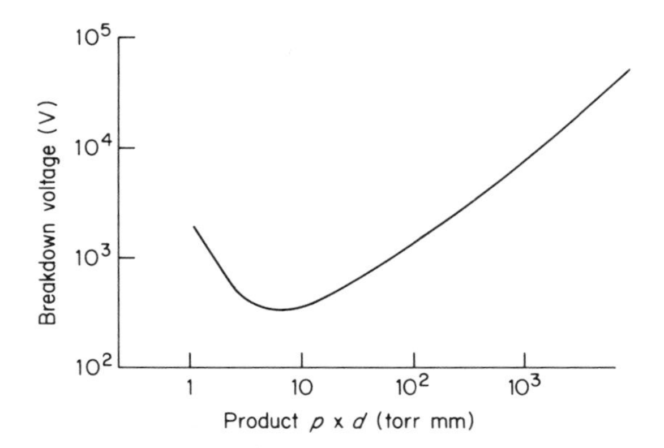

Fig. 3.4 Breakdown voltage as a function of gas pressure (p) and electrode spacing (d)

(b) Breakdown can still occur through the gas to the nearest earth point unless this point is sufficiently far removed from the ion source or an intervening restriction is present. The quenching of an electrical discharge at higher pressures ($>$ approximately 50 torr) created by the presence of some kind of capillary restriction is used in most CI inlet systems for magnetic sector instruments.

(c) Breakdown should not occur in the source housing since the pressures are so much lower unless the pressure is locally high or some sharp point produces a high field gradient.

3.2.3 Effect of operating conditions on CI spectra

The use of milder reagent gases such as isobutane or ammonia results in the formation of protonated molecular ions or addition products with very little excess energy. Under these conditions the input of additional thermal energy during sample vaporization has a much more noticeable effect on fragmentation than under methane chemical ionization or electron ionization conditions (e.g. Table 3.3) [39–42].

As a general rule, CI spectra should be recorded using the minimum source temperature compatible with maintaining the sample in the vapour phase (cf. section 2.3.2) although the use of a higher temperature can afford additional structural information [43] or prevent undesirable addition reactions [44] in some cases. In the special case of a method of sample introduction such as DCI (section 5.2.1), where a particularly labile sample is rapidly volatilized from a heated filament within the CI reagent gas ion plasma, operation without source heating can be employed. It should, however, be recognized that the use of a lower source temperature can eventually lead to problems of source contamination and memory from previous samples.

Table 3.3 Effect of temperature on CI spectra of geraniol. (Reproduced by permission of The Institute of Petroleum (ref. 19))

Ion source temperature (°C)	CH_4 CI ion mass (intensity)	NH_3 CI ion mass (intensity)
170	M − 1 (9) M − 17 (100) + other fragment ions	M + 18 (2) M (29) M − 17 (100)
135	M − 1 (7) M − 17 (100) + other fragment ions	M + 18 (7) M (56) M − 17 (100)
110	M − 1 (9) M − 17 (100) + other fragment ions	M = 18 (28) M (100) M − 17 (88)

An increase in reagent gas pressure has little effect on the quality of CI spectra generated by a more exothermic process such as proton transfer. Thus, there is little change in sample fragmentation at pressures above about 0.5 torr with pure methane since reactions involved in the production of reagent ions are essentially complete at that pressure (Fig. 3.5). With a less exothermic process, such as the formation of addition complexes, fragmentation is reduced at higher pressures through collisional stabilization of these complexes.

Experiments with methyl stearate confirm that fragmentation using pure methane has fallen to a minimum at a source pressure of 0.5 torr and that fragmentation using

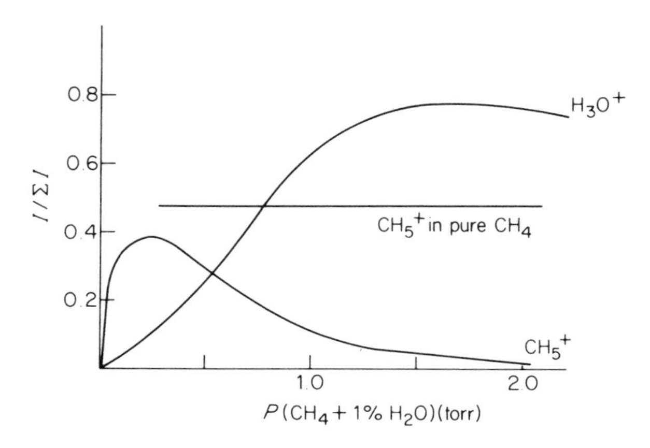

Fig. 3.5 Relative concentrations of CH_5^+ and H_3O^+ ions versus pressure of a mixture of CH_4 + 1% H_2O. (Reprinted with permission from M.S.B. Munson and F.H. Field, *J. Am. Chem. Soc.*, **87**, 4242 [48]. Copyright (1965) American Chemical Society

isobutane, which is already much less extensive, reaches a minimum at a source pressure of 0.25 torr. Measurement of absolute sensitivity using the $(M+H)^+$ ion at m/z 299, however, shows that a maximum sensitivity is achieved at pressures of approximately 0.2 torr for methane (a similar optimum is recorded for ammonia [45]) and 0.1 torr for isobutane. This disparity between the pressure for optimum sensitivity and that for minimum fragmentation is quite common and should be recognised, particularly when experiments requiring the best detection limits in CI are being set up. Small but consistent differences in sensitivity are also observed by using different reagent gases.

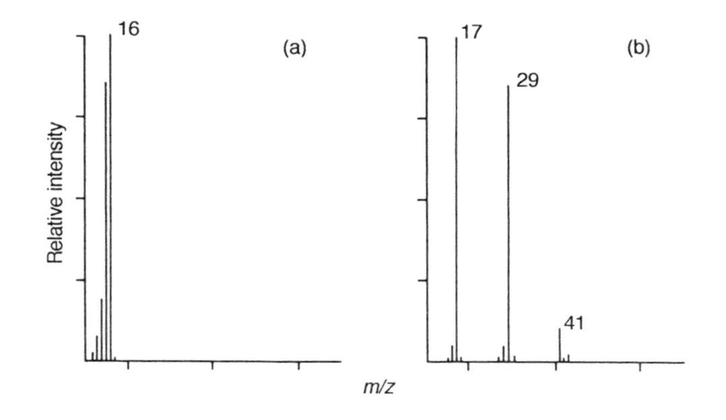

Fig. 3.6 Mass spectrum of methane at different pressures: (a) approximately 10^{-5} torr; (b) approximately 1 torr

The spectrum of a reagent gas such as methane undergoes a complete transformation in passing from low to high pressure conditions (Fig. 3.6). As an indication of the attainment of CI conditions using methane we expect to see the appearance of approximately equally abundant major ions at m/z 17 (CH_5^+) and m/z 29 ($C_2H_5^+$) and the disappearance of typical EI ions such as those at m/z 15 (CH_3^+) and m/z 16 (CH_4^+). A relative intensity at m/z 17 compared with m/z 16 in excess of 10 to 1 would be regarded as satisfactory for CI operation. In the case of isobutane reagent gas, reactions producing the reagent gas ion at m/z 57 proceed as shown:

$$(CH_3)_3CH^{+\cdot} \rightarrow (CH_3)_2CH^+ + CH_3^{\cdot} \quad (3.17)$$

$$(CH_3)_2CH^+ + (CH_3)_3CH \rightarrow (CH_3)_2CH_2 + (CH_3)_3C^+ \quad (3.18)$$

Harrison and Cotter [46] have reported the following test for the establishment of CI conditions using xylene as a test compound. With methane, both the intensity ratios m/z 107:m/z 106 and m/z 106:m/z 91 should be 10:1. With isobutane, *m/z* 107:*m/z* 106 should be 3:1 while *m/z* 106:*m/z* 91 should be 10:1. With ammonia, reagent gas ions are produced as follows

$$NH_3^{+\cdot} + NH_3 \rightarrow NH_4^+ + NH_2^{\cdot} \quad (3.19)$$

$$NH_4^+ + NH_3 \rightarrow (NH_3)_2H^+ \quad (3.20)$$

Relative intensities for NH_4^+ and $(NH_3)_2H^+$ of 12:1 have been reported for a typical source pressure of 0.2 torr[18]. Reagent gas spectra for H_2O and CH_3OH (both at 0.2 torr) have also been published [44].

The presence of even a low level of impurity in a reagent gas introduces the possibility of changes in the reagent gas spectrum as the overall pressure is altered. This is shown in a dramatic manner in Fig. 3.5 where the reagent ions and their relative abundances in pure methane are compared with those present in methane containing 1% water. Water, because of its high proton affinity, depletes the CH_5^+ ions

$$CH_5^+ + H_2O \rightarrow CH_4 + H_3O^+ \quad (3.21)$$

so that wet methane behaves more as a source of the gentle protonating agent H_3O^+ than of the strong proton donor CH_5^+, particularly at higher pressures [47,48]. The effect of impurities is greatly lessened when the main reagent gas is a weaker proton donor, e.g. isobutane or ammonia, that will not protonate common impurity molecules. An excessive pressure of the sample itself will have a marked effect on the CI spectrum under most conditions [12].

A sensitivity test may be carried out under CI conditions with methyl stearate using the basic methods described in sections 1.6.1 and 1.6.2 but monitoring *m/z* 299 (*m/z* 316 with ammonia). Alternatively, benzophenone (*m/z* 183) may be used as a test sample for chromatographic introduction. Bear in mind the effect on sensitivity of reagent gas composition and pressure and of temperature.

3.2.4 Chemical ionization with trapping instruments

With trapping instruments, much lower reagent gas pressures can be used for CI studies since the reaction time of reagent gas ions and analyte molecules can be made correspondingly longer. Typical operating pressures are 10^{-5} torr with ion trap instruments and 10^{-6} torr with FTMS instruments.

In practice, CI with the ion trap is effected by the use of a preliminary ionization period during which the reagent ion population is established. This is then followed by a reaction period of some tens of milliseconds, at a slightly higher r.f. potential, during which analyte ions are formed and stored. The r.f. potential is then ramped to effect a mass scan as usual. With an FTMS instrument, reagent gas may be introduced as a pulse so that it has been substantially pumped away before ion analysis takes place [49]. Alternatively, reagent ions may be generated in an external ion source.

There are a number of other differences between chemical ionization with trapping and scanning instruments. For example, trapping instruments offer the possibility of more selective chemical ionization reactions through the preliminary isolation of a single species of reagent ion [50]. On the other hand, since thermal electrons and heavy negative ions cannot be held in an ion trap at the same time, electron capture reactions (section 4.2.1) are generally not observed [51]. In addition, increased fragmentation has been observed in ion trap chemical ionization spectra so that, in some cases, the spectra recorded are significantly different from those obtained with scanning instruments [52,53].

3.3 Analytical Applications of CI

Two of the principal applications of chemical ionization, i.e. the determination of molecular weight information and the determination of structural and stereochemical information not normally available through electron impact ionization, are dealt with in sections 3.3.1 and 3.3.2 respectively. Chemical ionization is also widely used as a soft ionization technique preceding collision-induced decomposition in mixture analysis. This aspect is introduced briefly in section 3.3.3 and discussed at greater length in Chapter 7. Chemical ionization is also widely used in quantitative analysis. This topic is covered in section 8.3.3. References to earlier published CI data classified by reagent gas and compound type are listed in the appendices at the end of the chapter.

3.3.1 Molecular weight determination using CI

Many compounds fail to give a useful molecular ion under electron impact conditions whereas CI affords an intense quasi-molecular ion (e.g. $(M+H)^+$, $(M+NH_4)^+$) with the same sample (Fig. 3.7). Molecular weight information is an ideal complement to characteristic fragmentation patterns observed under electron impact conditions in the elucidation of the structure of unknowns. Both sets of data may conveniently be obtained, even for components eluted from high resolution capillary columns by rapid switching from EI to CI and vice versa between repetitive scans using a combined source of fixed geometry. On the other hand, many more complex and labile molecules will provide both quasi-molecular and fragment ion information from the CI spectrum alone, even under the mildest ionization conditions (Fig. 3.8). The use of ammonia as a reagent gas has been particularly successful as a method of providing molecular weight information and a large number of examples are described in ref. 18.

Reagent gases for CI may be specifically designed to give both molecular weight and structural information from a wide range of compounds. Mixed reagent gas systems that give this type of information (Fig. 3.2) are the argon–water and

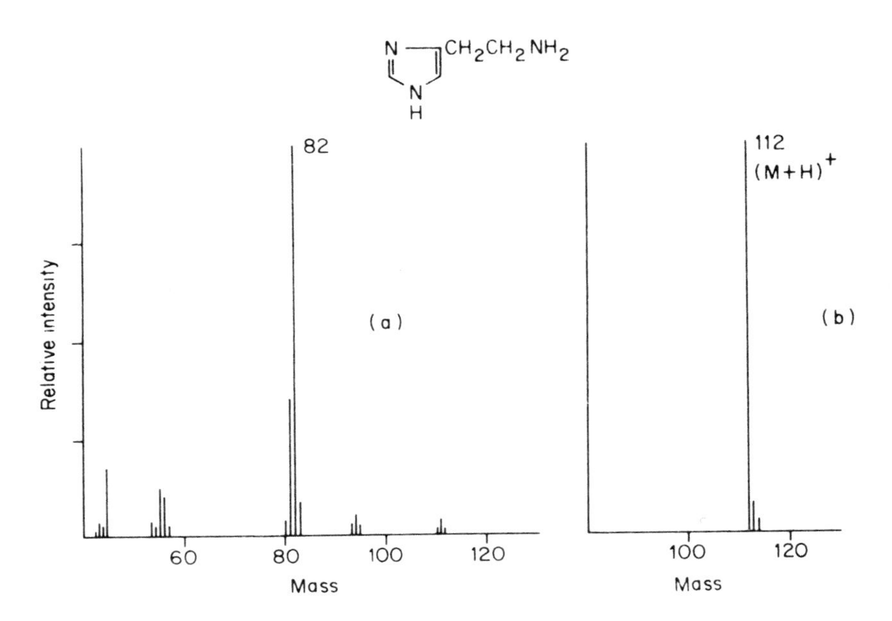

Fig. 3.7 Mass spectra of histamine. (a) Electron impact; (b) isobutane chemical ionization

nitrogen–nitric oxide mixtures discussed in section 3.1.1.2. Another interesting application in this area is the analysis of branched-chain fatty acids using ammonia to give molecular weight information followed by helium to give chain branching information [54].

In other cases, the use of a single standard reagent gas has been found to give all the information required. Thus, permethoxylated derivatives of polyunsaturated fatty acids analysed by combined GC–CI mass spectrometry using isobutane reagent gas give spectra containing both molecular ions and fragment ions which allow the location of the original double bond positions (Fig. 3.9) [55]. Standard CI operation has proved similarly advantageous compared with EI operation in other situations, such as the analysis of oil-shale hydrocarbons [56] and the sequencing of small peptides [57].

3.3.2 Structural and stereochemical information from CI

Chemical ionization may be used to provide structural information that is unavailable from a representative EI spectrum. For example, since isomerization between molecular ions prior to fragmentation is less likely under mild ionization conditions, the further fragmentation of isomeric and stereoisomeric compounds following chemical ionization is strongly influenced by molecular structure [12,18,56–68]. Thus, benzyl chloride and *p*-chlorotoluene are readily distinguished under methane CI conditions (Fig. 3.10), whereas the corresponding EI spectra show only minor intensity differences [66].

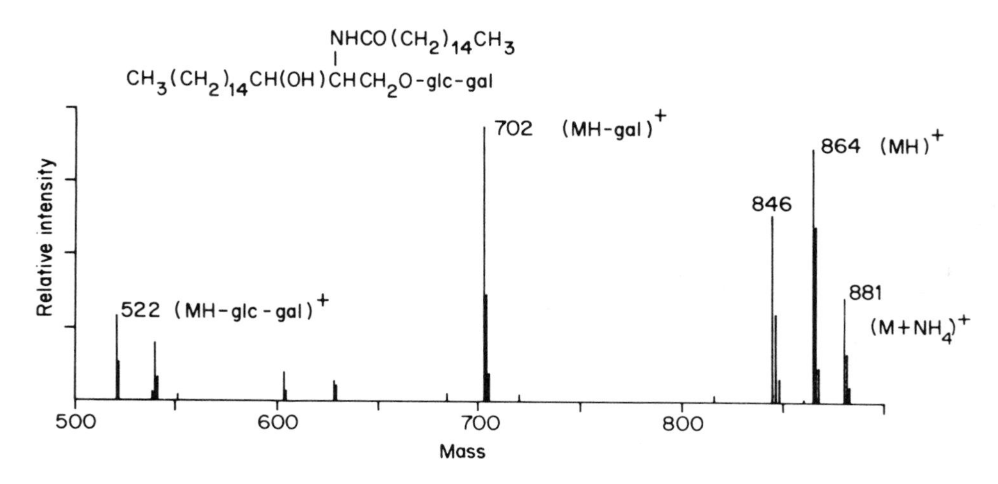

Fig. 3.8 Part DCI spectrum of *N*-palmitoyl dihydrolactocerebroside. Ammonia reagent gas. Sample evaporation from a polyimide coated wire (section 5.3.1.1). (Reprinted with permission from V.N. Reinhold and S.A. Carr, *Anal. Chem.*, **54**, 449 [80]. Copyright (1982) American Chemical Society)

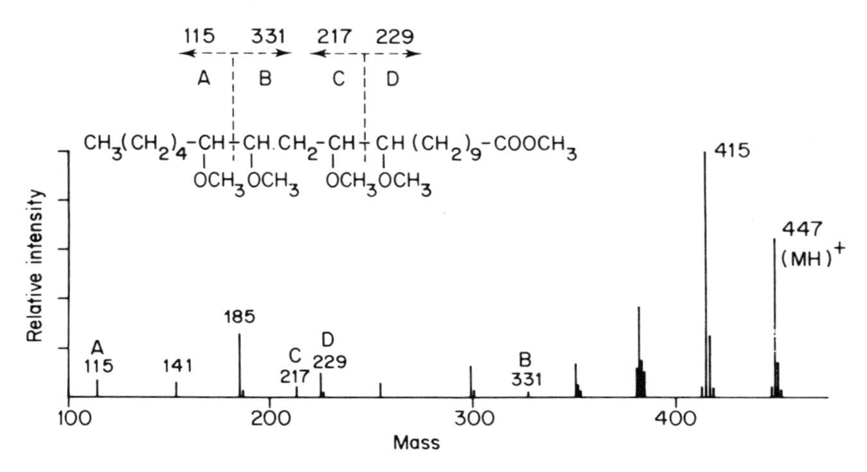

Fig. 3.9 Isobutane CI spectrum of methyl 11,12,14,15-tetramethoxy-eicosonate. (Reprinted with permission from M. Suzuki, T. Ariga, M. Sekine, E. Araki, and T. Miyatake, *Anal. Chem.*, **53**, 985 [55]. Copyright (1981) American Chemical Society)

Longevialle *et al.* [58,59] have reported that the stereoisomeric β-amino alcohols *1* and *2* show different degrees of dehydration when ionized under identical conditions (isobutane 0.5 torr, temperature $150\pm10°C$). The proximity of the amino group in isomer *1* means that proton transfer from the hydroxyl group to the amino group is highly competitive with dehydration so that the $(M+H-H_2O)^+$ ion has a relative intensity of only 0.5% in this case. In isomer *2* the two groups are further

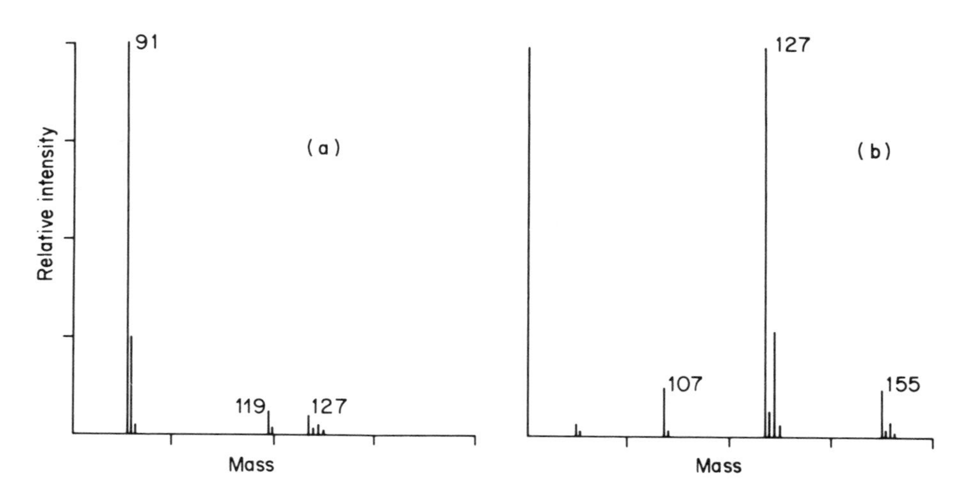

Fig. 3.10 Methane CI spectra of (a) benzyl chloride and (b) *p*-chlorotoluene (both molecular weight 126). (Reproduced with permission from ref. 81)

apart so that the same ion now has a relative intensity of 4.5%.

Munson [12] has observed that the isobutane CI spectra of endo-norborneol *3* and exo-norborneol *4* differ significantly in the relative abundances of $(M-H)^+$ ions. These differences result from the greater accessibility of the α-hydrogen in endo-norborneol but are emphasized by the use of bulkier, low energy reactant ions so that the methane CI spectra of the two isomers are almost identical.

CI spectra determined with the appropriate reagent gas may also be used to carry out functional group analysis. For example, the major ion in the nitric oxide CI spectra of ketones, esters, and carboxylic acids is $(M+NO)^+$, whereas the most abundant ion in the nitric oxide CI spectra of aldehydes and ethers is an $(M - H)^+$ ion formed by hydride abstraction [23].

Deuterated CI reagent gases such as D_2O, ND_3, CH_3OD, and C_2H_5OD may be used for the determination of active hydrogens in a number of specific organic structural environments through isotope exchange reactions [69–71]. Such determinations can also form the basis of a method of functional group analysis. For example, deuterated ammonia, ND_3, reacts with simple amines to produce $(M+D)^+$ and $(M+ND_4)^+$ ions in which other exchangeable hydrogens attached to the nitrogen have been replaced by deuterium. Thus, primary amines produce (M+4)

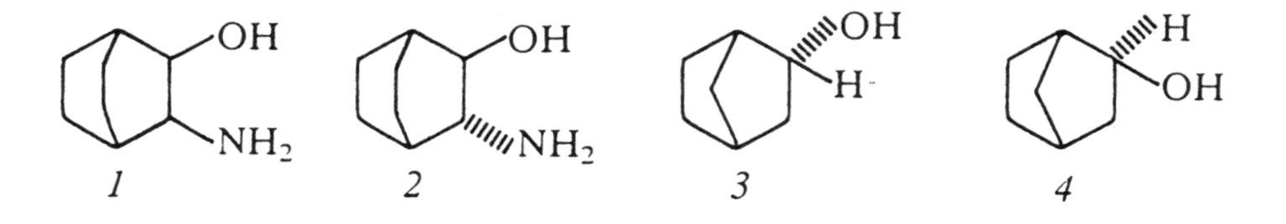

and (M+24) ions. Secondary amines produce (M+3) and (M+23) ions and tertiary amines produce (M+2) and (M+22) ions [72]. Similarly, spectra recorded with deutero-methanol (CH_3OD) have been used, in conjunction with EI spectra, to assign the compound class to heteroatom-containing aromatics by observation of deuterium exchange reactions [73].

Water chemical ionization of simple aliphatic and aromatic carbonyl compounds provides spectra that exhibit abundant $(M+H)^+$ ions as well as structurally informative fragment ions [44]. The use of an isobutane/vinyl methyl ether mixed reagent gas allows cyclopropenoid and mono-unsaturated fatty acid methyl esters to be distinguished [74]. The mono-unsaturated esters form vinyl methyl ether adduct ions whereas the cyclopropenoid esters do not (Fig. 3.11). Dimethyl ether has been proposed as another reagent gas for functional group analysis [75].

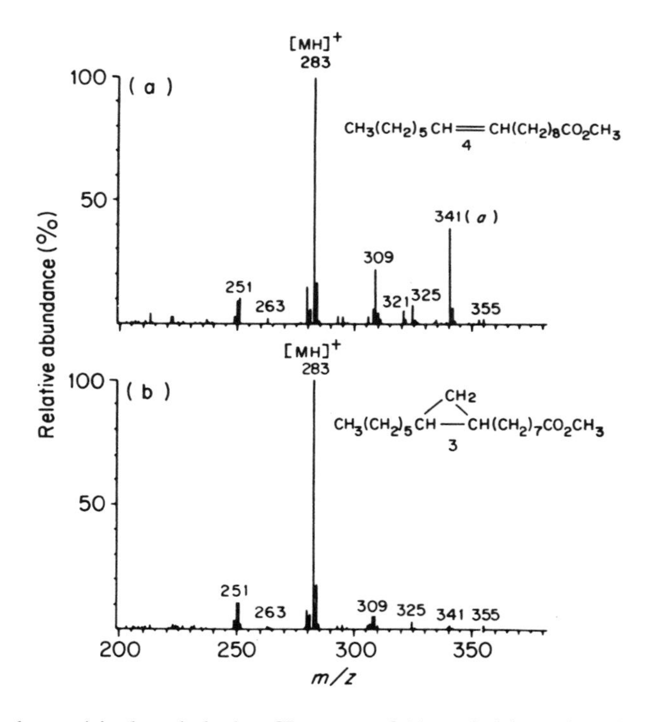

Fig. 3.11 Isobutane/vinyl methyl ether CI spectra of (a) methyl heptadec-10-enoate and (b) methyl cis-9,10-methylenehexadecanoate. (Reproduced with permission from ref. 74)

3.3.3 Mixture analysis using CI

CI spectra are sometimes sufficiently simple to allow a direct analysis of mixtures without prior fractionation. Mixtures of cholesteryl esters may be analysed in this manner since their methane or isobutane spectra contain only protonated acid ions and an ion at m/z 369 $(M+H - RCO_2H)^+$ [76]. Isobutane CI has also been applied to the direct analysis of amino and carboxylic acids [77,78] in biological fluids since such compounds afford mainly quasi-molecular ions under these conditions. Similarly, ammonia, rather than a more energetic reagent gas such as methane, is preferred for the direct analysis of additives in polypropylene since the ensuing spectra of the mixture are considerably simpler [79]. Further discussion of the role of CI in mixture analysis will be found in section 7.4.1 dealing with collision-induced decomposition following chemical ionization.

References

1. Munson, M.S.B., and Field, F.H. (1966), *J. Am. Chem. Soc.*, **88**, 2621.
2. Harrison, A.G. (1983), Chemical Ionization Mass Spectrometry, CRC Press, Boca Raton, Fla..
3. Field, F.H., and Munson, M.S.B. (1965), *J. Am. Chem. Soc.*, **87**, 3289.
4. Lias, S.G., Bartmess, J.E., Liebmann, J.F., Holmes, J.L., Levin, R.D., and Mallard, W.G. (1988), *J. Phys. Chem. Ref. Data*, **17**, Suppl.1.
5. Yinon, J., and Cohen, A. (1983), *Org. Mass Spectrom.*, **18**, 47.
6. Shaw, G.J., Eglinton, G., and Quirke, J.M.E. (1981), *Anal. Chem.*, **53**, 2014.
7. Harrison, A.G., Onuska, F.I., and Tsang, C.W. (1981), *Anal. Chem.*, **53**, 1183.
8. Brophy, J.J., Nelson, D., Shannon, J.S., and Middleton, S. (1979), *Org. Mass Spectrom.*, **14**, 379.
9. Keough, T., and DeStefano, A.J. (1981), *Org. Mass Spectrom.*, **16**, 527.
10. Einolf, N., and Munson, B, (1972), *Int. J. Mass Spectrom. Ion Processes*, **9**, 141.
11. Hatch, F., and Munson, B. (1977), *Anal. Chem.*, **49**, 169.
12. Munson, B. (1977), *Anal. Chem.*, **49**, 772A.
13. Hunt, D.F., and Ryan, J.F., III (1972), *Anal. Chem.*, **44**, 1306.
14. Jelus, B.L., Munson, B., and Fenselau, C. (1974), *Biomed. Mass Spectrom.*, **1**, 96.
15. Keough, T., Mihelich, E.D., and Eickhoff, D.A. (1984), *Anal. Chem.*, **56**, 1849.
16. Subba Rao, S.C., and Fenselau, C. (1978), *Anal. Chem.*, **50**, 511.
17. Sieck, L.W. (1983), *Anal. Chem.*, **55**, 38.
18. Westmore, J.B., and Alauddin M.M. (1986), *Mass Spectrom. Rev.*, **5**, 381.
19. Knight, D.W., Knight, M.E., Rossiter, M., Jones, A., and Games, D.E. (1980), in *Advances in Mass Spectrometry*, Vol. 8B (Ed. E.Quayle), p. 1116, Heyden & Son, London.
20. Rudewicz, R., and Munson, B. (1986), *Anal. Chem.*, **58**, 2903.
21. Bowen, D.V., and Field, F.H. (1974), *Org. Mass Spectrom.*, **9**, 195.
22. Vouros, P., and Carpino, L.A. (1974), *J. Org. Chem.*, **39**, 3777.
23. Hunt, D.F., and Ryan, J.F. (1972), *J. Chem. Soc. Chem. Commun.*, **1972**, 620.
24. Chai, R., and Harrison, A.G. (1981), *Anal. Chem.*, **53**, 34.
25. Greathead, R.J., and Jennings, K.R. (1980), *Org. Mass Spectrom.*, **15**, 431.
26. Budzikiewicz, H., Schneider, B., Busker, E., Boland, W., and Francke, W. (1987), *Org. Mass Spectrom.* **22**, 458.
27. Schneider, B., Breuer, M., Hartmann, H., and Budzikiewicz, H. (1989), *Org. Mass Spectrom.*, **24**, 216.
28. Jennings, K.R. (1979), in *Gas Phase Ion Chemistry*, Vol.2 (Ed. M.T.Bowers), p. 124, Academic Press.
29. Hunt, D.F., Harvey, T.M., Brumley, W.C., Ryan, J.F.,III, and Russell, J.W. (1982), *Anal. Chem.*, **54**, 492.

30. Chai, R., and Harrison, A.G. (1983), *Anal. Chem.*, **55**, 969.
31. Scheibel, R.C., Tanner, O.P., and Wood, K.V. (1981), *Anal. Chem.*, **53**, 550.
32. Hunt, D.F., McEwen, C.N., and Harvey, T.M. (1975), *Anal. Chem.*, **47**, 1730.
33. McLuckey, S.A., Glish, G.L., Asano, K.G., and Grant, B.C. (1988), *Anal. Chem.*, **60**, 2220.
34. Sofer, I., Zhu, J., Lee, H-S., Antos, W., and Lubman, D.M. (1990), *Appl. Spectrosc.*, **44**, 1391.
35. Huynh, C.K., and Vu Duc, T. (1989), Spectrosc. Int. J., **7**, 207.
36. Blum, W., and Richter, W.J. (1973), Tetrahedron Lett., **1973**, 835.
37. Beaune, P., Pileire, B., Rocchiccioli, F., Hardy, M., Leroux, J.P., and Cartier, P. (1980), *Anal. Chem. Symp. Ser.*, 4 (*Recent Dev. Mass Spectrom. Biochem. Med.*, **6**), 1.
38. Bose, A.K., Pramanik, B.N., and Bartner, P.L. (1982), *J. Org. Chem.*, **47**, 4008.
39. Munson, M.S.B., and Field, F.H. (1966), *J. Am. Chem. Soc.*, **88**, 4337.
40. Field, F.H. (1969), *J. Am. Chem. Soc.*, **91**, 6334.
41. Field, F.H. (1969), *J. Am. Chem. Soc.*, **91**, 2827.
42. Field, F.H., Munson, M.S.B., and Becker, D.A. (1966), *Adv. Chem. Ser.*, **58**, 167.
43. Fales, H.M., Milne, G.W.A., and Nicholson, R.S. (1971), *Anal. Chem.*, **43**, 1785.
44. Hawthorne, S.B., and Miller, D.J. (1986), *Appl. Spectrosc.*, **40**, 1200.
45. Morgan, R.P., Hayward, E.J., and Steel, G. (1979), *Org. Mass Spectrum.*, **14**, 627.
46. Harrison, A.G., and Cotter, R.J. (1990) in *Methods in Enzymology*, Vol. 193: *Mass Spectrometry* (Ed. J.A. McCloskey), Ch.1, Academic Press, San Diego.
47. Michnowicz, J., and Munson, B. (1972), *Org. Mass Spectrom.*, **6**, 283.
48. Munson, M.S.B., and Field, F.H. (1965), *J. Am. Chem. Soc.*, **87**, 4242.
49. Laukien, F.H., Alleman, M., Bischofberger, P., Grossman, P., Kellerhals, Hp., and Kofer, P. (1987), in *Fourier Transform Mass Spectrometry—Evolution, Innovation and Applications* (Ed. M.V. Buchanan), Ch.5. American Chemical Society, Washington, DC.
50. Strife, R.J., and Keller, P.R. (1989), *Org. Mass Spectrom.*, **24**, 201.
51. Todd, J.F.J., and Penman, A.D. (1991), *Int. J. Mass Spectrom. Ion Processes*, **106**, 1.
52. Dorey, R.C. (1989), *Org.Mass Spectrom.*, **24**, 973.
53. Boswell, S.M., Mather, R.E., and Todd J.F.J. (1990), *Int. J. Mass Spectrom. Ion Processes*, **99**, 139.
54. Vine, J. (1980), *J. Chromatogr.*, **196**, 415.
55. Suzuki, M., Ariga, T., Sekine, M., Araki, E., and Miyatake, T. (1981), *Anal. Chem.*, **53**, 985.
56. Bayer, W., Bayer, E., Heller, W., and Schmidt, K.-H. (1981), Z. *Anal. Chem.*, **305**, 15.
57. Gray, W.R., Wojcik, L.H., and Futrell, J.H. (1970), *Biochem. Biophys. Res. Commun.*, **41**, 1111.
58. Longevialle, P., Girard, J.-P., Rossi, J.-C., and Tichy, M. (1980), *Org. Mass Spectrom.*, **15**, 268.
59. Longevialle, P., Girard, J.-P., Rossi, J.-C., and Tichy, M. (1980), in *Advances in Mass Spectrometry*, Vol.8A (Ed. E. Quayle), p. 705, Heyden & Son, London.
60. Blanc-Muesser, M., Defaye, J., Foltz, R.L., and Horton, D. (1980), *Org. Mass Spectrom.*, **15**, 317.
61. Vermeulen, N.P.E., Cauvet, J., Luijten, W.C.M.M., and Bladeren, P.J. (1980), *Biomed. Mass Spectrom.*, **7**, 413.
62. McCloskey, J.A., Futrell, J.H., Elwood, T.A., Schram, K.H., Panzica, R.P., and Townsend, L.B. (1973), *J. Am. Chem. Soc.*, **95**, 5762.
63. DeJong, E.G., Heerma, W., and Sicherer, C.A.X.G.F. (1979), *Biomed. Mass Spectrom.*, **6**, 242.

64. Winkler, F.J., and Stahl, D. (1979), *J. Am. Chem. Soc.*, **101**, 3685.
65. Van Gaever, F., Monstrey, J., and Van de Sande, C.C. (1977), *Org. Mass Spectrom.*, **12**, 200.
66. Leung, H.-W., and Harrison, A.G. (1976), *Can. J. Chem.*, **54**, 3439.
67. Bastard, J., Do Khac Manh, D., Fetizon, M., Tabet, J.C., and Fraisse, D. (1981), *J. Chem. Soc., Perkin Trans.*, **2**, 1591.
68. Nagasawa, H.T., Magnan, S.D.J., and Foltz, R.L. (1982), *Biomed. Mass Spectrom.*, **9**, 252.
69. Hunt, D.F., and Sethi, S.K. (1980), *J. Am. Chem. Soc.*, **102**, 6953.
70. Lin, Y.Y., and Smith, L.L. (1979), *Biomed. Mass Spectrom.*, **6**, 15.
71. Hunt, D.F., McEwen, C.N., and Upham, R.A. (1972), *Anal. Chem.*, **44**, 1292.
72. Hunt, D.F., McEwan, C.N., and Upham, R.A. (1971), *Tetrahedron Lett.*, **1971**, 4539.
73. Miller, D.J., and Hawthorne, S.B. (1989), *Fuel*, **68**, 105.
74. Cristopher, R.K., and Duffield, A.M. (1980), *Biomed. Mass Spectrom.*, **7**, 429.
75. Keough, T. (1982), *Anal. Chem.*, **54**, 2540.
76. Murata, T., Takahashi, S., and Takeda, T. (1975), *Anal. Chem.*, **47**, 577.
77. Mee, J.M.L. (1980), *Am. Lab.*, **12**, 55.
78. Issachar, D., and Yinon, J. (1980), in *Advances in Mass Spectrometry*, Vol. 8B (Ed. A. Quayle), p. 1321, Heyden & Son, London.
79. Rudewicz, P., and Munson, B. (1986), *Anal. Chem.*, **58**, 358.
80. Reinhold, V.N., and Carr, S.A. (1982), *Anal. Chem.*, **54**, 499.
81. Howe, I., Williams, D.H., and Bowen, R.D. (1981), *Mass Spectrometry, Principles and Applications* (Second Edition), McGraw-Hill, New York.

Appendix 3.1 Published applications of positive ion chemical ionization

References are subdivided first by reagent gas and then into major categories of compounds analysed (categories printed in upper case). This table covers the most common reagent gases. Only references that offer detailed spectral data have been included. References cover the period up to the end of 1982.

Methane (including CH_4/argon)

ALICYCLICS. Cycloalkanes (37, 186). Cyclopentane diol derivatives (84). Cyclic diols (231). Cycloalkane esters (199). ALIPHATICS, MISCELLANEOUS. α-Mono- and α,ω-di-substituted alkanes (86). ALKALOIDS (191). Harringtonine, homoharringtonine TMS (265). Opium alkaloids (5). Quinine metabolites (62). AMINES. Polytertiary amines (202). n-Decylamine (184). Decamethylenediamine (184). Trimethylamine (202). Tri-n-butylamine (248). Polyamines (176). Cysteamine derivatives (177). Histamine TMS (38). Melatonin derivative (275). AMINO ACIDS and DERIVATIVES (93, 101, 105, 192). Amino acid fluorescamine derivatives (111). ANTIOBIOTICS. Botryodiplodin (227). AROMATICS, MISCELLANEOUS. Halobenzenes (122). BARBITURATES (75, 262). Barbiturate *N*-glucosides (141). BILE ACIDS and DERIVATIVES (7,

146). BIOCIDES, MISCELLANEOUS. Barnon (99). 5,6-Dihydro-2-methyl-1,4-oxathiin-3-carboxanilide and metabolites (168). CARBAMATE BIOCIDES (152). CARBONYL COMPOUNDS. Aldehydes (89). Cyclic- and α,β-unsaturated-ketone photodimers (218). *N*-alkyl- and *N*-aryl-aminoketones (88). Aryl ketones (198). Chloroacetophenone (262). Benzophenones (223). CARBOXYLIC ACIDS and DERIVATIVES. Esters (190). Fatty acid methyl esters (15). Unsaturated methyl esters (16). Hydroxylated fatty acid methyl esters, TMS (60). Propyl esters (78). n-Heptyl propionate (248). Methoxymethyl-acetate and -formate (194). Dicarboxylic acids and esters (200). Dicarboxylic acids, anhydrides and imides (23). Benzoic acids (80). Benzyl acetate (250). CROWN ETHERS (81). DRUGS and METABOLITES, MISCELLANEOUS (67). Phencyclidine and metabolites (162). Methylated sulphonylurea drugs (165). 1-(2-Chloroethyl)-3-cyclohexyl-1-nitrosourea (272). EPOXIDES (12). 3-Chloro-1,2-epoxypropane (248). FLAVONES (79). HETEROCYCLICS, MISCELLANEOUS. 2-Aryl-1,3-dithianes (25). Oxathiins (77). Isoxazoles (35). 3,5-Diaryl-1,2,4-oxadiazoles (54). HYDROCARBONS. ALIPHATIC. n-Alkanes (37). n-Hexadecane (246, 248). 7-n-Propyltridecane (248). Alkenes and alkynes (188). HYDROCARBONS. AROMATIC. Alkylbenzenes (187, 254). Alkylnaphthalenes (187). Toluene, cycloheptatriene and norbornadiene (193). Polycyclics (106). HYDROXY COMPOUNDS. n-Decanol (184, 248). 4-Dodecene-1-ol derivatives (41). Decamethylene diol (184). Aliphatic diols, TMS (70). Phenol TMS (238). Benzyl alcohols (80). ISOPRENOIDS. Hydrocarbons (37). Mono- and sesqui-terpenes (259). Abscisic acids (49). Bruceolides (47). LIPIDS. Sphingolipids (136). MYCOTOXINS. Aflatoxin B1 (245). NITROCOMPOUNDS. Nitroalkanes (83). Nitrobenzenes (19). 2,4,6-Trinitroaromatics (87). 1,3,5-Trinitro-1,3,5-triazacyclohexane (267). NUCLEOTIDES and RELATED COMPOUNDS. Nucleosides (108, 229). ORGANOMETALLICS (14, 72). PEPTIDES (74, 94). Permethylated peptides (163). Peracetylated, permethylated peptides (228). Dipeptide derivatives (115). Peptide amides (264). Polyamino-alcohols derived from peptides (226). Peptide intermediate (262). PHOSPHATE ESTERS. Organophosphorus insecticides (129, 151). Organophosphonates (3). PHTHALATES (140). Phthalate metabolites (135). Di-(2-ethylhexyl) phthalate metabolites (234). POLYCHLORINATED COMPOUNDS. Polychlorobiphenyls (34, 42). Toxaphene (154, 251). Polycyclic chlorinated insecticides and metabolites (98). Kepone and degradation products (175). PROSTAGLANDINS (133). Prostaglandin TMS (117). SACCHARIDES. Permethylated disaccharides (57, 183). Methylated alditol acetates of neutral and amino sugars (44). Amino sugar derivatives (180). Dianhydro hexopyranose derivatives (58). Glucose derivatives (145). STEROIDS (159). 17-Hydroxy-steroids (96). Monohydroxysteroids (63). Sterodial ketones (92). Estrogens (36). C_{27} Steroids (179). Steroidal spirolactones TMS (178). SULPHUR COMPOUNDS, MISCELLANEOUS. n-Decanethiol (184). Decamethylenedithiol (184). 2,3-Diphenyl thiirene-1,1-dioxide (230).

Isobutane

ALICYCLICS. Decaline (113). Tetrahydropyrans (113). Cyclohexanol (236). Cyclopentane diol derivatives (84). Cyclic diols (231). Cyclic diol stereoisomers (253). Cyclohexane esters (113). Cycloalkane esters (199). ALKALOIDS. Ergot alkaloids (27, 95). Heroin (149, 212). Acetylcodeine, caffeine, quinine (212). AMINES (69). Polytertiary amines (202). Trimethylamine (202). Ephedrine (208). 1-(2,5-Dimethoxy-4-methylphenyl)-2-amino propane and related compounds (241). AMINO ACIDS and DERIVATIVES (208). Methionine and analogs (39). Amino acids TMS (82). Amino acid thiohydantoins (100, 121). ANTIBIOTICS. Gentamicins (167). Anthracycline antibiotic (196). Carbohydrate antibiotics (119, 120). Macrolide antibiotics (33, 50, 156). AROMATICS, MISCELLANEOUS. 4-Substituted *N*-ethyl, *N*-methylaniline-*N*-oxides (97). Dibenzyl ethers (258). BARBITURATES (208). BILE ACIDS and DERIVATIVES (138). CARBAMATE BIOCIDES, Temik (208). Aldicarb and metabolites (270). CARBONYL COMPOUNDS. Benzophenones (223). CARBOXYLIC ACIDS and DERIVATIVES. Carboxylic acids (182, 214). Saturated and unsaturated wax esters (1). Mono-unsaturated and mono-epoxy fatty acid methyl esters (185). Unsaturated fatty acid methyl esters (double bond positions) (116, 169). Carboxylic acid TMS derivatives (9, 132). Methoxymethyl-acetate and -formate (194). Methyl-thiomethyl-acetate and -propionate (195). *p*-Fluoro- and *p*-nitro-benzyl acetate (249). Benzyl- and *t*-amyl-acetate (250). Dicarboxylic acids and esters (200). Di- and tri-carboxylic esters (203). CAROTENOIDS (181). CROWN ETHERS (81). DRUGS and METABOLITES, MISCELLANEOUS (139, 153, 237). Dicarbamates (150). Phenoprocoumon and metabolites (148). Warfarin and metabolites (160). Procaine, methapyrilene (212). 1-(2-Chloroethyl)-3-cyclohexyl-1-nitrosourea (272). EPOXIDES. Glycidyl ethers (26). Aliphatic epoxides (239). FLAVONES (79, 155). GLUCURONIDES and GLYCOSIDES. Drug glucuronide (144). Cannabinol glucuronide TMS (171). Desulphoglucosinolates TMS (224). HETEROCYCLICS, MISCELLANEOUS. 2-Aryl-1,3-dithianes (25). HYDROCARBONS, ALIPHATIC. Alkenes (Double bond position) (13, 28). Alkynes (2). HYDROXY COMPOUNDS. Saturated aliphatic alcohols (189). Aliphatic diols TMS (30). Catechols (24). ISOPRENOIDS. Phorbol esters (174). LIPIDS. Phospholipids (22). Sphingosine boronates (85). Cerebroside TMS (137). Triglycerides (208). NITRO-COMPOUNDS. Nitro-aromatics (45). 2,4,6-Trinitro-aromatics (87). 1,3,5-Trinitro-1,3,5-triazacyclohexane (267). NUCLEOTIDES AND RELATED COMPOUNDS. Zeatins TMS (6). ORGANOMETALLICS. Lanthanide chelates (204, 205). PEPTIDES. Dipeptide derivatives (157). Dipeptides TMS (114). Permethylated peptides (163, 170). Peptides TMS (82). Peptides, methyl ester, TFA (56). Peptide amides (264). Polyamino-alcohols derived from peptides (226). PHOSPHATE ESTERS. Organophosphorus insecticides (129, 208). Organophosphonates (3). Glyphosphate and metabolite derivatives (271). Tri-n-butyl phosphate (203). PHTHALATES (140). POLYCHLORINATED COMPOUNDS. Aromatic chlorinated pesticides (207). PROSTAGLANDINS (133). Prostaglandin cyclic boronates (172). Prostaglandin TMS (117). QUATERNARY AMINO COMPOUNDS (161). SACCHARIDES. Permethylated disaccharides (57, 183). Saccharides TMS (4, 9, 132). Saccharide dimethylacetals and thioacetals (20). Permethylated glycosylalditols (102). Aldose derivatives (124). Methylated hexitol acetates (127). Monosaccharide derivatives (128). Dianhydro-hexopyranose derivatives (58). Mono- and disaccharides (212). Sucrose (243). Aldobiouronic acids (242). Dialdose dianhydrides (242). Permethylated reduced oligosaccharides (17). STEROIDS (112, 208). Steroid TMS ethers (166). Mono-hydroxy cholesterols (63). Steroidal amino alcohols (255). Sterol esters (213). Aldosterone cyclic boronate (130). C_{27}-steroids (179). Brassinolide (143). SULPHUR COMPOUNDS, MISCELLANEOUS. Thiosulphinates, thiosulphonates and sulphinyl sulphones (268). 2,3-Diphenyl thiirene-1,1-dioxide (230).

Ammonia (including ND_3, NH_3/isobutane and NH_3/argon)

ALICYCLICS. Cyclic diol stereoisomers (253). ALIPHATICS, MISCELLANEOUS. α-Mono- and α,ω-di-substituted alkanes (86). ALKALOIDS. Isoquinoline alkaloids (261). Morphine (257). AMINES (73). Aromatic amines (64). Diaryl amines (52). *p*-Ethylaniline, *N*-Ethylaniline (232). Biogenic amines (261). Biogenic amine derivatives (134). Phenylephrine (257). Taurine (243). AMINO ACIDS and DERIVATIVES (107, 110). Amino acid thiohydantoins (100). Amino acid fluorescamine derivatives (111). Amino acids containing hydroxyl or thiol groups (257). ANTIBIOTICS. Antimycin antibiotics (221). Anthracycline antibiotic (222). Carbohydrate antibiotics (119, 120). Macrolide antibiotics (33, 50). AROMATICS, MISCELLANEOUS. Anisole (232). Benzonitrile (232). 4-Substituted *N*-ethyl,*N*-methylaniline-*N*-oxides (97). BILE ACIDS and DERIVATIVES (53). Bile acid ester acetates (273). Bile alcohol (276). CARBONYL COMPOUNDS. Ketones (235). Conjugated ketones (225). Aldehyde oximes (66). Decanal di-butyl acetal (232). Benzaldehyde, acetophenone (232). Anthrones (10). CARBOXYLIC ACIDS and DERIVATIVES. Oleic acid (257). Fatty acid methyl esters (15). Dicarboxylic acids and esters (200). Penta-erythritol ester (51). DRUGS and METABOLITES, MISCELLANEOUS (52). 5-Fluoro-uracil derivatives (29). Warfarin metabolites (160). Procaine (257). Phenacetin (257). 1-(2-Chloroethyl)-3-cyclohexyl-1-nitrosourea (272). GLUCURONIDES and GLYCOSIDES (11, 32, 274). Amygdalin (263, 99). Steroid glucosides (61). Δ^6-Tetrahydrocannabinol glucuronide (142). Desulphoglucosinolates TMS (48). HETEROCYCLICS, MISCELLANEOUS. *N*-Heterocycles (64). Imidazoles (46). Phenothiazines (51). HYDROCARBONS, AROMATIC. Benzene, toluene, trimethylbenzene (232). HYDROXY COMPOUNDS, Hydroxy polycyclic aromatics (257). Phenol

(232). ISOPRENOIDS. Mono- and sesquiterpenes (259). LIPIDS. Diglycerides (147). Triglycerides (211). Lysophosphatidic acids and esters (118). NITRO-COMPOUNDS. 1,3,5-Trinitro-1,3,5-triazacyclohexane (267). NUCLEOTIDES and RELATED COMPOUNDS. Nucleosides (108, 229). Ribofuranosylpurines (252). PEPTIDES (110). Proline dipeptides (221). Peptide amides (264). PHOSPHATE ESTERS (51). Organophosphorus insecticides (151). POLYCHLORINATED COMPOUNDS. Mirex (266). PROSTAGLANDINS. Prostaglandin TMS (117, 131). Thromboxane (131). QUATERNARY AMINO COMPOUNDS. Betaine (243). QUINONES. Anthraquinones (10). SACCHARIDES (32, 257). Oligosaccharide peracetates (103, 220). Permethylated disaccharides (57, 183). Saccharide dimethyl- and thio-acetals (20). Permethylated glycosylalditols (102). Aldose derivatives (125). Monosaccharides TMS (126). Monosaccharide derivatives (128). STEROIDS. Hydroxy- and keto-steroids (257). Mono-hydroxycholesterols (63). Steroidal spirolactone TMS (178). C_{27}-Steroids (179). SULPHUR COMPOUNDS, MISCELLANEOUS. 2,3-Diphenyl thiirene-1,1-dioxide (230). Sulphonamides (257).

Nitric oxide (including NO/nitrogen)

ALKALOIDS. Heroin (149). Morphine and tropane alkaloids (215). BILE ACIDS and DERIVATIVES. Bile acid methyl esters TMS (206). CARBONYL COMPOUNDS. Aldehydes (89). Ketones (256). CARBOXYLIC ACIDS and ESTERS (256). Palmitic acid (244). HYDROCARBONS, ALIPHATIC (246). Alkanes (209). Alkenes (216). Alkenes (double bond position) (13). Alkynes (2). Alkynes (triple bond position) (55). Conjugated trienes and tetraenes (269). HYDROCARBONS, AROMATIC (246). HYDROXY COMPOUNDS (65, 256). PHOSPHATE ESTERS. Parathion (244). STEROIDS. Hydroxysteroids TMS (68). Testosterone, cholesterol (244).

Hydrogen (including H_2/nitrogen)

AROMATICS, MISCELLANEOUS. Halobenzenes (104, 122). Dibenzyl ethers (258). *N,N′*-diaryl ureas (8). CARBOXYLIC ACIDS and ESTERS. Unsaturated methyl esters (16). Propyl esters (78). Dicarboxylic acids, anhydrides and imides (23). Benzoic acids (80). CAROTENOIDS (181). FLAVANOIDS (155). HETEROCYCLICS, MISCELLANEOUS. Isoxazoles (35). Oxathiins (77). HYDROCARBONS, AROMATIC. Alkylbenzenes (254). HYDROXY COMPOUNDS. C_5,C_6 alcohols (40). Benzyl alcohols (80). NITRO-COMPOUNDS. Nitrobenzenes (19). Nitroexplosives (91). 1,3,5-Trinitro-1,3,5-triazacyclohexane (267). PEPTIDES. Dipeptide derivatives (157). POLYCHLORINATED COMPOUNDS. Polychloro-biphenyls (34). PORPHYRINS (43). STEROIDS. Steroidal ketones (92). 17-Hydroxy steroids (96).

Appendix 3.2 Published applications of positive ion chemical ionization

References are subdivided first by reagent gas and then by compound analysed. This table covers less common reagent gases. Full details of the reagent gas are given in parentheses. Only references that offer detailed spectral data have been included. References cover the period up to the end of 1982.

Water

α,ω-Dicyanoalkanes (H_2O) (201). Valeronitrile (H_2O) (240). Acetone (H_2O) (240). Diethyl ether (H_2O) (240). Dicyanobenzenes (H_2O) (201). Aniline (H_2O) (240). Glutathione (H_2O) (210). Barbiturates (H_2O/Ar) (71). Aryl ketones (H_2O/CH_4 (198). Fatty acid methyl esters (H_2O/He) (109). Dicarboxylic acids and esters (H_2O) (200). Nitro explosives (H_2O) (158). 1,3,5-Trinitro-1,3,5-triazacyclohexane (H_2O) (267). Adenosine (D_2O) (247). Oligopeptide, *N*-acetyl, methyl ester (H_2O/CH_3CN) (173). 6-Ketoestradiol (D_2O) (247).

Alcohols

Dicarboxylic acids and esters (CH_3OH) (200). Organophosphorus insecticides (CH_3OH) (129).

Amines

Amines ($NH_2CH_2CH_2OH$/isobutane) (90). Amines ($NH_2CH_2CH_2NH_2$/isobutane) (90). Glucuronides TMS (pyridine/CH_4) (217). Olefins, double bond position (CH_3NH_2) (13). Polyhydroxy compounds, polyethers ($NH_2CH_2CH_2OH$/isobutane) (90). Polyhydroxy compounds, polyethers ($NH_2CH_2CH_2NH_2$/isobutane) (90). Nucleosides ($(CH_3)_2NH$) (108). Peptides ($NH_2CH_2CH_2OH$/isobutane) (90). Peptides ($NH_2CH_2CH_2OH$/isobutane) (90). Peptides ($NH_2CH_2CH_2NH_2$/isobutane (90). 2,3-Diphenyl thiirene-1,1-dioxide ($(CH_3)_2NH$) (230).

Inert gases

Decalins (He) (113). Tetrahydropyrans (He) (113). Cyclohexane esters (He) (113). Fatty acid methyl esters (He) (15, 109). Dipeptide derivatives (He) (123, 164).

Vinyl methyl ether mixtures

Olefins, double bond position (VME/CS_2/N_2) (18, 21). Olefins, double bond position (VME) (197). Mono-unsaturated and cyclopropanoid fatty acid methyl esters (VME/i-C_4H_{10}) (31).

Aliphatic hydrocarbons

Benzophenones (C_2H_6) (223). Benzophenones (C_3H_8) (223). Organophosphonates (C_2H_4) (3).

Miscellaneous

Tryptophan methyl ester (C_6H_6) (233). Unsaturated fatty acid esters (C_6H_6) (233). Fatty acids (HCO_2H) (59). C_5,C_6 alkanols (N_2O/H_2) (40). Dicarboxylic acids and esters ($(CH_3)_2O$) (200). Nalorphine (C_6H_6) (233). Dibenzothiophene (O_2/H_2) (245). Alkylbenzenes (N_2O/H_2) (254). Alkylbenzenes (CO/H_2) (254). Aromatic hydrocarbons (C_6H_5Cl) (260). Coronene (O_2/H_2 (245). Alcohols ($(CH_3)_2CO/CH_4$) (76). Alcohols (CH_3CHO/CH_4) (76). Alcohols TMS ($(CH_3)_4Si$) (219). Cholesterol (C_6H_6) (233).

References for Appendices 3.1 and 3.2

1. Gronneberg, T. O. (1979), *Chem. Scr.*, **13**, 56.
2. Busker, E., and Budzikiewicz, H. (1979), *Org. Mass Spectrom.*, **14**, 222.
3. Sass, S., and Fischer, T. L. (1979), *Org. Mass Spectrom.*, **14**, 257.
4. Schoots, A. C., Mikkers, F. E. P., Cramers, C. A. M. G., and Ringoir, S. (1979), *J. Chromatogr.*, **164**, 1.
5. Cone. E. J., Gorodetsky, C. W., Yeh, S. Y., Darwin, W. D., and Buchwald, W. F. (1982), *J. Chromatogr.*, **230**, 571.
6. Summons, R. E., Entsch, B., Letham, D. S., Gollnow, B. I., and MacLeod, J. K. (1980) *Planta*, **147**, 422.
7. Muschik, G. M., Wright, L. H., and Schroer, J. A. (1979), *Biomed. Mass Spectrom.*, **6**, 266.
8. Brophy, J. J., Nelson, D., and Shannon, J. S. (1979), *Org. Mass Spectrom.*, **14**, 379.
9. Schoots, A. C., and Leclerq, P. A. (1979), *Biomed. Mass Spectrom.*, **6**, 502.
10. Evans, F. J., Lee, M. G., and Games, D. E. (1979), *Biomed. Mass Spectrom.*, **6**, 374.
11. Vine, J., Brown, L., Boutagy, J., Thomas, R., and Nelson, D. (1979), *Biomed. Mass Spectrom*, **6**, 415.
12. Suzuki, S., Hori, Y., Das, R. C., and Koga, O. (1980), *Bull. Chem. Soc. Jpn.*, **53**, 1451.
13. Budzikiewicz, H., and Busker, E. (1980), *Tetrahedron*, **36**, 255.
14. Vandenheuvel, W. J. A. (1980), *J. Organomet. Chem.*, **190**, 73.
15. Vine, J. (1980), *J. Chromatogr.*, **196**, 415.
16. Harrison, A. G., and Ichikawa, H. (1980), *Org. Mass Spectrom.*, **15**, 244.
17. Fournet, B., Dhalluin, J. M., Strecker, G., and Montreuil, J. (1980), *Anal. Biochem.*, **108**, 35.
18. Chai, R., and Harrison, A. G. (1981), *Anal. Chem.*, **53**, 34.
19. Harrison, A. G., and Kallury, R. K. M. R. (1980), *Org. Mass Spectrom.*, **15**, 284.
20. Blanc-Meusser, M., Defaye, J., Foltz, R. L., and Horton, D. (1980), *Org. Mass Spectrom.*, **15**, 317.
21. Greathead, R. J., and Jennings, K. R. (1980), *Org. Mass Spectrom.*, **15**, 431.
22. Polonsky, J., Tence, M., Varenne, P., Das, B. C., Lunel, J., and Benveniste, J. (1980) *Proc. Natl. Acad. Sci. U.S.A.*, **77**, 7019.
23. Harrison, A. G., and Kallury, R. K. M. R. (1980), *Org. Mass Spectrom.*, **15**, 277.

24. Roepstorff, P., and Andersen, S. O. (1980), *Biomed. Mass Spectrom.*, **7**, 317.
25. Shaw, C. J., and Kane, V. V. (1980), *Org. Mass Spectrom.*, **15**, 502.
27. Porter, J. K., and Betowski, D. (1981), *J. Agric. Food Chem.*, **29**, 650.
28. Suzuki, M., Ariga, T., Sekine, M., Araki, E., and Miyatake, T. (1981), *Anal. Chem.*, **53**, 985.
29. Marunaka, T., Umeno, Y., Minami, Y., and Shibata, T. (1980), *Biomed. Mass Spectrom.*, **7**, 331.
30. Wolfschuetz, R., Schwarz, H., Blum, W., and Richter, W. (1981), *Org. Mass Spectrom.*, **16**, 37.
31. Christopher, R. K., and Duffield, A. M. (1980), *Biomed. Mass Spectrom.*, **7**, 429.
32. Games, D. E., and Lewis, E. (1980), *Biomed. Mass Spectrom.*, **7**, 433.
33. Suzuki, M., Harada, K., Takeda, N., and Tatematsu, A. (1981), *Heterocycles*, **15**, 1123.
34. Harrison, A. G., Onuska, F. I., and Tsang, C. W. (1981), *Anal. Chem.*, **53**, 1183.
35. Kallury, R. K. M. R., and Hemalatha, J. (1980), *Org. Mass Spectrom.*, **15**, 659.
36. Cairns, T., Siegmund, E. G., and Rader, B. (1981), *Anal. Chem.*, **53**, 1217.
37. Bayer, W., Bayer, E., Heller, W., and Schmidt, K. H. (1981), *Fresenius' Z. Anal. Chem.*, **305**, 15.
38. Henion, J. D., Nosanchuk, J. S., and Bilder, B. M. (1981), *J. Chromatogr.*, **213**, 475.
39. Cooper, A. J. L., Griffith, O. W., Meister, A., and Field, F. H. (1981), *Biomed. Mass Spectrom.*, **8**, 95.
40. Herman, J. A., and Harrison, A. G. (1981), *Canad. J. Chem.*, **59**, 2125, 2133.
41. Hogge, L. R., and Olson, D. J. H. (1982), *J. Chromatogr. Sci.*, **20**, 109.
42. Cairns, T., and Siegmund, E. G. (1981), *Anal. Chem.*, **53**, 1599.
43. Shaw, G. J., Eglinton, G., and Quirke, J. M. E. (1981), *Anal, Chem.*, **53**, 2014.
44. Laine, R. A. (1981), *Anal. Chem.*, **116**, 383.
45. Gielsdorf, W. (1981), *Fresenius' Z. Anal. Chem.*, **308**, 123.
46. Haskins, N. J., Ford, G. C., Waddell, K. A., Dickens, J. P., Dyer, R. L., Hamill, B. J., and Harrow, T. A. (1981), *Biomed. Mass Spectrum.*, **8**, 351.
47. Baldwin, M. A., Carter, D. M., Darwish, F. A., and Philipson, J. S., (1981), *Biomed. Mass Spectrom.*, **8**, 362.
48. Eagles, J., Fenwick, G. R., and Heaney, R. K. (1981), *Biomed. Mass Spectrom.*, **8**, 265.
49. Netting, A. G., Millborrow, B. V., and Duffield, A. M. (1982), *Phytochemistry*, **21**, 385.
50. Suzuki, M., Harada, K., Takeda, N., and Tatematsu, A. (1981), *Biomed. Mass Spectrom.*, **8**, 332.
51. Zeman, A. (1982), *Fresenius' Z. Anal. Chem.*, **310**, 243.
52. Kauert, G., Drasch, G., and Von Meyer, L. (1979), *Beitr. Gerichtl. Med.*, **37**, 329.
53. Kuriyama, K., Ban, Y., Nakashima, T., and Murata, T. (1979), *Steroids*, **34**, 717.
54. Selva, A., and Facchetti, S. (1980), *Adv. Mass Spectrom.*, **8A**, 723.
55. Budzikiewicz, H., Busker, E., and Brauner, A. (1980), *Adv. Mass Spectrom.*, **8A**, 713.
56. Koenig, W. A., Krohn, H., Greiner, M., Brueckner, H., and Jung, G. (1980), *Adv. Mass Spectrom.*, **8B**, 1109.
57. DeJong, E. G., Heerma, W., and Dijkstra, G. (1980), *Adv. Mass Spectrom.*, **8B**, 1314.
58. Kadenstev, V. I., Kaymarazov, A. G., Chizov, O. S., Cerny, M., Trnka, T., and Turecek, F. (1982), *Biomed. Mass Spectrom.*, **9**, 130.
59. Beaune, P., Pileire, B., Rocchiccioli, F., Hardy, M., Leroux, J. P., and Cartier, P. (1980), *Anal. Chem. Symp. Ser.*, **4**, (*Recent Dev. Mass Spectrom. Biochem. Med* **6**), 1
60. Scheutwinkel-Reich, M., and Stan, H. J. (1980), *Anal. Chem. Symp. Ser.*, **4** (*Recent Dev. Mass Spectrom. Biochem. Med.*, **6**), 45.
61. Kastelic-Suhadolc, T. (1980), *Eur. J. Mass Spectrom., Biochem. Med. Environ. Res.*, **1**, 193.

62. Liddle, C., Graham, G. G., Christopher, R. K., Bhuwapathanapun, S., and Duffield, A. M. (1981), *Xenobiotica*, **11**, 81.
63. Lin, Y. Y., Low, C.-E., and Smith, L. L. (1981), *J. Steroid Biochem.*, **14**, 563.
64. Bucanan, M. V. (1982), *Anal. Chem.*, **54**, 570.
65. Hunt, D. F., Harvey, T. M., Brumley, W. C., Ryan, J. F., III, and Russell, J. W. (1982), *Anal. Chem.*, **54**, 492.
66. Levine, S. P., Harvey, T. M., Waeghe, T. J., and Shapiro, R. H. (1981), *Anal. Chem.*, **53**, 805.
67. Finkle, B. S., Foltz, R. L., and Taylor, D. M. (1974), *J. Chromatogr. Sci.*, **12**, 304.
68. Jelus, B. L., Munson, B., and Fenselau, C. (1974), *Biomed. Mass Spectrom.*, **1**, 96.
69. Milne, G. W. A., Fales, H. M., and Colburn, R. W. (1973), *Anal. Chem.*, **45**, 1952.
70. Blum, W., and Richter, W. J. (1973), *Tetrahedron Lett.*, **1973**, 835.
71. Hunt, D. F., and Ryan, J. F., III (1973), *Anal. Chem.*, **45**, 1306.
72. Hunt, D. F., Russell, J. W., and Torian, R. L. (1972), *J. Organometal. Chem.*, **43**, 175.
73. Hunt, D. F., McEwen, C. N., and Upham, R. A. (1971), *Tetrahedron Lett.*, **1971**, 4539.
74. Kiryushkin, A. A., Fales, H. M., Axenrod, T., Gilbert, E. J., and Milne, G. W. A. (1971), *Org. Mass Spectrom.*, **5**, 19.
75. Fales, H. M., Milne, G. W. A., and Axenrod, T. (1970), *Anal. Chem.*, **42**, 1432 (correction in 1971, *Anal. Chem.*, **43**, 1461).
76. Hunt, D. F., and Ryan, J. F., III (1971), *Tetrahedron Lett.*, **1971**, 4535.
77. Harrison, A. G., and Onuska, F. I. (1978), *Org. Mass Spectrom.*, **13**, 35.
78. Benoit, F. M., and Harrison, A. G. (1978), *Org. Mass Spectrom.*, **13**, 128.
79. Yinon, J., Issachar, D., and Boettger, H. G. (1978), *Org. Mass Spectrom.*, **13**, 167.
80. Ichikawa, H., and Harrison, A. G. (1978), *Org. Mass Spectrom.*, **13**, 389.
81. Van Gaever, F., Van de Sande, C. C., Bucquoye, M., and Goethals, E. J. (1978), *Org. Mass Spectrom.*, **13**, 486.
82. Budzikiewicz, H., and Meissner, G. (1978), *Org. Mass Spectrom.*, **13**, 608.
83. Chizhov, O. S., Kadenstev, V. I., Palmbach, G. G., Burstein, K. Ia., Shevelev, S. A., and Feinsilberg, A. A. (1978), *Org. Mass Spectrom.*, **13**, 611.
84. Claeys, M., and Van Haver, D. (1977), *Org. Mass Spectrom.*, **12**, 531.
85. Gaskell, S. J., and Brooks, C. J. W. (1977), *Org. Mass Spectrom.*, **12**, 651.
86. Weinkam, R. J., and Gal, J. (1976), *Org. Mass Spectrom.*, **11**, 188.
87. Zitrin, S., and Yinon, J. (1976), *Org. Mass Spectrom.*, **11**, 388.
88. Audoye, P., Gaset, A., and Lattes, A. (1975), *Org. Mass Spectrom.*, **10**, 669.
89. Jardine, I., and Fenselau, C. (1975), *Org. Mass Spectrom.*, **10**, 748.
90. Bowen, D. V., and Field, F. H. (1974), *Org. Mass Spectrom.*, **9**, 195.
91. Gillis, R. G., Lacey, M. J., and Shannon, J. S. (1974), *Org. Mass Spectrom.*, **9**, 359.
92. Michnowicz, J., and Munson, B. (1974), *Org. Mass Spectrom.*, **8**, 49.
93. Leclerq, P. A., and Desiderio, D. M. (1973), *Org. Mass Spectrom.*, **7**, 515.
94. Baldwin, M. A., and McLafferty, F. W. (1973), *Org. Mass Spectrom.*, **7**, 1353.
95. Eckers, C., Games, D. E., Mallen, D. N. B., and Swann, B. P. (1982), *Biomed. Mass Spectrom.*, **9**, 162.
96. Michnowicz, J., and Munson, B. (1972), *Org. Mass Spectrom.*, **6**, 765.
97. Cowan, D. A., Patterson, L. H., Damani, L. A., and Gorrod, J. W. (1982), *Biomed. Mass Spectrom.*, **9**, 233.
98. Biros, F. J., Dougherty, R. C., and Dalton, J. (1972), *Org. Mass Spectrom.*, **6**, 1161.
99. Page, J. A. (1982), *Anal. Proc.*, **19**, 307.
100. Okada, K., and Sakuno, A. (1978), *Org. Mass Spectrom.*, **13**, 535.
101. Weinkam, R. J. (1978), *J. Org. Chem.*, **43**, 2581.
102. Chizov, O. S., Kadentsev, V. I., Solov'yov, A. A., Levonowich, P. F., and Dougherty, R. C. (1976), **41**, 3425.

103. Dougherty, R. C., Roberts, J. D., Binkley, W. W., Chizov, O. S., Kadentsev, V. I., and Solov'yov, A. A. (1974), *J. Org. Chem.*, **39**, 451.
104. Leung, H. -W., and Harrison, A. G. (1979), *J. Am. Chem. Soc.*, **101**, 3168.
105. Tsang, C. W., and Harrison, A. G. (1976), *J. Am. Chem. Soc.*, **98**, 1301.
106. Lee, M. L., and Hites, R. A. (1977), *J. Am. Chem. Soc.*, **99**, 2008.
107. Gaffney, J. S., Pierce, R. C., and Friedman, L. (1977), *J. Am. Chem. Soc.*, **99**, 4293.
108. Wilson, M. S., and McCloskey, J. A. (1975), *J. Am. Chem. Soc.*, **97**, 3436.
109. Arsenault, G. P. (1972), *J. Am. Chem. Soc.*, **94**, 8241.
110. Gaffney, J. S., Pierce, R. C., and Friedman, L. (1977), *Int. J. Mass Spec. Ion Phys.*, **25**, 439.
111. Shieh, J. -J., Leung, K., and Desiderio, D. M. (1977), *Anal. Lett.*, **10**, 575.
112. Nowlin, J. G., Carroll, D. I., Dzidie, I., Horning, M. G., Stillwell, R. N., and Horning, E. C. (1979), *Anal. Lett.*, **12**, 573.
113. Van de Sande, C. C., van Gaever, F., Sandra, P., and Monstrey, J. (1977), *Z. Naturforsch.*, **32B**, 573.
114. Krutzsch, H. C., and Kindt, T. J. (1979), *Anal. Biochem.*, **92**, 525.
115. Seifert, W. E. Jr., McKee, R. E., Beckner, C. F., and Caprioli, R. M. (1978), *Anal. Biochem.*, **88**, 149.
116. Ariga, T., Araki, E., and Murata, T. (1977), *Anal. Biochem.*, **83**, 474.
117. Ariga, T., Suzuki, M., Morita, I., Murota, S. -I., and Miyatake, T. (1978), *Anal. Biochem.*, **90**, 174.
118. Tokumura, A., Handa, Y., Yoshioka, Y., Higashimoto, M., and Tuskatani, H. (1982), *Chem. and Pharm. Bull.*, **30**, 2119.
119. Horton, D., Wander, J. D., and Foltz, R. L. (1974), *Anal. Biochem.*, **59**, 452.
120. Horton, D., Wander, J. D., and Foltz, R. L. (1973), *Anal. Biochem.*, **55**, 123.
121. Fales, H. M., Nagai, Y., Milne, G. W. A., Brewer, H. B., Jr, Bronzert, T. J., and Piscano, J. J. (1971), *Anal. Biochem.*, **43**, 288.
122. Leung, H. -W., and Harrison, A. G. (1976), *Can. J. Chem.*, **54**, 3439.
123. Faull, K. F., Schier, G. M., Schlesinger, P., and Halpern, B. (1976), *Clin. Chim. Acta.*, **70**, 313.
124. Li, B. W., Cochran, T. W., and Verrellotti, J. R. (1977), *Carbohydr. Res.*, **59**, 567.
125. Seymour, F. R., Chen, E. C. M., and Bishop, S. H. (1979), *Carbohydr. Res.*, **73**, 19.
126. Murata, T., and Takahashi, S. (1978), *Carbohydr. Res.*, **62**, 1.
127. McNeil, M., and Albersheim, P. (1977), *Carbohydr. Res.*, **56**, 239.
128. Horton, D., Wander, J. D., and Foltz, R. L. (1974), *Carbohydr. Res.*, **36**, 75.
129. Stan, H. -J. (1977), *Fresenius' Z. Anal. Chem.*, **287**, 104.
130. Gaskell, S. J., and Brooks, C. J. W. (1978), *J. Chromatogr.*, **158**, 331.
131. Morita, I., Murota, S. -I., Suzuki, M., Ariga, T., and Miyatake, T. (1978), *J. Chromatogr.*, **154**, 285.
132. Schoots, A. C., Mikkers, F. E. P., Cramers, C. A. M. G., and Ringoir, S. (1979), *J. Chromatogr.*, **164**, 1.
133. Oswald, E. O., Parks, D., Eling, T., and Corbett, B. J. (1974), *J. Chromatogr.*, **93**, 47.
134. Miyazaki, H., Hashimoto, Y., Iwanaga, M., and Kubodera, T. (1974), *J. Chromatogr.*, **99**, 575.
135. Albro, P. W., Thomas, R., and Fishbein, L. (1973), *J. Chromatogr.*, **76**, 321.
136. Markey, S. P., and Wenger, D. A. (1974), *Chem. Phys. Lipids*, **12**, 182.
137. Murata, T., Ariga, T., Oshima, M., and Miyatake, T. (1978), *J. Lipid. Res.*, **19**, 370.
138. Cowen, A. E., Hofmann, A. F., Hachey, D. L., Thomas, P. J., Belobaba, D. T. E., Klein, P. D., and Tökes, L. (1976), *J. Lip. Res.*, **17**, 231.
139. Saferstein, R., Manura, J. J., Brettell, T. A., and De, P. K. (1978), *J. Anal. Toxicol.*, **2**, 245.

140. Addison, J. B. (1979), *Analyst*, **104**, 846.
141. Tang, B. K., Kalow, W., and Grey, A. A. (1979), *Drug Metab. Dispos.*, **7**, 315.
142. Pallante, S., Lyle, M. A., and Fenselau, C. (1978), *Drug Metab. Dispos.*, **6**, 389.
143. Grove, M. D., Spencer, G. F., Rohwedder, W. K., Mandava, N., Worley, J. F., Warthen, J. D., Jr, Steffens, G. L., Flippen-Anderson, J. L., and Cook, J. C., Jr (1979), *Nature*, **281**, 216.
144. Chang, T. T. L., Kuhlmann, Ch. F., Schillings, R. T., Sisenwine, S. F., Tio, C. O., and Ruelius, H. W. (1973), *Experientia*, **29**, 653.
145. Hogg, A. M., and Nagabhushan, T. L. (1972), *Tetrahedron Lett.*, **1972**, 4827.
146. Kelsey, M. I., Muschik, G. M., and Sexton, S. A. (1978), *Lipids*, **13**, 152.
147. Kagawa, Y., and Ariga, T. (1977), *J. Biochem. (Tokyo)*, **81**, 1161.
148. Pohl, L. R., Haddock, R., Garland, W. A., and Trager, W. F. (1975), *J. Med. Chem.*, **18**, 513.
149. Jardine, I., and Fenselau, C. (1975), *J. Forensic Sci.*, **20**, 373.
150. Nelson, G. L., Kuhlmann, C. F., and Chang, T. L. (1974), *J. Pharm. Sci.*, **63**, 1959.
151. Holmstead, R. L., and Casida, J. E. (1974), *J. Ass. Offic. Anal. Chem.*, **57**, 1050.
152. Holmstead, R. L., and Casida, J. E. (1975), *J. Ass. Offic. Anal. Chem.*, **58**, 541.
153. Saferstein, R., and Chao, J. -M. (1973), *J. Ass. Offic. Anal. Chem.*, **56**, 1234.
154. Holmstead, R. L., Khalifa, S., and Casida, J. E. (1974), *J. Agric. Food Chem.*, **22**, 939.
155. Clark-Lewis, J. W., Harwood, C. N., Lacey, M. J., and Shannon, J. S. (1973), *Aust. J. Chem.*, **26**, 1577.
156. Mitscher, L. A., Showalter, H. D. H., and Foltz, R. L. (1972), *J. Chem. Soc. Chem. Commun.*, **1972**, 796.
157. Schier, G. M., Halpern, B., and Milne, G. W. A. (1974), *Biomed. Mass Spectrom.*, **1**, 212.
158. Yinon, J. (1974), *Biomed. Mass Spectrom.*, **1**, 393.
159. Haskins, N. J., Games, D. E., and Taylor, K. T. (1974), *Biomed. Mass Spectrom.*, **1**, 423.
160. Pohl, L. R., Nelson, S. D., Garland, W. A., and Trager, W. F. (1975), *Biomed. Mass Spectrom.*, **2**, 23.
161. Shabanowitz, J., Byrnes, P., Maelicke, A., Bowen, D. V., and Field, F. H. (1975), *Biomed. Mass Spectrom.*, **2**, 164.
162. Lin, D. C. K., Fentiman, A. F., Jr, Foltz, R. L., Forney, R. D., Jr, and Sunshine, I. (1975), *Biomed. Mass Spectrom.*, **2**, 206.
163. Mudgett, M., Bowen, D. V., Kindt, T. J., and Field, F. H. (1975), *Biomed. Mass Spectrom.*, **2**, 254.
164. Schier, G. M., Bolton, P. D., and Halpern, B. (1976), *Biomed. Mass Spectrom.*, **3**, 32.
165. Midha, K. K., Awang, D. V. C., McGilveray, I. J., and Kleber, J. (1976), *Biomed. Mass Spectrom.*, **3**, 100.
166. Smith, A. G., Gaskell, S. J., and Brooks, C. J. W. (1976), *Biomed. Mass Spectrom.*, **3**, 161.
167. Parfitt, R. T., Games, D. E., Rossiter, M., Rogers, M. S., and Weston, A. (1976), *Biomed. Mass Spectrom.*, **3**, 232.
168. Onuska, F. I., Comba, M. E., and Harrison, A. G. (1976), *Biomed. Mass Spectrom.*, **3**, 248.
169. Araki, E., Ariga, T., and Murata, T. (1976), *Biomed. Mass Spectrom.*, **3**, 261.
170. Mudgett, M., Bowen, D. V., Field, F. H., and Kindt, T. J. (1977), *Biomed. Mass Spectrom.*, **4**, 159.
171. Lyle, M. A., Pallante, S., Head, K., and Fenselau, C. (1977), *Biomed. Mass Spectrom.*, **4**, 190.
172. Smith, A. G., and Brooks, C. J. W. (1977), *Biomed. Mass Spectrom.*, **4**, 258.

173. Dawkins, B. G., Arpino, P. J., and McLafferty, F. W. (1978), *Biomed. Mass Spectrom.*, **5**, 1.
174. Solomon, J. J., Van Duuren, B. L., and Tseng, S. -S. (1978), *Biomed. Mass Spectrom.*, **5**, 164.
175. Harless, R. L., Harris, D. E., Sovocool, G. W., Zehr, R. D., Wilson, N. K., and Oswald, E. O. (1978), *Biomed. Mass Spectrom.*, **5**, 232.
176. Weinkam, R. J. (1978), *Biomed. Mass Spectrom.*, **5**, 334.
177. Seyama, Y., Kawaguchi, A., Kasama, T., Sasaki, K., Arai, K., Okuda, S., and Yamakawa, T. (1978), *Biomed. Mass Spectrom.*, **5**, 357.
178. Boreham, D. R., Ford, G. C., Haskins, N. J., Vose, C. W., and Palmer, R. F. (1978), *Biomed. Mass Spectrom.*, **5**, 524.
179. Lin, Y. Y., and Smith, L. L. (1978), *Biomed. Mass Spectrom.*, **5**, 604.
180. Bowser, D. V., Teece, R. C., and Somani, S. M. (1978), *Biomed. Mass Spectrom.*, **5**, 627.
181. Carnevale, J., Cole, E. R., Nelson, D., and Shannon, J. S. (1978), *Biomed. Mass Spectrom.*, **5**, 641.
182. Issachar, D., and Yinon, J. (1979), *Biomed. Mass Spectrom.*, **6**, 47.
183. De Jong, E. G., Heerma, W., and Sicherer, C. A. X. G. F. (1979), *Biomed. Mass Spectrom.*, **6**, 242.
184. Dzidic, I., and McCloskey, J. A. (1971), *J. Am. Chem. Soc.*, **93**, 4955.
185. Weinkam, R. J. (1974), *J. Am. Chem. Soc.*, **96**, 1032.
186. Field, F. H., and Munson, M. S. B. (1967), *J. Am. Chem. Soc.*, **89**, 4272.
187. Munson, M. S. B., and Field, F. H. (1967), *J. Am. Chem. Soc.*, **89**, 1047.
188. Field, F. H. (1968), *J. Am. Chem. Soc.*, **90**, 5649.
189. Field, F. H. (1970), *J. Am. Chem. Soc.*, **92**, 2672.
190. Munson, M. S. B., and Field, F. H. (1966), *J. Am. Chem. Soc.*, **88**, 4337.
191. Fales, H. M., Lloyd, H. A., and Milne, G. W. A. (1970), *J. Am. Chem. Soc.*, **92**, 1590.
192. Milne, G. W. A., Axenrod, T., and Fales, H. M. (1970), *J. Am. Chem. Soc.*, **92**, 5170.
193. Field, F. H. (1967), *J. Am. Chem. Soc.*, **89**, 5328.
194. Weeks, D. P., and Field, F. H. (1970), *J. Am. Chem. Soc.*, **92**, 1600.
195. Field, F. H., and Weeks, D. P. (1970), *J. Am. Chem. Soc.*, **92**, 6521.
196. David, L., Scanzi, E., Fraisse, D., and Tabet, J. C. (1982), *Tetrahedron*, **38**, 1619.
197. Ferrer-Correia, A. J. V., Jennings, K. R., and Sen Sharma, D. K. (1976), *Org. Mass Spectrom.*, **11**, 867.
198. Michnowicz, J., and Munson, B. (1972), *Org. Mass Spectrom.*, **6**, 283.
199. Van Gaever, F., Monstrey, J., and Van de Sande, C. C. (1977), *Org. Mass Spectrom.*, **12**, 200.
200. Weinkam, R. J., and Gal, J. (1976), *Org. Mass Spectrom.*, **11**, 197.
201. Price, P., Swofford, H. S., Jr, and Buttrill, S. E., Jr (1976), *Anal. Chem.*, **48**, 494.
202. Whitney, T. A., Klemann, L. P., and Field, F. H. (1971), *Anal. Chem.*, **43**, 1048.
203. Fales, H. M., Milne, G. W. A., and Nicholson, R. S. (1971), *Anal. Chem.*, **43**, 1785.
204. Risby, T. H., Jurs, P. C., Lampe, F. W., and Yergey, A. L. (1974), *Anal. Chem.*, **46**, 161.
205. Risby, T. H., Jurs, P. C., Lampe, F. W., and Yergey, A. L. (1974), *Anal. Chem.*, **46**, 726.
206. Jelus, B., Munson, B., and Fenselau, C. (1974), *Anal. Chem.*, **46**, 729.
207. Dougherty, R. C., Roberts, J. D., and Biros, F. J. (1975), *Anal. Chem.*, **47**, 54.
208. Fales, H. M., Milne, G. W. A., WInkler, H. U., Beckey, H. D., Damico, J. N., and Barron, R. (1975), *Anal. Chem.*, **47**, 207.
209. Hunt, D. F., and Harvey, T. M. (1975), *Anal. Chem.*, **47**, 1965.
210. Buttrill, S. E., Jr, and Findeis, A. F. (1976), *Anal. Chem.*, **48**, 626.

211. Murata, T., and Takahashi, S. (1977), *Anal. Chem.*, **49**, 728.
212. Chao, J. -M., Saferstein, R., and Manura, J. (1974), *Anal. Chem.*, **46**, 296.
213. Murata, T., Takahashi, S., and Takeda, T. (1975), *Anal. Chem.*, **47**, 577.
214. Murata, T., Takahashi, S., and Takeda, T. (1975), *Anal. Chem.*, **47**, 573.
215. Jardine, I., and Fenselau, C. (1975), *Anal. Chem.*, **47**, 730.
216. Hunt, D. F., and Harvey, T. M. (1975), *Anal. Chem.*, **47**, 2136.
217. Johnson, L. P., Subba Rao, S. C., and Fenselau, C. (1978), *Anal. Chem.*, **50**, 2022.
218. Ziffer, H., Fales, H. M., Milne, G. W. A., and Field, F. H. (1970), *J. Am. Chem. Soc.*, **92**, 1597.
219. Odiorne, T. J., Harvey, D. J., and Vouros, P. (1973), *J. Org. Chem.*, **38**, 4274.
220. Dougherty, R. C., Roberts, J. D., Binkley, W. W., Chizov, O. S., Kadentsev, V. I., and Solov'yov, A. A. (1974), *J. Org. Chem.*, **39**, 451.
221. Haegele, K. D., and Desiderio, D. M., Jr (1974), *J. Org. Chem.*, **39**, 1078.
222. David, L., Scanzi, E., Fraisse, D., Beloeil, J. C., and Tabet, J. C. (1982), *Tetrahedron*, **38**, 1631.
223. Michnowicz, J., and Munson, B. (1970), *Org. Mass Spectrom.*, **4**, 481.
224. Christensen, B. W., Kjaer, A., Madsen, J. O., Olsen, C. E., Olson, O., and Sørensen, H. (1982), *Tetrahedron*, **38**, 353.
225. Dzidic, I., and McCloskey, J. A. (1972), *Org. Mass Spectrom.*, **6**, 939.
226. Stan, H. -J., and Banz, M. (1980), *Anal. Chem. Symp. Ser.*, **4**, *(Recent Dev. Mass Spectrom. Med.*, **6**), 53.
227. Arsenault, G. P., Althaus, J. R., and Divekar, P. V. (1969), *J. Chem. Soc. Chem. Comm.*, **1969**, 1414.
228. Gray, W. R., Wojcik, L. H., and Futrell, J. H. (1970), *Biochem. Biophys. Res. Comm.*, **41**, 1111.
229. Wilson, M. S., Dzidic, I., and McCloskey, J. A. (1971), *Biochem. Biophys. Acta*, **240**, 623.
230. Vouros, P., and Carpino, L. A. (1974), *J. Org. Chem.*, **39**, 3777.
231. Winkler, J., and McLafferty, F. W. (1974), *Tetrahedron*, **30**, 2971.
232. Keough, T., and Stefano, A. J. (1981), *Org. Mass Spectrom.*, **16**, 527.
233. Subba Rao, S. C., and Fenselau, C. (1978), *Anal. Chem.*, **50**, 511.
234. Harvan, D. J., Hass, J. R., Albro, P. W., and Friesen, M. D. (1980), *Biomed. Mass Spectrom.*, **6**, 242.
235. Macquestiau, A.,, Flammang, B., and Nielsen, L. (1980), *Org. Mass Spectrom.*, **15**, 376.
236. Jelus, B. L., Murray, R. K., Jr, and Munson, B. (1975), *J. Am. Chem. Soc.*, **97**, 2362.
237. Milne, G. W. A., Fales, H. M., and Axenrod, T. (1971), *Anal. Chem.*, **43**, 1815.
238. Fentiman, A. F., Jr, Foltz, R. L., and Kinzer, G. W. (1973), *Anal. Chem.*, **45**, 580.
239. Tumlinson, J. H., Heath, R. R., and Doolittle, R. E. (1974), *Anal. Chem.*, **46**, 1309.
240. Price, P., Martinsen, D. P., Upham, R. A., Swofford, H. S., Jr, and Buttrill, S. E., Jr, (1975), *Anal. Chem.*, **47**, 190.
241. Weinkam, R. J., Gal, J., Callery, P., and Castagnoli, N., Jr (1976), *Anal. Chem.*, **48**, 203.
242. Fujiwara, T., and Arai, K. (1982), *Carbohydr. Res.*, **101**, 287.
243. Carroll, D. I., Dzidic, I., Horning, M. G., Montgomery, F. E., Nowlin, J. G., Stillwell, R. N., Thenot, J. -P., and Horning, E. C. (1979), *Anal. Chem.*, **51**, 1858.
244. Einolf, N., and Munson, B. (1972), *Int. J. Mass Spec. Ion. Phys.*, **9**, 141.
245. Hunt, D. F., Stafford, G. C., Jr, Crow, F. W., and Russell, J. W. (1976), *Anal. Chem.*, **48**, 2098.
246. Hunt, D. F., McEwen, C. N., and Harvey, T. M. (1975), *Anal. Chem.*, **47**, 1730.
247. Hunt, D. F., McEwen, C. N., and Upham, R. A. (1972), *Anal. Chem.*, **44**, 1292.

248. Munson, M. S. B., and Field, F. H. (1966), *J. Am. Chem. Soc.*, **88**, 2621.
249. Field, F. H. (1969), *J. Am. Chem. Soc.*, **91**, 6334.
250. Field, F. H. (1969), *J. Am. Chem. Soc.*, **91**, 2827.
251. Cairns, T., Siegmund, E. G., and Froberg, J. E. (1981), *Biomed. Mass Spectrom.*, **8**, 569.
252. McCloskey, J. A., Futrell, J. H., Elwood, T. A., Schram, K. H., Panzica, R. P., and Townsend, L. B. (1973), *J. Am. Chem. Soc.*, **95**, 5762.
253. Winkler, F. J., and Stahl, D. (1979), *J. Am. Chem. Soc.*, **101**, 3685.
254. Herman, J. A., and Harrison, A. G. (1981), *Org. Mass Spectrom.*, **16**, 423.
255. Longevialle, P., Milne, G. W. A., and Fales, H. M. (1973), *J. Am. Chem. Soc.*, **95**, 6666.
256. Hunt, D. F., and Ryan, J. F. (1972), *J. Chem. Soc. Chem. Comm.*, **1972**, 620.
257. Lin, Y. Y., and Smith, L. L. (1979), *Biomed. Mass Spectrom.*, **6**, 15.
258. Kingston, E. E., Shannon, J. S., Diakiw, V., and Lacey, M. J., *Org. Mass Spectrom.*, **16**, 428.
259. Knight, D. W., Knight, M. E., Rossiter, M., Jones, A., and Games, D. E. (1980), *Adv. Mass Spectrom.*, **8B**, 1116.
260. Sieck, L. W. (1980), *Chem. Brit.*, **16**, 38.
261. Wood, G. W., Mak, N., and Hogg, A. M. (1976), *Anal. Chem.*, **48**, 981.
262. Brandenberger, H. (1980), *Anal. Chem. Symp. Ser.*, **4**, *(Recent Dev. Mass Spectrom. Biochem. Med.*, **6**), 391.
263. Cairns, T., and Siegmund, E. G. (1982), *Biomed. Mass Spectrom.*, **9**, 307.
264. Björkman, S. (1982), *Biomed. Mass Spectrom.*, **9**, 315.
265. Roboz, J., Greaves, J., Jui, H., and Holland, J. F. (1982), *Biomed. Mass Spectrom.*, **9**, 510.
266. Cairns, T., and Siegmund, E. G. (1981), *Org. Mass Spectrom.*, **16**, 555.
267. Zitrin, S. (1982), *Org. Mass Spectrom.*, **17**, 74.
268. Freeman, F., and Angeletakis, C. N. (1982), *Org. Mass Spectrom..*, **17**, 114.
269. Brauner, A., Budzikiewicz, H., and Boland, W. (1982), *Org. Mass Spectrom.*, **17**, 161.
270. Muszkat, L., and Aharonson, N. (1982), *J. Agric. Food Chem.*, **30**, 613.
271. Guinivan, R. A., Thompson, N. P., and Wheeler, W. B. (1982), *J. Agric. Food Chem.*, **30**, 977.
272. Smith, R. G., and Cheung, L. K. (1982), *J. Chromatogr.*, **229**, 464.
273. Murata, T., Takahashi, S., Ohnishi, S., Hosoi, K., Nakashima, T., Ban, Y., and Kuriyama, K. (1982), *J. Chromatogr.*, **239**, 571.
274. Cairns, T., and Siegmund, E. G. (1982), *Anal. Chem.*, **54**, 2456.
275. Lewy, A. J. and Markey, S. P. (1978), **201**, 741.
276. Shefer, S., Salen, G., Cheng, F. W., Dayal, B., Batta, A. K., Tint, G. S., Bose, A. K., and Pramanik, B. N. (1982), *Anal. Biochem.*, **121**, 23.

Chapter 4

Negative Ion Chemical Ionization

4.1 Introduction

During an ionization process such as electron impact ionization, negatively as well as positively charged ions are formed. Normally, the negative ions remain undetected because ion source and focusing potentials allow only the extraction of positive ions from the source. Other changes are often necessary since, for example, magnetic analysers do not transmit negative ions unless the magnet current is reversed, and a standard electron multiplier does not usually detect negative ions without some modification. These, and other modifications, are discussed in section 4.3.

Even when the appropriate instrumental modifications are made, further limitations to the analytical usefulness of negative ions formed by electron impact are found. First, many organic compounds do not form molecular or quasi-molecular anions under conventional EI conditions but instead provide spectra dominated by structurally insignificant low mass ions such as CN^- and F^- [1]. Second, the sensitivity for negative ion production under electron impact conditions is several orders of magnitude lower than that for positive ion production.

In contrast to the case of ionization under conventional electron impact conditions, low energy electrons are readily captured by many organic compounds without inducing extensive fragmentation (section 4.2.1). A resurgence of interest in negative ions began with the development of a simple and convenient means of generating a large population of electrons with near thermal energy in the ion source. This was achieved by Hunt *et al.* [2] following earlier work by Von Ardenne *et al.* [3] and Dougherty and Weisenberger [4] by the simple expedient of using the reagent gas in a conventional chemical ionization source as a moderator for the initially energetic electrons. Anionic reagent gas ions may also be generated in a chemical ionization source (section 4.2.2), providing a further extension to the experimental control of analytical information available under CI conditions.

Negative ion formation is not limited to CI operation since most of the newer ionization techniques, developed for the analysis of labile molecules (Chapters 5 and 6), also operate readily in this mode. With techniques such as fast atom bombardment (Chapter 5) and thermospray ionization (Chapter 6), ion-molecule reactions are involved in the formation of analyte-related ions in the negative ion mode just as in the positive ion mode (section 3.1.1). An ability to form negative ions is useful in any analytical technique since the molecular anion may have greater stability than the corresponding positively charged ion and may also provide complementary structural information through different fragmentation processes.

4.2 Reactions in Negative Ion Chemical Ionization [5]

Under CI conditions, the reagent gas affords positive reagent ions as detailed in Chapter 3, together with electrons of near thermal energy and negatively charged reagent ions in the case where stable anions can be formed from the reagent gas. Thus, further reactions to produce negative ions from the sample fall into two categories: capture of thermal electrons by the sample molecules and ion–molecule reactions between sample and reagent gas ions. These two categories are now discussed in some detail.

4.2.1 Ion formation by electron capture

Formation of negative ions by the interaction of electrons and molecules can occur by three mechanisms formalized as follows:

$$AB + e^- \rightarrow AB^{-\cdot} \text{ (associative resonance capture)} \quad (4.1)$$

$$AB + e^- \rightarrow A^- + B^\cdot \text{ (dissociative resonance capture)} \quad (4.2)$$

$$AB + e^- \rightarrow A^+ + B^- + e^- \text{ (ion-pair production)} \quad (4.3)$$

Under CI conditions the reagent gas acts both as a means of producing thermal electrons and as a source of molecules for collisional stabilization of the ions formed. For example, the electron bombardment of methane at a pressure of approximately 1 torr and containing a low level of an electron capturing material, such as nitrobenzene, results in the following reactions:

$$e^- \xrightarrow{CH_4} e^-_{th} \quad (4.4)$$

$$e^-_{th} + C_6H_5NO_2 \rightarrow C_6H_5NO_2{}^{-*} \quad (4.5)$$

$$C_6H_5NO_2{}^{-*} \xrightarrow{CH_4} C_6H_5NO_2{}^{-} \quad (4.6)$$

where equation (4.6) represents collisional stabilization. Equation (4.5) is the associative resonance capture process (electron energy 0–2 eV) which is widely used analytically to form molecular anions from compounds with a positive electron affinity, i.e. that are electron capturing (Fig. 4.1).

The rate constant for this reaction (attachment rate) shows very large variations amongst compounds so that different compounds show very different sensitivities under electron capture conditions [6,7]. In favourable cases the attachment rate is

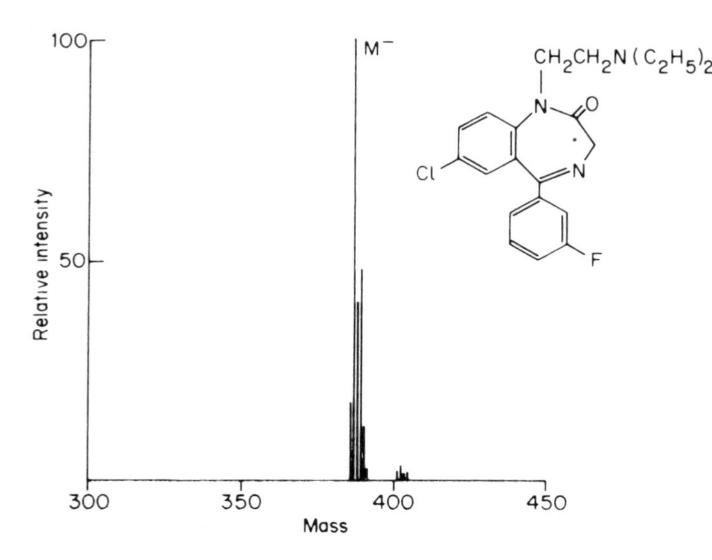

Fig. 4.1 Electron capture spectrum of flurazepam. Methane reagent gas. (Reprinted with permission from B.J. Miwa, W.A. Garland, and P. Blumenthal, *Anal. Chem.*, **53**, 793 [37]. Copyright (1981) American Chemical Society)

considerably greater than the rate of ion formation in the positive ion chemical ionization mode so that enhanced sensitivity results (section 4.2.3) [2,8].

Since the electron affinity (defined as the lowest energy required to remove an electron from the molecular anion) for most organic molecules does not exceed 2.5 eV [9,10], any excess energy from the electron attachment process must be removed by collisional stabilization otherwise ions are lost by electron auto-detachment:

$$AB^{-*} \rightarrow AB + e^- \qquad (4.7)$$

A relatively high source pressure is therefore necessary in many cases to minimize loss of sensitivity (section 4.3.1).

Dissociative resonance capture is observed with electrons in the energy range 0–15 eV. For example, the primary reaction producing negative ions from the electron bombardment of methylene chloride at 1 torr [11] is a dissociative capture reaction:

$$CH_2Cl_2 + e^- \rightarrow Cl^- + CH_2Cl_2^{\cdot} \qquad (4.8)$$

Both dissociative and associative capture processes are enhanced under CI conditions because of the increased abundance of low energy electrons.

Ion pair production occurs over a wide range of electron energies above 10 eV and is the principal method of negative ion formation under conventional electron impact conditions.

4.2.2 Ion formation by ion–molecule reactions [12]

These reactions can conveniently be grouped into four major categories, viz.

(a) Proton transfer $M + X^- \rightarrow (M - H)^- + XH$
(b) Charge exchange $M + X^- \rightarrow M^- + X$
(c) Nucleophilic addition $M + X^- \rightarrow MX^-$
(d) Nucleophilic displacement $AB + X^- \rightarrow BX + A^-$

4.2.2.1 Proton transfer

The tendency for an anion X^- to accept a proton from a particular analyte molecule may be assessed from a knowledge of proton affinity values for the anions [13]. The proton affinity of anion X^- is the negative of the change in enthalpy or heat content accompanying the reaction

$$X^- + H^+ \rightarrow XH \qquad PA(X^-) = -\Delta H \tag{4.9}$$

Thus, observation of the reaction

$$X^- + HY \rightarrow XH + Y^- \tag{4.10}$$

implies that $PA(X^-) > PA(Y^-)$, i.e. the reaction is exothermic.

Table 4.1 Proton affinities of selected anions

Anion	$PA(X^-)$ (kJ mole^{-1})
NH_2^-	1689
H^-	1675
OH^-	1635
$O^{-\cdot}$	1599
CH_3O^-	1592
$(CH_3)_2CHO^-$	1571
CH_2CN^-	1560
F^-	1554
$C_5H_5^-$	1481
$O_2^{-\cdot}$	1476
CN^-	1469
$CH_3CO_2^-$	1459
Cl^-	1395

Representative proton affinity values for anions are listed in Table 4.1. From these data we see that an anion such as CH_3O^- may be used to generate other anions with a smaller proton affinity by the addition of the corresponding neutral.

For example, CH_3O^- generated by the addition of approximately 0.1% methyl nitrite [14] to methane at 1 torr,

$$CH_3ONO + e^- \rightarrow CH_3O^- + NO \quad (4.11)$$

can be used to form the cyclopentadienyl anion,

$$CH_3O^- + C_5H_6 \rightarrow C_5H_5^- + CH_3OH \quad (4.12)$$

by adding cyclopentadiene until the methoxyl ion is exhausted [2].

Analyte anions produced by proton transfer have very little excess internal energy. As can be seen from equation (4.10), no new bonds are involved in the formation of the analyte anion so that most of the excess energy resides in the neutral derived from the reagent gas ion. As a result, negative ion proton transfer spectra generally contain abundant quasi-molecular ions but very few fragment ions.

The OH^- ion, conveniently formed by electron bombardment of a mixture of approximately equal amounts of methane and nitrous oxide [15,16],

$$N_2O + e^- \rightarrow N_2 + O^{-\cdot} \quad (4.13)$$

$$O^{-\cdot} + CH_4 \rightarrow OH^- + CH_3^{\cdot} \quad (4.14)$$

is a strong base that will abstract protons from a wide range of compounds and has been extensively studied as an analytical reagent (see Appendix 4.1). Anions with a lower proton affinity such as Cl^- are of limited value for the production of analyte anions by proton transfer, although they will form hydrogen-bonded complexes with more acidic molecules (section 4.2.2.3). These same hydrogen-bonded intermediates may alternatively react by elimination of a small neutral molecule such as HF, HCl, or H_2O [54]:

$$X— + H—C—C—A \longrightarrow [X.....H.....A]^- + C{=}C \quad (4.15)$$

Such an elimination reaction can occur even when the anion has too low a proton affinity to promote a proton transfer reaction such as CN^-:

$$CN^- + CF_2H\text{-}CF_2H \rightarrow CNHF— + CF_2{=}CHF \quad (4.16)$$

4.2.2.2 *Charge exchange*

The charge exchange reaction

$$M + X^{-\cdot} \rightarrow M^{-\cdot} + X \quad (4.17)$$

will occur if the electron affinity (section 4.2.1) of molecule M is greater than that of molecule X. Thus, the relatively low electron affinity of the oxygen molecule means that O_2^- is able to ionize samples with higher electron affinities by charge exchange:

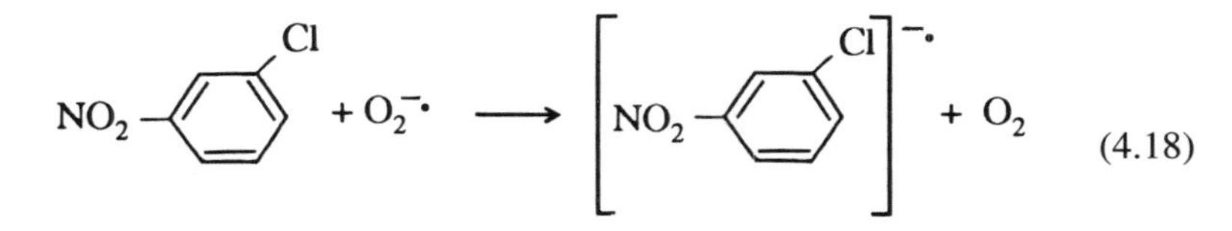

(4.18)

Solvated $O_2^{-\cdot}$ is commonly encountered as a reagent ion in LC–MS with a corona discharge atmospheric pressure ionization source (section 2.4.6).

Charge exchange also provides a mechanism for the loss of analyte ions [17]. For example, fluorine radicals, formed when Freon reagent gas is used as a source of chloride ions, have such a high electron affinity that they readily remove analyte ions, resulting in a reduced sensitivity for electron capturing materials:

$$M^{-\cdot} + F^{\cdot} \rightarrow M + F^- \quad (4.19)$$

4.2.2.3 Nucleophilic addition

Stable addition complexes may be formed by nucleophilic attack with anions of low proton affinity, such as Cl^- and $O_2^{-\cdot}$, which cannot usually react by proton transfer. The chloride anion formed by electron bombardment of methylene dichloride [11,18], methylene dichloride–methane mixtures [17,19], or dichlorodifluoromethane (Freon 12) [20] forms stable hydrogen-bonded complexes which account for the majority of the ionization in the spectra of a number of acids, amides, aromatic amines, and phenols [11]. The chloride anion also forms stable $(M+Cl)^-$ complexes with a range of other compound types (Fig. 4.2) and has been used in the analysis of polychlorinated aliphatic compounds [21–23], organophosphorus pesticides [18], and oligosaccharides [20,24] (section 4.4).

The $O_2^{-\cdot}$ ion formed from oxygen, and implicated in atmospheric pressure ionization processes, is a stronger base than Cl^- and can react by proton transfer with strongly acidic materials such as nitrophenols [25,26]. With less acidic materials such as aliphatic alcohols, nucleophilic addition to give $(M+O_2)^-$ ions is the major reaction [27].

Under some circumstances the reagent gas itself is susceptible to nucleophilic attack by anions originating from the analyte molecule. For example, polycyclic aromatic hydrocarbons [27] or poly-chlorinated aromatics [17,25] ionized by electron capture will react with oxygen to form addition complexes which then rearrange to form diagnostic phenoxide ions (section 4.4):

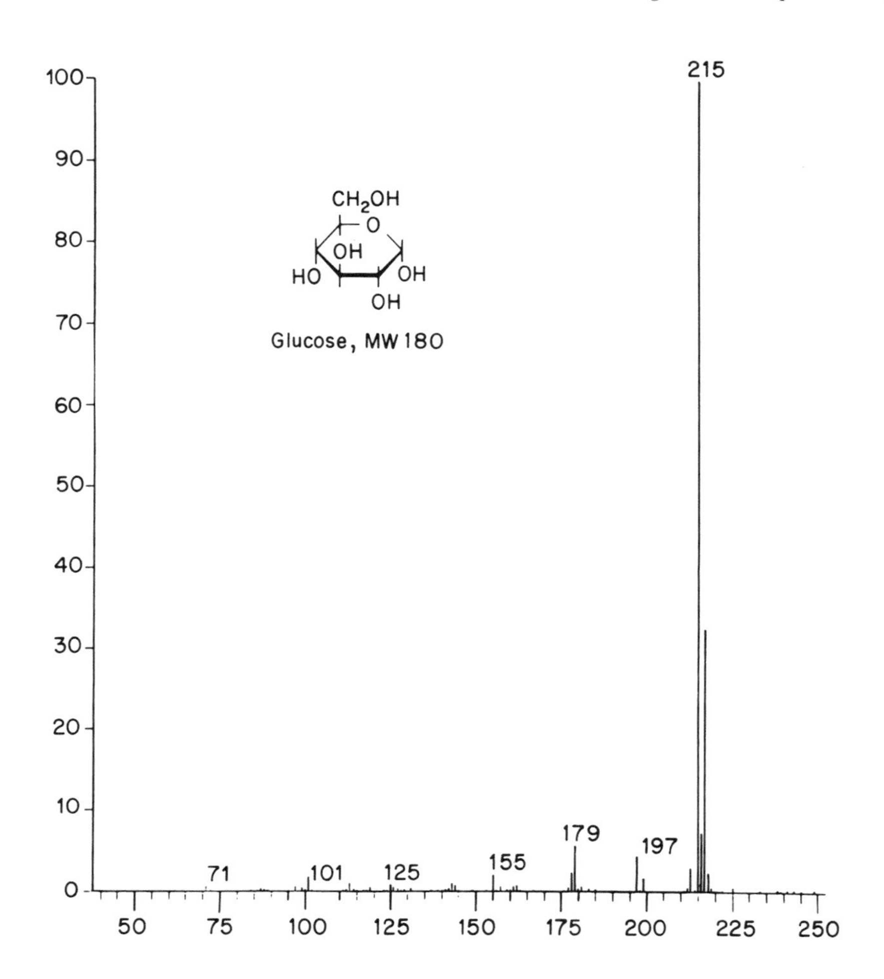

Fig. 4.2 Chloride addition spectrum of glucose. (Courtesy Kratos Analytical Instruments)

$$(Ar{-}X)^{-} + O_2 \longrightarrow \left[Ar \begin{matrix} \cdot\cdot X \\ \cdot\cdot O{-}O \end{matrix} \right]^{-} \longrightarrow ArO^{-} + OX\cdot \qquad (4.20)$$

where X = H, Cl

Again, the $(M - H)^-$ ions formed from some steroid alcohols and unsaturated steroids by proton transfer to OH^- will react with N_2O present in the gas mixture (equations 4.13 and 4.14) to form addition products with the composition $(M+N_2O - H)^-$ [28–30].

4.2.2.4 Nucleophilic displacement

OH^- and $O^{-\cdot}$ (formed by electron bombardment of nitrous oxide) are both strongly basic anions so that a proton transfer reaction leading to the formation of an $(M - H)^-$ ion is the major reaction with many compounds in each case. Both ions can, however, participate in nucleophilic displacement reactions with suitable compounds [6,28,54] so that, for example, the spectra of aliphatic esters (RCO_2R^1) also contain carboxylate anions (RCO_2^-) in each case. $O^{-\cdot}$ is a particularly strong nucleophile and will also displace hydrogen atoms from aromatic compounds [54]:

$$ArH + O^{-\cdot} \rightarrow ArO^- + H^\cdot \qquad (4.21)$$

4.2.3 Sensitivity under electron capture conditions

A study of the relative sensitivities of positive ion (PI) and negative ion (NI) chemical ionization and electron impact ionization gave the results listed in Table 4.2 [31]. The most instructive comparison available from these data is that provided by the total ionization figures. From these it can be seen that while the total ionization under electron capture conditions can equal or exceed that recorded under electron impact conditions, the total ionization in positive ion chemical ionization is at least an order of magnitude less.

Theoretical treatments by Siegel [32] and Field [6] lead to the conclusion that under equivalent chemical ionization conditions the ratio of negative to positive ion current is equal to the ratio of the rate constants for electron capture by sample molecules (K^-) and for the reaction of sample molecules with positive reagent ions (K^+). While values of (K^+) do not exceed $3–4 \times 10^{-9}\,cm^3\,mol^{-1}\,s^{-1}$, the rate constant ($K^-$) may well be larger because of the high mobility of electrons in the chemical ionization plasma.

Rate constants for electron capture vary considerably with molecular structure so that sensitivity under these conditions is very dependent on the choice of sample [6,7]. In some cases, as reported by Hunt *et al.* [2], the electron capture sensitivity can be orders of magnitude greater than that observed in positive ion chemical ionization. The relative sensitivity figure for trinitrotoluene, for example, is 1042.

Hunt and Crow [8] have, however, also pointed out that these high sensitivities cannot be maintained with higher sample flow rates since the rate of supply of electrons then becomes limiting. This saturation effect was observed in practice when more than 20 ng of the pentafluoropropionyl derivative of an amine was introduced via a GC column [33]. Other electron capturing materials introduced into the ion source at the same time will have the same effect so that, for example, chlorinated solvents would not be used in this type of work.

Table 4.2 Sensitivities in different ionization modes

Compound	MW	Base peak sensitivity (C μg^{-1})			Total ionization (C μg^{-1})		
		PICI	NICI	EI	PICI	NICI	EI
17-hydroxy-1,4,6-androstatriene-3-one	326	5.8×10^{-11}	2.9×10^{-9}	No data	1.7×10^{-10}	4.2×10^{-9}	No data
Anthracene	154	5.1×10^{-11}	3.1×10^{-9}	1.3×10^{-9}	1.1×10^{-10}	3.8×10^{-9}	2.6×10^{-9}
Benzil	210	1.2×10^{-11}	1.3×10^{-9}	No data	3.4×10^{-11}	1.5×10^{-9}	No data
Methyl stearate	298	1.0×10^{-11}	—	5.0×10^{-10}	2.7×10^{-10}	—	2.5×10^{-9}

4.3 Instrumentation for Negative Ion Operation

The formation of a beam of negative ions for analysis in quadrupole, time-of-flight, or magnetic sector instruments demands that both ion source and associated lens potentials be reversed from positive ion operation. A quadrupole analyser is then impartial to the sign of the charge on the ions, as is a time-of-flight analyser. A magnetic analyser is not impartial. Thus, the magnet current must be reversed, as must the voltages on the electrostatic analyser, to deal with negative ions. With an FT–MS instrument, positive or negative ion operation is selected by applying an appropriate potential to the trapping plates. Ions of the opposite polarity are not trapped. The use of an r.f. field in an ion trap mass spectrometer, on the other hand, means that both positive and negative ions can be trapped simultaneously [34].

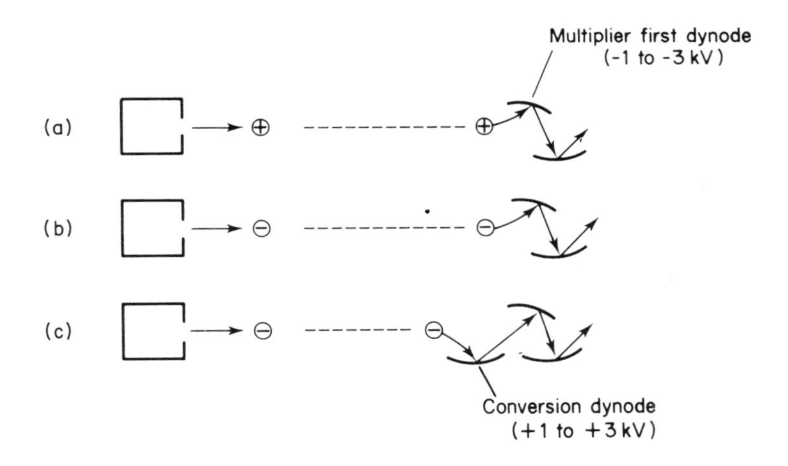

Fig. 4.3 Ion detection using an electron multiplier: (a) for positive ions; (b) for negative ions; (c) using an extra conversion dynode

Operation of an electron multiplier (section 1.4.1) for the detection of positive ions is illustrated in Fig. 4.3(a). Under these conditions, positive ions are accelerated towards the first dynode which is held at a high negative potential (1–3 kV) depending on the gain required. When the detection of negative ions is required, this situation is no longer satisfactory unless the ions originate from, for example, a magnetic sector instrument operated with a higher ion source potential (Fig. 4.3b). With lower source potentials, negative ions either no longer reach the electron multiplier or impact on the first dynode with insufficient energy to cause secondary electron emission.

A convenient solution to this problem that leaves the electron multiplier output at earth potential is to add a separate conversion dynode, held at a positive potential, prior to the multiplier [35] (Fig. 4.3c). The impact of energetic negative ions on this plate results in the emission of positive ions which are then detected by the electron multiplier in the normal manner. In positive ion operation, the conversion dynode may be operated at a high negative potential as a post-acceleration detector (section 1.4.3), or ions may be allowed to reach the multiplier directly without being deflected to a separate dynode.

4.3.1 Effect of operating conditions on negative ion CI spectra

Negative ion chemical ionization spectra can show a substantial temperature dependence for a number of reasons. For example, the input of thermal energy may activate dissociative electron capture reactions [36,37], as is shown in the data of Table 4.3. In practice this aflatoxin analysis was carried out at an ion source temperature of 130°C to prevent undue source contamination. The formation of anions by nucleophilic addition can likewise be very temperature dependent in cases where only weakly bound complexes are formed [11,18]. Thermally induced fragmentation is also always likely where an ionization technique is used for the analysis of more labile materials.

Table 4.3 Effect of temperature on NICI spectrum of aflatoxin b_1. (Reproduced with permission from W.C. Brumley *et al.*, *Anal. Chem.*, **53**, 2003[36]. Copyright (1981) American Chemical Society)

Ion source temperature (°C)	Relative intensity		
	M^-	$(M - H)^-$	$(M - CH_3)^-$
60	59.0	22.0	3.2
100	47.8	22.4	14.9
130	37.3	17.9	30.6
150	15.2	6.0	63.4
200	2.2	4.1	86.7

In addition, reactions leading to the production of reagent ions are themselves temperature dependent. For example, the concentration of chloride ions in the plasma produced from methylene dichloride increases with temperature, as does the abundance of chloride attachment ions $(M+Cl)^-$ in the absence of competing reactions [11]. In the case of OH^- chemical ionization using N_2O/CH_4 mixtures, a minimum source temperature of 200°C is necessary for the efficient production of hydroxyl ions [6]. Methods for the preparation of other anionic reagent gas mixtures may be found in the literature cited in Appendix 4.1.

An alternative method in which the solid sample is mixed with a salt that can act as a source of reagent ions on heating has also been described [38]. The use of ammonium chloride provides chloride ions whereas sodium bicarbonate serves as a source of hydroxyl ions. This technique can be used to provide CI spectra in a conventional EI source (cf. sections 3.2.1 and 5.3.1.2).

Since the formation of stable negative ions by non-dissociative electron capture involves collisional stabilization with another reagent gas molecule subsequent to ionization (section 4.2.1) it follows that electron capture ionization is particularly strongly dependent on source pressure [6,17]. Furthermore, this dependence varies from sample to sample since the lifetimes of excited negative molecular ions vary over a wide range [6,10]. For example, Miwa *et al.* [37] determined the relationship between the molecular anion intensity for flurazepam and methane pressure during the development of an assay method for this material. From a source housing pressure of 4×10^{-4} torr at which virtually no molecular anion was observed, a maximum sensitivity was achieved at 6×10^{-4} torr.

For electron capture work, high pressure source conditions can be established with methane etc., as described in section 3.2.3. Gases such as methane, isobutane, argon, and nitrogen are transparent in the electron capture mode so that on switching to negative ion operation the only ions present in the background will be due to impurities (e.g. OH^-, Cl^-, or the isotope clusters of ReO_3^- and ReO_4^- between m/z 237 and 251, when a rhenium filament is used). If ammonia is used as a reagent gas, a reasonably intense peak will be seen at m/z 16 (NH_2^-).

Analyte sensitivity with different reagent gases depends on the ability of the gas to thermalize electrons and its ability to collisionally stabilize the analyte ions. The following ranking of reagent gases, in order of increasing effectiveness for electron capture, has been established: $He < Ne = N_2 < CH_4 = Ar = Kr < Xe = i\text{-}Bu < CO_2$, although it should be borne in mind that these different gases can require very different source pressures for optimization[39]. Again, reactions that are competitive with electron capture may occur with one gas but not with another [5,40]. Impurities in the reagent gas can be an additional contribution to the variability of electron capture spectra. For example, oxygen may react with some ions from polycyclic aromatic hydrocarbons by gas-phase and wall-catalysed reactions [41]. Mechanisms for the formation of artefact ions under electron capture conditions have been considered in some detail [42].

Background peaks may be sufficient for an initial tuning of the instrument, otherwise a sample eventually to be used for sensitivity determination (e.g. anthraquinone or anthracene) or a low pressure of any fluorinated reference compound (see Appendix A at the end of the book) may be used. The reagent gas pressure and source controls, including electron volts, should all be adjusted for maximum sensitivity at this stage. Temperature should be maintained at the minimum level necessary for sample volatilization. The use of small magnets positioned on the vacuum housing above the conversion dynode and the ion source has been advocated by some authors [37] as a means of improving sensitivity and signal-to-noise ratio. A number of recent publications have commented on the dependence of electron capture negative ion CI spectra on experimental conditions and have emphasized and have emphasized the lack of reproducibility from instrument to instrument, both in terms of relative abundance and absolute sensitivity [43,44].

A sensitivity test can be carried out under electron capture CI conditions using the method described in section 1.6.1. With anthraquinone as the test compound, monitor M^- at m/z 208. Care should be taken to limit the amount of anthraquinone used so that source saturation does not occur (section 4.2.3). In addition, the use of electron-capturing solvents should be avoided in all sample handling procedures. Further adjustment of source temperature pressure and tuning may well be necessary to provide optimum conditions for any subsequent sample analysis.

The quality of spectra recorded using anionic reagent ions may be assessed with suitable test compounds, e.g. glucose for Cl^- addition. It should be noted, however, that while the very low level of fragmentation in Fig. 4.2 is typical of what can be achieved, the spectrum quality, as is the case with all less energetic reagent gas systems, is as much a test of sample introduction technique as of CI source performance.

4.4 Analytical Applications of Negative Ion CI [5]

Negative ion chemical ionization provides further ion–molecule reactions which complement those available in positive ion chemical ionization and which similarly may be selected so as to provide the information required in a particular analysis. For example, the commonly used hydroxyl ion is a strong base which reacts, principally by proton abstraction and with little fragmentation, with a wide range of compounds containing an acidic hydrogen. Thus, tertiary alcohols readily provide molecular weight information using OH^- as a reactant ion [6,16], whereas they are difficult to analyse directly by most other mass spectrometric methods (cf. section 3.1.1.4). The hydroxyl ion has also been used as a reagent ion under desorption chemical ionization conditions (Chapter 5) for the analysis of high molecular weight polar materials such as the tetraglycoside purpureaglycoside which provides an intense $(M - H)^-$ ion (m/z 925) as well as fragment ion information under these conditions (Fig. 4.4) [12].

The stability of ions formed by proton abstraction can be very sensitive to stereochemical differences (cf. section 3.3.2) so that, for example, *cis*- and *trans*-cyclic diols may be distinguished since the alkoxide ions of the *cis* configuration are preferentially stabilized by hydrogen bonding [45]. Enhanced selectivity may be obtained for these compounds by the use of F^- (prepared from CHF_3) as a reagent ion since the proton affinities of fluoride and secondary alkoxide ions are very similar (Table 4.1):

$$>COH + F^- \rightleftharpoons >COH\cdots\cdots F^- \rightleftharpoons >CO^- + HF \qquad (4.22)$$

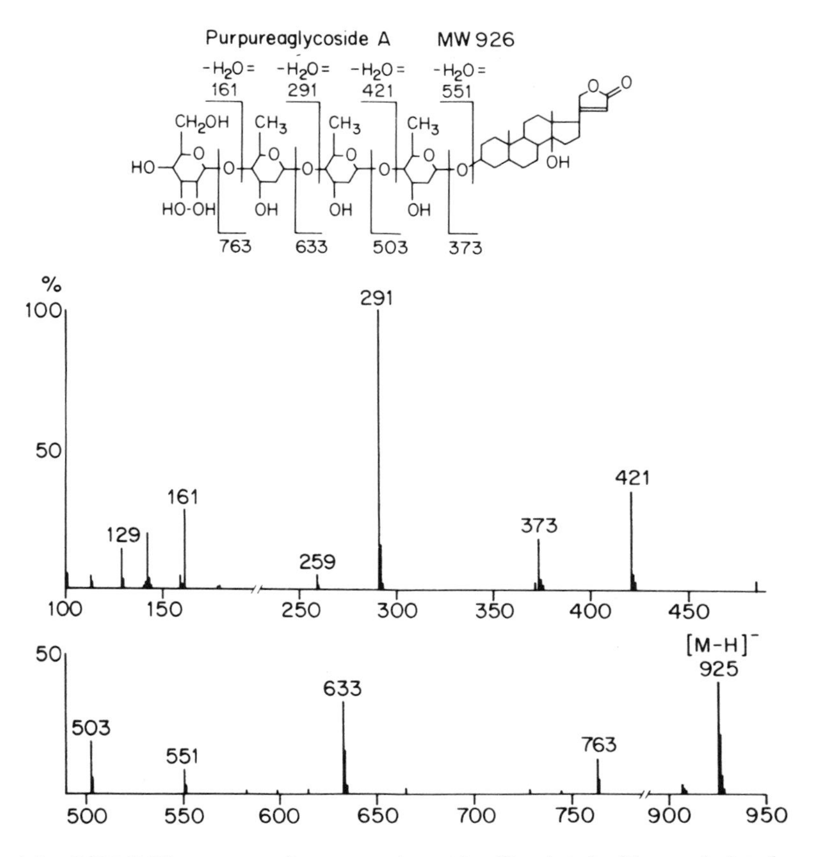

Fig. 4.4 OH^- DCI spectrum of purpureaglycoside. (Reprinted with permission from A.P. Bruins, *Anal. Chem.*, **52**, 605[12]. Copyright (1980) American Chemical Society)

In this case the base peak in the spectra of the *trans* diols is an $(M+F)^-$ ion formed by fluoride attachment, whereas the decrease in proton affinity due to hydrogen bonding in the alkoxide ion means that the spectra of the *cis* diols show an $(M - H)^-$ ion as base peak [45] (Table 4.4).

While the hydroxyl ion will abstract a proton from a wide range of compounds, the attachment reaction with chloride ions is much more restricted in its applicability. This selectivity has been used to good effect, particularly by Dougherty and co-workers, in the development of screening methods for polychlorinated aliphatics [22,23,46,47] and organophosphorus pesticides [18] by the recognition of $(M + Cl)^-$ ions. Such screening methods can be simple to implement since many co-occurring materials do not react with the chloride ion and therefore little pre-purification is required. The chloride ion also undergoes an attachment reaction with zinc dithiophosphate esters used to increase the performance of lubricating oils, and in this case an oil may be analysed directly for these additives since the majority of its constituents are also transparent to the reagent ion [48]. Similar analyses have been proposed for the direct analysis of phenols in coal-derived liquids using Cl^- CI [49] and of naphthenic acids in crude oils using F^- CI [50].

Table 4.4 Partial NICI spectra of cyclic diols (F^- reagent ion). (Reproduced with permission from ref. 45)

	Relative intensity			
Compound	$(M - H)^-$	$(M + F)^-$	$(M_2 - H)^-$	$(M_2 + F)^-$
cis-1,3-Cyclohexanediol	59	7	28	0.6
trans-1,3-Cyclohexanediol	24	47	15	3
cis-1,4-Cyclohexanediol	90	7	2	—
trans-1,4-Cyclohexanediol	33	56	6	0.7

Phenoxide ions formed by reaction with oxygen may be used in the same way for the recognition of polychlorinated aromatics. Thus, the use of a controlled mixture of isobutane, methylene dichloride, and oxygen has been recommended as a method of providing characteristic ions from a wide range of polychlorinated materials in environmental screening [17,21,23,47,51] (Fig. 4.5). Even the more generally reactive hydroxyl ion displays some selectivity by ionizing alkylated aromatic and heteroaromatic compounds, but not alkanes or cycloalkanes in crude oils [29,30].

Electron capture ionization also displays considerable selectivity amongst compound types and can be used to differentiate isomeric species [52]. Structural information, although sometimes available under electron capture conditions, is, however, limited and difficult to predict, so that the principal application of the technique is as a highly sensitive method for quantitative analysis (Chapter 8) of electron capturing compounds. Currently this one area of application accounts for the great majority of published work in the field of negative ion CI [5].

Compound types that show a high electron affinity and are therefore suitable for electron capture analysis include halogenated compounds, nitro-compounds, phosphate esters and polycyclic aromatic hydrocarbons [5]—a remarkable number of compounds of environmental interest. Another approach to electron capture negative ion CI analysis is to prepare an electron capturing derivative of the analyte. Perfluoropropionyl, perfluorobenzyl, and perfluorobenzoyl derivatives, amongst others, have been used for this purpose [5,53]. In this case, it is necessary that ions that are characteristic of the analyte, rather than solely of the derivatizing group, are available to be monitored.

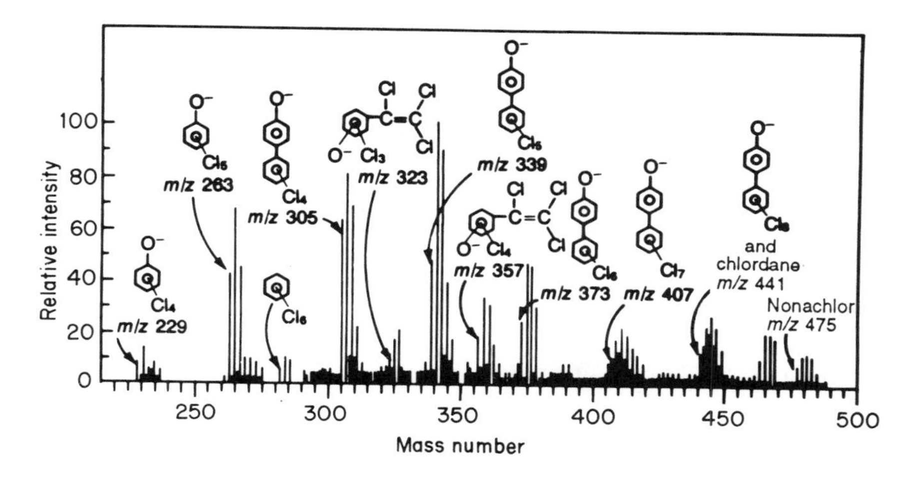

Fig. 4.5 Negative ion CI spectrum of an extract of Lake Ontario trout. Only negative mass defect ions are plotted. Reagent gases isobutane methylene dichloride, and oxygen. (Reprinted with permission from R.C. Dougherty, *Anal. Chem.*, **53**, 625A[17]. Copyright (1981) American Chemical Society)

Earlier references to published negative ion CI data are listed in Appendix 4.1.

References

1. Bowie, J.H., and Williams, B.D. (1975), in *MTP International Review of Science*, Physical Chemistry Series Two, Vol. 5 (Ed. A. Maccoll), p. 89, Butterworths, London.
2. Hunt, D.F., Stafford, G.C., Jr, Crow, F.W., and Russell, J.W. (1976), *Anal. Chem.*, **48**, 2098.
3. Von Ardenne, M., Steinfelder, K., and Tuemmler, R. (1971), *Electronen-anlagerungs-Massenspektrographie Organischer Substanzen*, Springer-Verlag, New York.
4. Dougherty, R.C., and Weisenberger, C.R. (1968), *J. Am. Chem. Soc.*, **90**, 6570.
5. Budzikiewicz, H. (1986), *Mass Spectrom. Rev.*, 5, 345.
6. Field, F.H. (1980), *28th Annual Conference on Mass Spectrometry and Allied Topics*, New York, p.2.
7. Massey, H. (1970), *Negative Ions*, Cambridge University Press, Cambridge.
8. Hunt, D.F., and Crow, F.W. (1978), *Anal.Chem.*, **50**, 1781.
9. Janousek, B.K., and Brauman, J.I. (1979), in *Gas Phase Ion Chemistry*, Vol. 2 (Ed.M.T.Bowers), p. 53, Academic Press.
10. Christophorou, L.G. (1971), *Atomic and Molecular Radiation Physics*, Wiley–Interscience, New York.
11. Tannenbaum, H.P., Roberts, J.D., and Dougherty, R.C. (1975), *Anal. Chem.*, **47**, 49.
12. Bruins, A.P. (1980), *Anal. Chem.*, **52**, 605.
13. Lias, S.G., Bartmess, J.E., Liebmann, J.F., Holmes, J.L., Levin, R.D., and Mallard, W.G. (1988), *J. Phys. Chem. Ref. Data*, **17**, Suppl.1.
14. Caldwell, G., and Bartmess, J.E. (1982), *Org. Mass Spectrom.*, **17**, 456.
15. Siegel, M.W., (1981), *Int. J. Mass Spectrom. Ion Processes*, **40**, 265.

16. Smit, A.L.C., and Field, F.H. (1977), *J. Am. Chem. Soc.*, **99**, 6471.
17. Dougherty, R.C. (1981), *Anal. Chem.*, **53**, 625A.
18. Dougherty, R.C., and Wander, J.D. (1980), *Biomed. Mass Spectrom.*, **7**, 401.
19. Crow, F.W., Bjorseth, A., Knapp, K.T., and Bennett, R. (1981), *Anal. Chem.*, **53**, 619.
20. Ganguly, A.L., Cappuccino, N.F., Fujiwara, H., and Bose, A.K. (1979), *J. Chem. Soc. Chem. Commun.*, **1979**, 148.
21. Dougherty, R.C., Whitaker, M.J., Smith, L.M., Stalling, D.L., and Kuehl, D.W. (1980), *EHP, Environ. Health Perspect.*, **36**, 103.
22. Dougherty, R.C. (1981), *Biomed. Mass Spectrom.*, **8**, 283.
23. Kuehl, D.W., Whittaker, M.J., and Dougherty, R.C. (1980), *Anal. Chem.*, **52**, 935.
24. Bose, A.K., Pramanik, B., Tabei, K., and Bates, A.D. (1980), *28th Annual Conference on Mass Spectrometry and Allied Topics*, New York, p.636.
25. Dzidic, I., Carroll, D.I., Stillwell, R.N., and Horning, E.C. (1975), *Anal. Chem.*, **47**, 1308.
26. Dzidic, I., Carroll, D.I., Stillwell, R.N., and Horning, E.C. (1974), *J. Am. Chem. Soc.*, **96**, 5258.
27. Hunt, D.F., McEwen, C.N., and Harvey, T.M. (1975), *Anal. Chem.*, **47**, 1730.
28. Roy, T.A., Field, F.H., Lin, Y.Y., and Smith, L.L. (1979), *Anal. Chem.*, **51**, 272.
29. Burke, P., Jennings, K.R., Morgan, R.P., and Gilchrist, C.A. (1982), *Anal. Chem.*, **54**, 1304.
30. Sieck, L.W., Burke, P., and Jennings, K.R. (1979), *Anal. Chem.*, **51**, 2232.
31. Chapman.J.R. (1980), *Pract. Spectrosc.*, **3B**, 364.
32. Siegel, M.W. (1980), *Pract. Spectrosc.*, **3B**, 297.
33. Williams, C.M., Crowley, J.R., and Crouch, M.W. (1981), *29th Annual Conference on Mass Spectrometry and Allied Topics*, Minneapolis, p. 685.
34. Mather, R.E., and Todd, J.F.J. (1980), *Int. J. Mass Spectrom. Ion Processes*, **33**, 159.
35. Stafford, G.C. (1980), *EHP, Environ. Health. Perspect.*, **36**, 85.
36. Brumley, W.C., Nesheim, S., Trucksess, M.W., Dreifuss, P.A., Roach, J.A.G., Andrzejewski, D., Eppley, R.M., Pohland, A.E., Thorpe, C.W., and Sphon, J.A. (1981), *Anal. Chem.*, **53**, 2003.
37. Miwa, B.J., Garland, W.A., and Blumenthal, P. (1981), *Anal. Chem.*, **53**, 793.
38. Madhusudanan, K.P. (1984), *Org. Mass Spectrom.*, **19**, 517.
39. Gregor, I.K., and Guilhaus, M. (1984), *Int. J. Mass Spectrom. Ion Processes*, **56**, 167.
40. Sears, L.J., and Grimsrud, E.P. (1989), *Anal. Chem.*, **61**, 2523.
41. Stemmler, E.A., and Buchanan, M.V. (1989), *Org. Mass Spectrom.*, **24**, 94.
42. Sears, L.J., Campbell, J.A., and Grimsrud, E.P. (1987), *Biomed. Environ. Mass Spectrom.*, **14**, 401.
43. Stan, H.-J., and Kellner, G. (1989), *Biomed. Environ. Mass Spectrom.*, **18**, 645.
44. Arbogast, B., Budde, W.L., Deinzer, M., Dougherty, R.C., Eichelberger, J., Foltz, R.D., Grimm, C.C., Hites, R.A., Sakashita, C., and Stemmler, E. (1990), *Org. Mass Spectrom.*, **25**, 191.
45. Houriet, R., Stahl, D., and Winkler, F.J. (1980), *EHP, Environ. Health Perspect.*, **36**, 63.
46. Hudec, T., Thean, J., Kuehl, D., and Dougherty, R.C. (1981), *Science*, **211**, 951.
47. Dougherty, R.C. (1980), *Pract. Spectrosc.*, **3B**, 306.
48. Morgan, R.P., Gilchrist, C.A., Jennings, K.R., and Gregor, I.K. (1983), *Int. J. Mass Spectrom. Ion Processes*, **46**, 309.
49. Anderson, G.B., Johns, R.B., Porter, Q.N., and Strachan, M.G. (1984), *Org. Mass Spectrom.*, **19**, 583.
50. Dzidic, I., Somerville, A.C., Raia, J.C., and Hart, H.V. (1988), *Anal. Chem.*, **60**, 1318.
51. Kuehl, D.W., and Dougherty, R.C. (1980), *Env. Sci. Technol.*, **14**, 447.
52. Stemmler, E.A., and Hites, R.A. (1988), *Org. Mass Spectrom.*, **17**, 311.

53. Harrison, A.G., and Cotter, R.J. (1990), in *Methods in Enzymology*, Vol. 193: *Mass Spectrometry* (Ed. J.A. McCloskey), Ch.1, Academic Press, San Diego.
54. Jennings, K.R. (1977), in *Specialist Periodical Report, Mass Spectrometry*, Vol. 4 (Ed. R.A.W. Johnstone), p. 203, Chemical Society, London.

Appendix 4.1 Published applications of negative ion chemical ionization

References are subdivided by reagent ion and then into major categories of compounds analyzed (categories are printed in upper case). Only references that offer detailed spectral data have been included. References cover the period up to the end of 1982.

Electron capture

ALIPHATICS, MISCELLANEOUS (74). Nitriles (14). AMINES. Amphetamines (38). Perfluorobenzoyl amphetamine (67). Perfluoro derivatives of amphetamine and dopamine (71). Fluorescamine derivatives of primary amines (17, 19). Melatonin, spirocyclic pentafluoropropionyl derivative (66). Indole-3-acetic acid, pentafluorobenzyl ether (16). AROMATICS, MISCELLANEOUS (50). Chloroacetophenone (78). Nitriles (14). Polycyclic aromatics, oxygenated (48). BARBITURATES (20). Phenobarbital (78). Amytal (67). BILE ACIDS AND DERIVATIVES (79). BIOCIDES, MISCELLANEOUS. Barnon (57). DRUGS, MISCELLANEOUS (65). Sulpha drugs (12). Phenothiazines (58). Δ^9-Tetrahydrocannabinol (71). HETEROCYCLICS, MISCELLANEOUS. Tetrahydroisoquinolines (4). HYDROCARBONS, AROMATIC (75, 50). HYDROXY COMPOUNDS. Dodec-4-en-1-ol derivatives (29). MYCOTOXINS (18). Aflatoxins (18). Aflatoxin B_1 (67). NITRO-COMPOUNDS. Nitro-explosives (52). Glyceryl trinitrate (24). Nitrate esters (51). Nitro-aromatics (31). Nitrated polycyclic aromatic hydrocarbons (62). ORGANOMETALLICS. Volatile metal chelates (1, 33, 41). PEPTIDES. Peptide intermediate (78). PHOSPHATE ESTERS. Organophosphorus pesticides (8). POLYCHLORINATED COMPOUNDS. Chlorinated pesticides, polycyclic (47). Toxaphene, octachlorodane (22). Chlorinated pesticides, aromatic (60). Polychloroanisoles (56). Polychlorinated 2-phenoxyphenols (27). Hydroxyoctachlorobiphenyl ether derivatives (7). Polychlorinated dibenzo-*p*-dioxins (35). Alkoxypolychlorodibenzofurans (55). Polyhalogenated hydrocarbons (13). STEROIDS. Conjugated unsaturated steroids (50). Anabolic steroid derivatives (64). Corticosteroid derivatives (3). SULPHUR COMPOUNDS, MISCELLANEOUS. Thiols (76). Thioethers (76). Aromatic thioethers (10). TERPENES. Sesquiterpenes (70)

Hydroxyl ion (OH^-)

ALIPHATICS, MISCELLANEOUS. Acid chlorides (6). Ethers (44). AMINES (44). AMINO ACIDS (44). AROMATICS, MISCELLANEOUS. Nitrophenols (80).

CARBONYL COMPOUNDS (30, 80). Ketones (21, 44). CARBOXYLIC ACIDS AND DERIVATIVES. Esters (80). DRUGS, MISCELLANEOUS. Methadone, acetylmethadol and metabolites (59). GLYCOSIDES. Steroid glycosides (53). HYDROCARBONS, AROMATIC (44, 63, 72). HYDROXY COMPOUNDS (28, 44). MYCOTOXINS. Tricothecenes (77). SACCHARIDES. Glucose (80). STEROIDS (34). SULPHUR COMPOUNDS, MISCELLANEOUS. Thiols (44). TERPENES. Terpene Esters (32). Bornyl esters (9). Terpene alcohols (32). Oxygenated terpenes (11, 32).

Chloride ion (Cl^-)

ALICYCLICS. Cyclic diol stereoisomers (69). ALIPHATICS, MISCELLANEOUS (61). ALKALOIDS. Reserpine (79). ANTIBIOTICS. TETRACYCLINE (79). Lasolocid (79). AROMATICS, MISCELLANEOUS (61). BILE ESTERS (79). BIOCIDES, MISCELLANEOUS. Barnon (57). NITRO-COMPOUNDS. Ethylene glycol dinitrate (23). Glyceryl trinitrate (23, 24). PHOSPHATE ESTERS. Organophosphate pesticides (15). POLYCHLORINATED COMPOUNDS. (49). Polychlorinated aromatics (39, 40). Polyhalogenated hydrocarbons (13). PROSTAGLANDINS. Prostaglandin $F_{2\alpha}$ (79). SACCHARIDES (36). STEROIDS. Testosterone (79).

Oxygen and mixtures containing oxygen (O_2^-)

AROMATICS, MISCELLANEOUS. Halobenzenes (46). HETEROCYCLICS. Dibenzothiophene (67). HYDROCARBONS, AROMATIC. Coronene (67). Polycyclic aromatic hydrocarbons (68). POLYCHLORINATED COMPOUNDS (49). Polychlorinated aromatics (26). Polychlorinated anisoles (56). Polychlorinated phenols (5). Pentachlorophenol (25). Polychlorobiphenyls (2, 5). Polychlorodibenzo-p-dioxins (35, 45, 54).

Miscellaneous

Aliphatics, miscellaneous (O^-) (42). Acid chlorides (O^-) (6). Barbiturates (O^-) (37). Carbonyl compounds (O^-) (43). Carboxylic acids (OCH_3^-) (21). Cyclic diol stereoisomers ((F^-) (69). Hydroxy compounds (O^-) (28). Peptide, permethylated (OCH_3^-) (67). Phenothiazines (O^-) (58). Pyridines, methyl (O^-) (73).

References for Appendix 4.1

1. Dakternieks, D. R., Fraser, I. W., Garnett, J. L., and Gregor, I. K. (1979), *Org. Mass Spectrom.*, **14**, 330.
2. Busch, K. L., Norstroem, A., Bursey, M. M., Hass, J. R., and Nilsson, C. A. (1979), *Biomed. Mass Spectrom.*, **6**, 157.

3. Houghton, E., Teale, P., Dumasia, M. C., and Wellby, J. K. (1982), *Biomed. Mass Spectrom.*, **9**, 459.
4. Murray, S., and Waddell, K. A. (1982), *Biomed. Mass Spectrom.*, **9**, 466.
5. Kuehl, D. W., Whitaker, M. J., and Dougherty, R. C. (1980), *Anal. Chem.*, **52**, 935.
6. Lloyd, J. R., Agosta, W. C., and Field, F. H. (1980), *J. Org. Chem.*, **45**, 1614.
7. Deinzer, M., Griffin, D., Miller, T., and Skinner, R. (1979), *Biomed. Mass Spectrom.*, **6**, 301.
8. Stan, H. -J. and Kellner, G. (1982), *Biomed. Mass Spectrom.*, **9**, 483.
9. Bambagiotti, M. A., Giannellini, V., Coran, S. A., Vincieri, F. F., Moneti, G., Selva, A., and Traldi, P. (1982), *Biomed. Mass Spectrom.*, **9**, 495.
10. Lewis, E., and Peach, M. E. (1982), *Org. Mass Spectrom.*, **17**, 597.
11. Hendriks, H., and Bruins, A. P. (1980), *J. Chromatogr.*, **190**, 321.
12. Roach, J. A. G., Sphon, J. A., Hunt, D. F., and Crow, F. W. (1980), *J. Assoc. Off. Anal. Chem.*, **63**, 452.
13. Crow, F. W., Bjorseth, A., Knapp, K. T., and Bennett, R. (1981), *Anal. Chem.*, **53**, 619.
14. Busch, K. L., Parker, C. E., Harvan, D. J., and Bursey, M. M. (1981), *Anal. Spectrosc.*, **35**, 85.
15. Dougherty, R. C., and Wander, J. D. (1980), *Biomed. Mass Spectrom.*, **7**, 401.
16. Epstein, E., and Cohen, J. D. (1981), *J. Chromatogr.*, **209**, 413.
17. Bergner, E. A., and Dougherty, R. C. (1981), *Biomed. Mass Spectrom.*, **8**, 204.
18. Brumley, W. C., Nesheim, S., Trucksess, M., Trucksess, E. W., Dreifuss, P. A., Roach, J. A. G., Andrzejewski, D., Eppley, R. M., and Pohland, A. E. (1981), *Anal. Chem.*, **53**, 2003.
19. Bergner, E. A., and Dougherty, R. C. (1981), *Biomed. Mass Spectrom.*, **8**, 208.
20. Jones, L. V., and Whitehouse, M. J. (1981), *Biomed. Mass Spectrom.*, **8**, 231.
21. Hunt, D. F., Giordani, A. B., Shabanowitz, J., and Rhodes, G. (1982), *J. Org. Chem.*, **47**, 738.
22. Ribick, M. A., Dubay, G. R., Petty, J. D., Stalling, D. L., and Schmitt, C. J. (1982), *Environ. Sci. Technol.*, **16**, 310.
23. Bouma, W. J., and Jennings, K. R. (1981), *Org. Mass Spectrom.*, **16**, 331.
24. Idzu, G., Ishibashi, M., Miyazaki, H., and Yamamoto, K. (1982), *J. Chromatogr.*, **229**, 327.
25. Kuehl, D. W. and Dougherty, R. C. (1980), *Environ. Sci. Technol.*, **14**, 447.
26. Dougherty, R. C., Whitaker, M. J., Smith, L. M., Stalling, D. L., and Kuehl, D. W. (1980), *EHP, Environ. Health Perspect.*, **36**, 103.
27. Busch, K. L., Norstroem, A., Nilsson, C. A., Bursey, M. M., and Hass, J. R. (1980), *EHP, Environ. Health Perspect.*, **36**, 125.
28. Houriet, R., Stahl, D., and Winkler, F. J. (1980), *EHP, Environ. Health Perspect.*, **36**, 63.
29. Hogge, L. R., and Olson, D. J. H. (1982), *J. Chromatogr. Sci.*, **20**, 109.
30. Hunt, D. F., Sethi, S. K., and Shabanowitz, J. (1980), *EHP, Environ. Health Perspect.*, **36**, 33.
31. Newton, D. L., Erickson, M. D., Tomer, K. B., Pellizzari, E. D., and Gentry, P. (1982), *Environ. Sci. Technol.*, **16**, 206.
32. Bruins, A. P. (1979), *Anal. Chem.*, **51**, 967.
33. Prescott, S. R., and Risby, T. H. (1978), *Anal. Chem.*, **50**, 562.
34. Roy, T. A., Field, F. H., Lin, Y. Y., and Smith, L. L. (1979), *Anal, Chem.*, **51**, 272.
35. Hass, J. R., Friesen, M. D., Harvan, D. J., and Parker, C. E. (1978), *Anal. Chem.*, **50**, 1474.
36. Ganguly, A. K., Cappuccino, N. F., Fujiwara, H., and Bose, A. K. (1979), *J. Chem. Soc. Chem. Commun.*, **1979**, 148.

37. Frangi-Schnyder, D., and Brandenberger, H. (1978), *Fresenius' Z. Anal. Chem.*, **290**, 153.
38. Mårde, Y., and Ryhage, (1978), *Clin. Chem.*, **24**, 1720.
39. Dougherty, R. C., and Piotrowska, K. (1976), *Proc. Natl. Acad. Sci. USA*. **73**, 1777.
40. Dougherty, R. C., and Piotrowska, K. (1976), *J. Assoc. Off. Anal. Chem.*, **59**, 1023.
41. Prescott, S. R., Campana, J. E., and Risby, T. H. (1977), *Anal. Chem.*, **49**, 1501.
42. Dawson, J. H. J., and Jennings, K. R. (1976), *J. Chem. Soc., Faraday Trans. 2.*, **72**, 700.
43. Harrison, A. G., and Jennings, K. R. (1976), *J. Chem. Soc., Faraday Trans. 2.*, **72**, 1601.
44. Smit, A. L. C., and Field, F. H. (1977), *J. Am. Chem. Soc.*, **99**, 6471.
45. Hunt, D. F., Harvey, T. M., and Russell, J. W. (1975), *J. Chem. Soc. Chem. Commun.*, **1975**, 151.
46. Levonowich, P. F., Tannenbaum, H. P., and Dougherty, R. C. (1975), *J. Chem. Soc. Chem. Commun.*, **1975**, 597.
47. Dougherty, R. C., Dalton, J., and Biros, F. J. (1972), *Org. Mass Spectrom.*, **6**, 1171.
48. Yergey, J. A., Risby, T. H., and Lestz, S. S. (1982), *Anal. Chem.*, **54**, 354.
49. Dougherty, R. C. (1981), *Anal. Chem.*, **53**, 625A.
50. Chapman, J. R. (1980), *Pract. Spectrosc.*, **3B**, 364.
51. Bignall, J. C., Davies, N. W., Power, M., Roberts, M. S., Cossum, P. A., and Boyd, G. W. (1981), *Anal. Chem. Symp. Ser.*, **7** *(Recent Dev. Mass Spectrom. Biochem. Med. Environ. Res.)*. 111.
52. Gielsdorf, W. (1981), *Fresenius' Z. Anal. Chem.*, **308**, 123.
53. Bruins, A. P. (1980), *Adv. Mass Spectrom.*, **8A**, 246.
54. Hass, J. R., Friesen, M. D., and Hoffman, M. K. (1979), *Org. Mass Spectrom.*, **14**, 9.
55. Deinzer, M., Griffin, D., Miller, T., Lamberton, J., Freeman, P., and Jonas, V. (1982), *Biomed. Mass Spectrom.*, **9**, 85.
56. Busch, K. L., Hass, J. R., and Bursey, M. M. (1978), *Org. Mass Spectrom.*, **13**, 604.
57. Page, J. A. (1982), *Anal. Proc.*, **19**, 307.
58. Ryhage, R., and Brandenberger, H. (1978), *Biomed. Mass Spectrom.*, **5**, 615.
59. Smit, A. L. C., and Field, F. H. (1978), *Biomed. Mass Spectrom.*, **5**, 572.
60. Dougherty, R. C., Roberts, J. D., and Biros, F. J. (1975), *Anal. Chem.*, **47**, 54.
61. Tannenbaum, H. P., Roberts, J. D., and Dougherty, R. C. (1975), *Anal. Chem.*, **47**, 49.
62. Ramdahl, T., and Urdal, K. (1982), *Anal. Chem.*, **54**, 2256.
63. Sieck, L. W., Jennings, K. R., and Burke, P. D. (1979), *Anal. Chem.*, **51**, 2232.
64. Stan, H. -J., Quantz, D., and Abraham, B. (1981), *Recent Dev. Mass Spectrom. Biochem. Med.*, **7**, 335.
65. Garland, W. A., and Min, B. H. (1979), *J. Chromatogr.*, **172**, 279.
66. Lewy, A. J., and Markey, S. P. (1978), *Science*, **201**, 741.
67. Hunt, D. F., Staford, G. C., Jr, Crow, F. W., and Russell, J. W. (1976), *Anal. Chem.*, **48**, 2098.
68. Hunt, D. F., McEwen, C. N., and Harvey, T. M. (1975), *Anal. Chem.*, **47**, 1730.
69. Winkler, F. J., and Stahl, D. (1979), *J. Am. Chem. Soc.*, **101**, 3685.
70. Knight, D. W., Knight, M. E., Rossiter, M., Jones, A., and Games, D. E. (1980), *Adv. Mass Spectrom.*, **8B**, 1116.
71. Hunt, D. F., and Crow, F. W. (1978), *Anal. Chem.*, **50**, 1781.
72. Sieck, L. W. (1980), *Chem. Brit.*, **16**, 38.
73. Bruins, A. P., Ferrer-Correia, A. J., Harrison, A. G., Jennings, K. R., and Mitchum, R. K. (1978), *Adv. Mass Spectrom.*, **7**, 355.
74. Large, R., and Knof, H. (1976), *Org. Mass Spectrom.*, **11**, 582.
75. Albers, G., and Knof, H. (1977), *Org. Mass Spectrom.*, **12**, 698.
76. Knof, H., Large, R., and Albers, G. (1976), *Anal. Chem.*, **48**, 2120.

77. Brumley, W. C., Andrzejewski, D., Trucksess, E. W., Dreifuss, P. A., Roach, J. A. G., Eppley, R. M., Thomas, F. S., THorpe, C. W., and Sphon, J. A. (1982), *Biomed. Mass Spectrom.*, **9**, 451.
78. Brandenberger, H. (1980), *Anal. Chem. Symp. Ser.*, **4**, *Recent Dev. Mass Spectrom. Biochem. Med.*, 6), 391.
79. Bose, A. K., Fujiwara, H., and Pramanik, B. N. (1979), *Tetrahedron Lett.*, **1979**, 4017.
80. Hunt, D. F., Shabanowitz, J., and Giordani, A. B. (1980), *Anal. Chem.*, **52**, 386.

Chapter 5

The Ionization of Labile Materials (Part I)

5.1 Introduction

The general application of electron impact ionization to a wide range of organic compounds is limited by two factors. First, the requirement for sample vaporization prior to ionization may lead to thermal decomposition of involatile or thermally unstable materials so that a useful spectrum is not obtained. Second, the electron energy of 70 eV usually employed leads to extensive fragmentation and perhaps undetectable molecular ions.

The second difficulty may often be overcome by the use of softer ionization techniques such as chemical ionization, which has already been described in Chapters 3 and 4. The first difficulty has encouraged the development of new ionization techniques in which ions are produced with minimum thermal damage. These methods, which have proliferated greatly in recent years, may, following Fenn *et al.* [1], be assigned to two general categories:

(a) 'energy-sudden' methods which involve very rapid vaporization, e.g. by rapid heating or sputtering by energetic particles;
(b) 'field desorption' methods which use very strong electrostatic fields to extract ions from a substrate.

Energy sudden methods will be described in the present chapter while a discussion of field desorption methods, which includes a number of methods for ionization of samples from solution, is deferred to Chapter 6.

5.1.1 Energy-sudden methods

It should be noted at the outset that while at least some energy-sudden methods are very simply effected in practice, in many cases they are not fully understood in theory and may bring about ion production by a number of different processes, depending on the sample and conditions employed.

Energy-sudden methods can be regarded to some extent as extensions of an idea originally proposed by Beuhler *et al.* [2]. These authors argued that since the rate constant for vaporization is higher than that for decomposition at higher temperatures (Fig. 5.1) the more rapidly these high temperatures could be achieved in the vicinity of the sample, the more vaporization would be favoured at the expense of decomposition. Thus, the idea of rapid sample heating gained currency and the DCI technique was born. Despite the undoubtedly different forms of interaction between the energetic particles used and the sample, all the other energy-sudden techniques to be discussed also achieve the equivalent of rapid heating during bombardment.

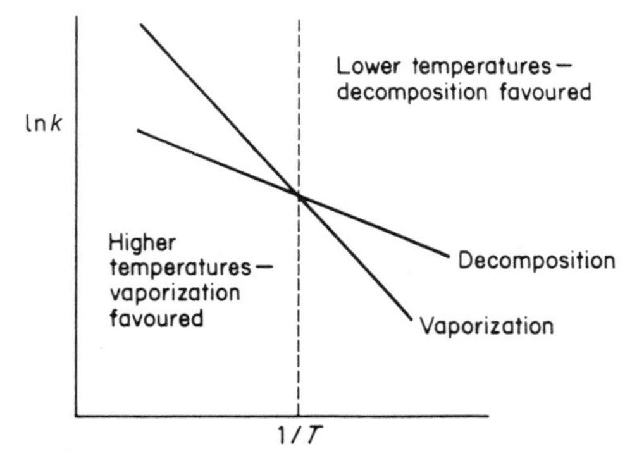

Fig. 5.1 Arrhenius plots for decomposition and vaporization reactions. Rate constant $k = A \exp(-E/RT)$ ($\ln k = \ln A - E/RT$). (Reprinted with permission from G. Doyle Daves Jr, *Acc. Chem. Res.*, **12**, 359[4]. Copyright (1979) American Chemical Society)

Another similarity between these methods is the pronounced influence of the matrix that surrounds the sample. Indeed, very notable improvements, principally to fast atom bombardment and laser desorption, have been obtained by the use of a specialized matrix in which the sample is dispersed. The matrix has a number of different functions and these will be discussed in detail under the appropriate technique. The use of a matrix is part of the sample preparation technology for these methods and the importance of sample preparation in all of these energy-sudden methods cannot be stressed too highly.

5.2 Theory

5.2.1 Desorption CI and 'in-beam' mass spectrometry [3,4]

It has long been recognized that a number of very simple strategies could be adopted to produce a spectrum of a labile sample without undue decomposition. For example, in 1958 Reed [5] showed that the evaporation of a solid sample placed close to the electron beam in an EI source produced spectra that were well suited to the determination of molecular weight.

An equivalent technique in which the sample is coated on the surface of an inert probe tip brought close to the electron beam was named 'in-beam' mass spectrometry [6] and has been used to provide molecular weight and fragment ion information from a number of complex natural products [7,8,9,10] (section 2.3.3). One advantage of this technique appears to derive from the close proximity of the sites of evaporation and ionization so that surface catalyzed thermal degradation is minimized. In addition the formation of protonated molecular ions indicates that the process may partly be one of self chemical ionization, perhaps due to a locally high vapour pressure.

A related technique, using sample deposited on the surface of an extended probe tip which was then introduced completely into the plasma contained in a CI source, was first described by Baldwin and McLafferty [11]. In this way, these authors obtained chemical ionization spectra from relatively involatile samples at lower temperatures than with a conventional probe. Further investigations have confirmed the general utility and simplicity of in-beam techniques and have also suggested other strategies which can be used to enhance quasi-molecular ion production. These strategies have included the minimization of sample–sample or sample–surface interaction by the use of thinly dispersed sample layers and inert probe tip surfaces and rapid heating of the sample probe [3,4,7,12–14].

A number of investigators have observed the time and temperature dependence of quasi-molecular ion production by EI and CI in-beam processes [8,15,16]. Figure 5.2 shows selected ion current profiles from underivatized sucrose introduced on a Vespel probe heated only indirectly by the ion source block whose temperature was adjusted prior to insertion [8]. At low heating rates (Fig. 5.2a) almost no quasi-molecular ion is observed. Higher heating rates (Fig. 5.2b, c) allow the temperature

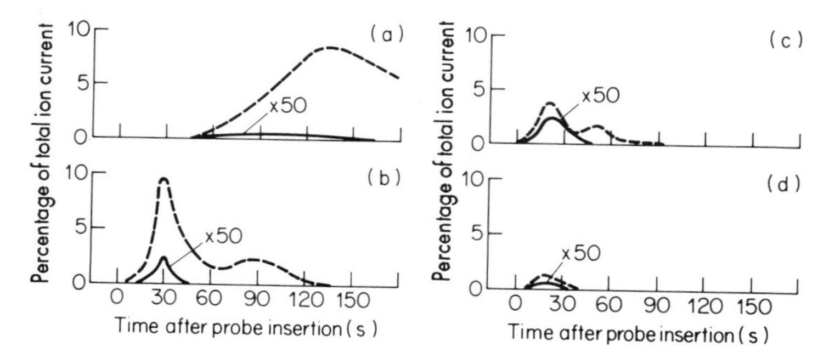

Fig. 5.2 Selected ion current profiles for the $(MH)^+$ ion (———) and the $(MH\text{-}H_2O)^+$ ion (— — — —) of sucrose at different ion source temperatures T_s: (a) 180°C; (b) 230°C; (c) 280°C; (d) 310°C. (Reprinted with permission from R.J. Cotter, *Anal. Chem.*, **52**, 1589A[3]. Copyright (1980) American Chemical Society)

to reach a value where vaporization is favoured compared with decomposition and the quasi-molecular ion is greatly enhanced.

The two maxima in the ion current traces seen in Fig. 5.2(b), (c) are taken to indicate the existence of two processes. i.e. vaporization of intact neutral molecules with some fragmentation, and vaporization of decomposition products formed on the surface. At a source temperature of 310°C (Fig. 5.2d) these two processes can no longer be distinguished.

A logical development of the discovery of the dependence of spectrum quality on heating rate was to provide a sample probe with an integral heater capable of very rapid heating rates. This approach, pioneered by Beuhler *et al.* [14] was taken a step further by Hunt [17] using an activated field desorption (FD) emitter in a CI

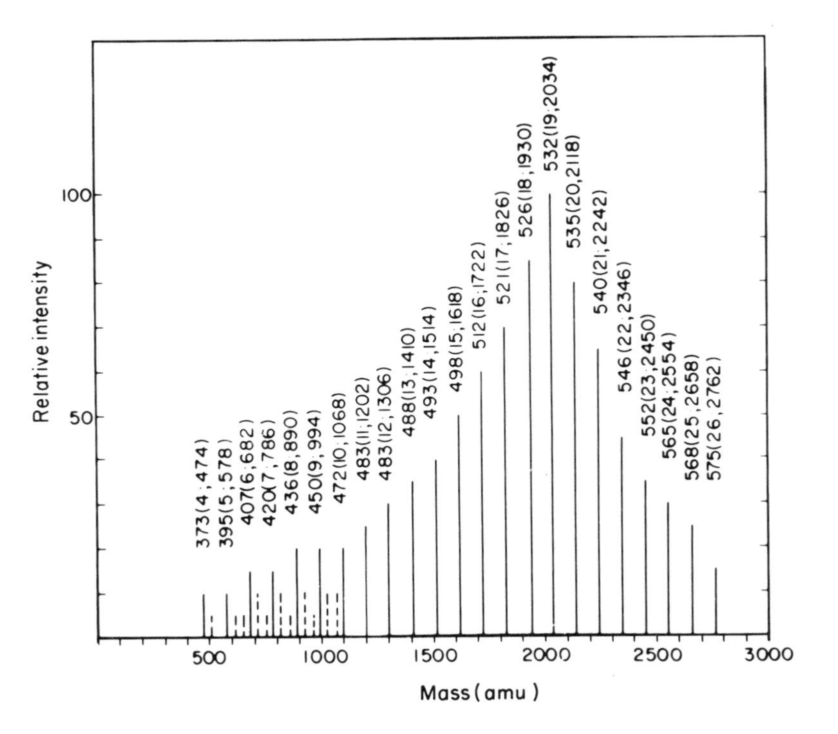

Fig. 5.3 Partial DCI spectrum of polystyrene (argon reagent gas). The solid lines are the nmer spectrum and the numbers in parentheses give first the number of monomer units and second the mono-isotopic mass. The numbers not in parentheses are the evaporation temperatures (in K). The dashed lines are the fragment peaks. (Reprinted with permission from H.R. Usdeth and L. Friedman, *Anal. Chem.*, **53**, 29[19]. Copyright (1981) American Chemical Society)

source. Not only does the electrically heated emitter wire permit very rapid heating rates but the microneedles provide a large surface area for dispersing the sample within the ionizing plasma. This technique, termed desorption chemical ionization (DCI), rapidly gained popularity, particularly in combination with the use of less energetic CI reagent gases.

With the replacement of activated emitters by electrically heated filaments (section 5.3.1) which require no complex preparation, DCI is now established as a simple but effective technique for dealing with labile samples. The same rapid evaporation technique may also be used with advantage in an EI source [10,18] and is known as desorption electron impact (DEI). The partial DCI spectrum obtained by the rapid evaporation of a styrene polymer from an electrically heated rhenium ribbon [19] is reproduced in Fig. 5.3.

5.2.2 Secondary ion mass spectrometry (SIMS)

Analysis of the secondary ions emitted when a surface is bombarded with an energetic primary ion beam is an established method for the analysis of the surface layers of inorganic materials such as metals, alloys and semiconductors [20,21]. Application of SIMS to the qualitative analysis of larger organic molecules deposited on a surface was pioneered by Benninghoven who introduced the technique known as 'static' SIMS [22]. The primary ion current densities of a few nanoamperes per square centimetre used in this technique cause less sample damage than is the case with higher current densities [23]. In principle, static SIMS can be used with scanning mass spectrometers. However, because it is a relatively low flux technique, there is considerable advantage in using a time-of-flight analyser to collect all the ions.

Mass spectra with abundant molecular ions such as $(M+H)^+$and $(M+Ag)^+$ are obtained by static SIMS from thin layers of many organic compounds deposited on a substrate such as Ag. The use of a nitrocellulose matrix was subsequently found to offer a considerable extension to the applicability of the SIMS technique. For example, bovine insulin (MW 5734) that could barely be detected when deposited on an Ag substrate could readily be detected when nitrocellulose was used [24]. As we shall see in the next section, the FAB technique introduced by Barber *et al.* [25] uses much more intense primary ion fluxes in conjunction with a liquid matrix, e.g. glycerol, which promotes surface renewal and limits surface damage effects. Under these conditions, both ion and neutral bombardment are practical techniques and are discussed together in the following section.

5.2.2.1 Fast atom bombardment(FAB)/liquid secondary ion mass spectrometry (LSIMS): ion formation

The principle of the FAB method [25], is illustrated in Fig. 5.4 and further details are given in section 5.3.2. A beam of energetic particles, directed to strike

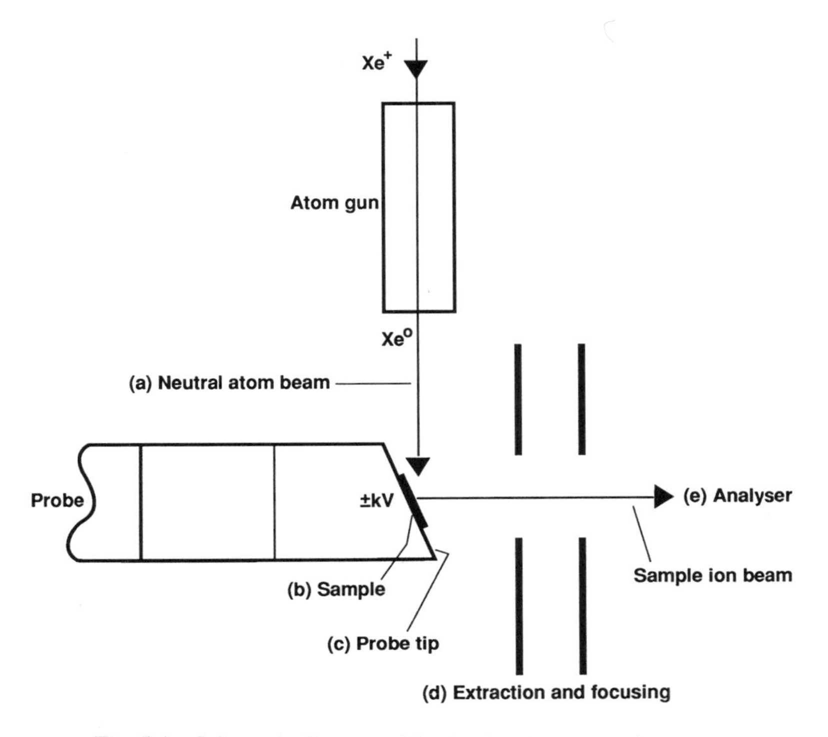

Fig. 5.4 Schematic diagram of fast ion bombardment ion source

a sample film carried on a clean metal support, produces an intense thermal spike whose energy is then dissipated through the outer layers of the sample surface. Barber's original work used fast neutral atoms, e.g. Ar^0 or Xe^0, as the sputtering agent. Later workers [26] who introduced the use of a primary ion beam, e.g. Cs^+, in place of the neutral beam, suggested the term liquid SIMS (LSIMS), rather than FAB, for this alternative mode of operation.

With a dry deposited sample there is a rapid decay of the secondary ion yield due to surface damage by the intense incident beam. However, these problems can be circumvented by the sample preparation method used in FAB/LSIMS, i.e. dissolving the sample in a liquid matrix such as glycerol [25,27]. The liquid medium is then able to provide continuous surface renewal so that intense primary beams may be used [28]. The overall result is that secondary ion beams with a useful intensity for scanning instruments, e.g. magnetic sector instruments, may be prolonged for periods of 20 min or more. These advantages were sufficient for FAB/LSIMS to become the standard method for the analysis of polar or labile samples, particularly those of higher molecular weight. In practice, an upper molecular weight limit of approximately 24 000 daltons with a 35 keV caesium ion gun was established [29].

The removal of damage created by the primary beam is very efficient in liquid glycerol. The most likely mechanisms for surface removal are volatilization of damage products together with matrix material and macroscopic flow within the matrix which occurs under ion bombardment conditions [30,31]. Experiments have shown that at -20°C glycerol is too viscous to allow effective renewal whereas at temperatures of 40°C or more, glycerol becomes too volatile and is rapidly lost. Optimum sensitivity is achieved at 25°C [32]. Sample mobility may occasionally be increased by gentle heating of the metal tip without the addition of glycerol, e.g. in the analysis of saccharides.

In addition to the use of a liquid matrix, the use of a neutral primary beam was initially introduced as another essential component of the FAB technique, both to avoid charging of insulating samples and to overcome difficulties associated with the introduction of a charged beam into the high voltage source of a magnetic sector mass spectrometer [33]. Subsequent experiments using samples prepared in a liquid matrix have, however, suggested that the charge or neutrality of similarly energetic primary beams does not affect either the intensity or the lifetime of the secondary ion beam or the quality of the secondary ion mass spectra obtained. Thus, a comparison of 6 keV Cs^+ ions and 8 keV Xe atoms as primary beams showed similar spectra with no evidence for sample charging using the Cs^+ beam [26] (Fig. 5.5).

On the other hand, major changes in primary particle mass and energy do affect the secondary ion yield in sputtering experiments [34]. In general, the secondary ion yield is increased by the use of more massive particles so that, for example, a primary beam of xenon neutrals is more effective than argon, which in turn is more effective than neon [35,36].

A further increase in secondary ion yield is observed in recent experiments in which organic targets were bombarded with energetic molecular rather than atomic ions. For example, yield increases up to a factor of 50 were observed when phenylalanine was bombarded with $Cs_3I_2^+$ instead of Cs^+, far more than would be expected on the basis of a change of momentum alone [37]. A comparison of the sputtering of organics with SF_6^0 (mass 146) and Cs^+ (mass 133) showed a factor of 9–24 increase in the intensity of the secondary ion beam when the molecular SF_6^0 beam was used [38]. Similarly, Barofsky [39] has shown much improved secondary ion yields from bombardment with bismuth and gold cluster ions.

The mechanism of ion formation under FAB/LSIMS conditions is by no means completely understood and appears to be a process of some complexity. Early speculation [40] attributed the spectrum to the desorption of pre-formed ions which could then undergo further ion–molecule reactions in the selvedge, or dense gas region, just above the liquid surface. While it is well established that pre-ionized molecules are preferentially desorbed [25,41], this picture is an over-simplification. The collision cascade that results from the impact of the fast atom or ion causes extensive ionization in the matrix. In a wider region of the matrix, rapid heating

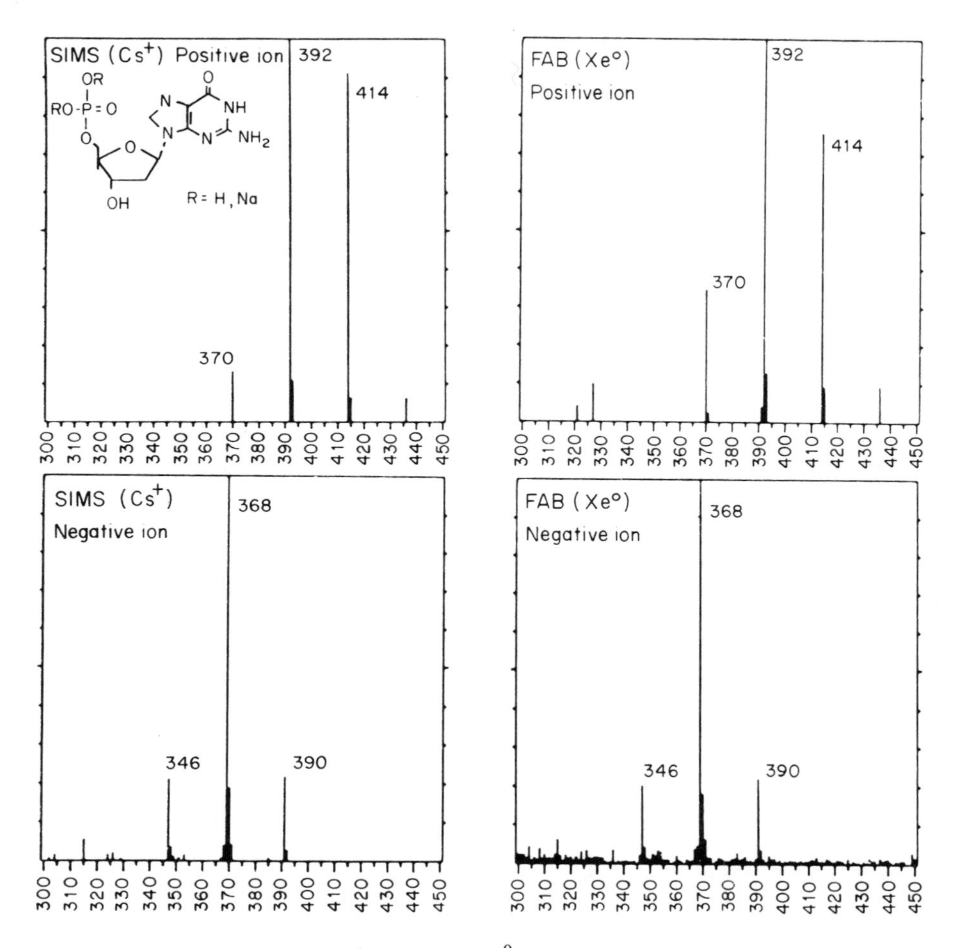

Fig. 5.5 Comparative SIMS(Cs^+) and FAB(Xe^0) spectra (positive and negative ion modes) of 5′-deoxyguanosine monophosphate. (Reprinted with permission from W. Aberth, K.M. Straub and A.L. Burlingame, *Anal. Chem.*, **54**, 2029[26]. Copyright (1982) American Chemical Society)

leads to an explosive transition from the liquid to a dense gas state in which analyte molecules may be substantially matrix free. The final FAB spectrum results from pre-formed sample ions, if such ions are present, and from the reactions of collision cascade ions and neutral analyte species in the dense gas state [42].

The expansion that results from this heating rapidly lowers the temperature again and the very short temperature maximum that results offers an explanation of how thermally labile molecules can survive the desorption process. For example, only unimolecular reactions with rates above *ca.* 10^{11} s^{-1}, at the very maximum temperature reached in the phase explosion, will lead to significant fragmentation of the analyte molecules.

5.2.2.2 *Fast atom bombardment(FAB)/liquid SIMS (LSIMS): appearance of spectra*

As previously mentioned, analyte ions may be pre-formed, or they may be generated as a result of energy deposition. For pre-formed ions, a number of authors have shown a direct correlation between ion equilibria in solution and ion abundances measured in the gas phase [43,44]. For example, Caprioli [44] found that ion abundances measured from FAB spectra could be used to calculate pK_a values for a variety of weak acids to within $\pm 0.04\,\mathrm{p}K_a$ units.

In cases where pre-formed ions are not involved, evidence for the importance of gas-phase equilibria has been gathered by examining compounds of known proton affinity. Sunner and co-workers [45–47] reported that the yields of protonated molecular ions from binary mixtures depend on the relative gas-phase basicities of the mixture components. Thus, the $(M+H)^+$ ion which corresponds to the compound with the lower gas-phase basicity is suppressed since proton transfer occurs preferentially to the more basic compound.

The general similarity of FAB/LSIMS and chemical ionization spectra which both contain even-electron ions such as $(M+H)^+$ or $(M - H)^-$ (Fig. 5.5) rather than $M^{+\cdot}$ or $M^{-\cdot}$ ions has been noted [33]. In part this may be a result of ion–molecule reactions which occur in the dense gas region above the sample surface, after the primary ionization event, and which almost certainly involve ions formed from the glycerol matrix.

Another example of the influence of gas-phase equilibria is the absence of $(M + H)^+$ ions in the FAB spectra of many neutral saccharides recorded using diethanolamine and triethanolamine matrices [48,49]. Both of these matrices have high proton affinities so that the equilibrium in the following equation

$$\text{(trehalose....H....matrix)}^+ \rightleftharpoons \text{(trehalose....H)}^+ + \text{matrix} \qquad (5.1)$$

favours formation of the solvated ion and a protonated ion is only observed when glycerol is used [48].

Factors other than gas- and liquid-phase equilibria are also important. Although gas-phase basicities correlated qualitatively with suppression in Sunner's work, it was felt that quantitative agreement was not as expected and it was suggested that this was due to surface enrichment of one component. Recent experiments with quaternary ammonium salts reported by Lacey and Keough [50] have confirmed that both surface activity and gas-phase basicity contribute to the overall appearance of FAB spectra from analyte mixtures. Surface activity also plays a vital role in determining the relative intensities of molecular ions when mixtures of peptides are analysed by FAB/LSIMS (see section 5.3.2.2).

A variety of chemical reactions can accompany FAB/LSIMS as well as other forms of 'energy-sudden' ionization methods and these have been comprehensively reviewed recently [51]. Details of beam-initiated reactions in FAB/LSIMS which involve the matrix will be found in section 5.3.2.2, while reactions such as cationization which involve matrix additives are discussed in section 5.3.2.3. Intermolecular reactions of analyte molecules [52,53] and reactions of analytes with the metal of the probe tip have also been reported [54]. Cluster ions, e.g. $(2M+H)^+$, and adduct ions, e.g. $(M+matrix+H)^+$ (cf. equation 5.1), are also commonly observed in FAB/LSIMS spectra.

Higher molecular weight analytes such as proteins generate multiply-charged ions as a result of the multiplicity of separate sites for protonation. In fact, the spectra of small proteins (MW 10 000–25 000) often show the doubly-charged ion as the most intense ion representative of the molecular mass [55].

Although the particular virtue of FAB/LSIMS is its ability to produce intact molecular ions from very large molecules, it is not a 'soft' ionization technique like field desorption (Chapter 6). Bombardment with highly energetic particles ensures that a wide range of energies is supplied to molecules throughout the surface lattice [56] so that a comprehensive range of fragment ions as well as molecular ions is formed [28,57].

The amount of fragmentation relative to the molecular ion in the spectra of peptides decreases as the amount of sample is decreased. Weaker fragment ions are often hidden by matrix-related peaks although the continuous-flow FAB technique offers advantages in this respect (section 2.4.4). Fragmentation also decreases with increasing molecular weight, and may be entirely absent for larger peptides, irrespective of the amount of sample analysed [58].

5.2.3 Plasma desorption mass spectrometry (PDMS) [59]

The principle of the ^{252}Cf plasma desorption experiment is shown later in Fig. 5.12 and further details are also given in section 5.3.3. The sample is deposited on a 0.5–1 μm aluminium or aluminized polyester foil. Fission fragments from a ^{252}Cf source placed just behind the sample foil are able to penetrate the thin foil and cause desorption of sample ions from the other side. These primary particles are highly energetic, typically approximately 100 MeV, compared with the kiloelectronvolt energy primary beams used in FAB/LSIMS.

Developed [60] somewhat earlier than FAB, PDMS has always been able to analyse somewhat higher molecular weight samples than FAB, principally because of the very much more energetic primary beam particles. Thus, until the introduction of electrospray ionization and matrix-assisted laser desorption/ionization, PDMS was the method of choice for the analysis of very high molecular weight samples.

As the energetic ions penetrate the solid surface, they form a fission track, a region of damage due to the deposition of energy in the solid lattice. Only low mass ions (e.g. H^+, Na^+, CH_3^+, CN^-) are sputtered from this region. It is believed

[61] that the sites of desorption of intact molecular species are located at some distance from the fission track. Experimental data suggest that this sputtering is due to a rapid collective motion of the lattice such as an expansion of the region surrounding the track [62] or a shock wave [63] which transfers momentum outward from the track. Energy transferred to molecules adsorbed on the surface distant from the fission track is then sufficient to overcome the weak interactions that hold molecules to surfaces and to each other, but does not break covalent bonds.

The introduction of nitrocellulose as a substrate for sample preparation [64] made it possible to record peptide and protein spectra of much higher quality than those obtained from samples prepared by electrospray deposition (section 5.3.3.1). Results from several groups [64–66] indicated that analyte molecules which are weakly bound to the nitrocellulose desorb with less fragmentation than those desorbed from bulk material.

The nitrocellulose technique also provided an improved molecular ion yield while requiring much less sample than the electrospray technique [67]. Carefully controlled experiments indicated that the maximum desorption yield per sample amount was, in fact, recorded with a five- to fiftyfold excess of sample over the calculated amount needed to provide a monolayer coverage of the nitrocellulose target [68]. It is believed that this is at least partly due to the fact that the actual surface area is much greater due to surface roughness of the nitrocellulose layer.

Observations of an increased ion yield, an increase in the relative abundance of multiply charged ions, and narrower mass spectral peaks using an nitrocellulose target have been taken to indicate a reduced binding energy between sample and substrate [69,70]. A similar relative increase in the intensity of multiply-charged ions as the sample loading on the nitrocellulose is reduced has been attributed to a reduction in the interaction with other sample molecules [68,71]. Experiments with beams of different energy from a tandem accelerator have also shown that less energy is needed to desorb insulin from a nitrocellulose surface than from multilayer electrosprayed sample preparations [70,72].

As a general comment on the plasma desorption process, Macfarlane *et al.* [73] have suggested that there may be domains in the solid sample layer where the molecules are perfectly oriented with the neighbouring molecules such that they form a highly ordered aggregate. These aggregates behave as a supra-molecule in that they can be collectively excited as if they were a single species. Energy is efficiently distributed between members of the aggregate so that a single molecule is not excited into dissociative states. When this aggregate is desorbed, it dissociates into its components but there is a high probability that a molecule can survive intact.

5.2.4 Laser desorption mass spectrometry (LDMS)

The first part of this discussion concerns laser desorption/ionization experiments in which both of these processes are initiated by the same laser pulse but do not involve the use of an absorbing matrix (cf. section 5.2.4.2). In many infrared laser experiments, which generally involve laser power densities $\leq 10^7$–10^8 W cm^{-2}, some features of the spectra obtained suggest the existence of a thermal process and it has been suggested that energy transfer is essentially achieved via substrate absorption, followed by conductive heating of the sample layer [74]. For example, unlike ultraviolet laser desorption, cationization is much more frequent than protonation. This cationization is almost certainly a result of gas-phase ion–molecule reactions that have been shown to involve ions with essentially thermal energies [75–78].

Cotter *et al.* [79] found that alkali metal ion desorption occurs directly after the laser shot and coincides with the highest temperatures in the sample. Neutral desorption occurs over a longer time period. It has been suggested [80] that intact neutrals are desorbed from cooler regions outside the area of laser impact whereas alkali ion emission takes place within the laser impact area where the temperature is much higher. It has been further proposed [81] that the weak bonds between the sample and substrate act as a 'bottleneck' which restricts energy transfer from the heated substrate to the internal degrees of freedom of the sample molecule.

Most UV laser experiments have used tightly focused laser beams and correspondingly higher power densities (*ca.* 10^{10} W cm^{-2}) [e.g. refs. 82 and 83]. While thermal processes have been demonstrated with a UV laser under some low power density conditions, conventional high power density conditions generate ions with properties that exclude the involvement of a thermal process [77,84]. For example, laser desorbed ions show kinetic energies of several to several tens of electronvolts [85], equivalent to a temperature that is much too high for the formation of stable molecular ions by a thermal evaporation process. On the other hand, the kinetic energy of the neutrals corresponds to a temperature that is much too low for such a process [86]. On this basis, Hillenkamp [85] has concluded that the process of desorption cannot be an equilibrium process.

The mechanisms that lead to desorption/ionization from the solid sample with a UV laser are believed to be collective excitation processes in the solid phase, which have already been discussed for sputtering in PDMS, as well as individual electronic excitation and photochemical reactions [87]. The main difference between PDMS and UV laser desorption/ionization is the much greater energy deposited in unit volume with the former technique.

5.2.4.1 Separation of desorption and ionization processes

A lack of reproducibility in laser desorption/ionization experiments, perhaps due to matrix effects, led several workers to separate the desorption and ionization processes. Post-ionization of the more abundant neutral molecules in the gas-phase is also a promising method for increasing instrumental sensitivity since a typical ion-to-neutral ratio in laser desorption of organics is about 10^{-3} [88,89]. Again, the power density for simple desorption has been found to be less critical than that

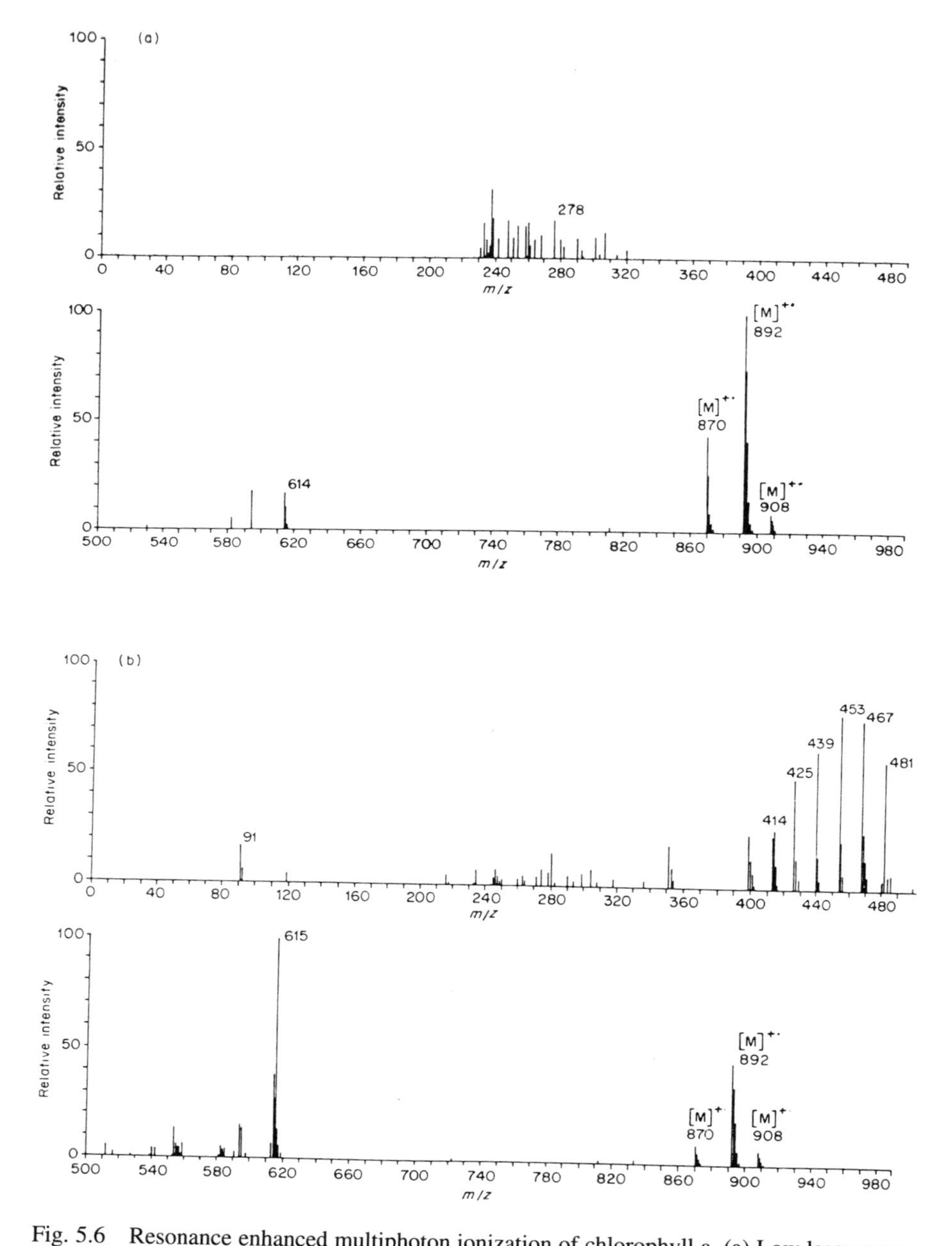

Fig. 5.6 Resonance enhanced multiphoton ionization of chlorophyll a. (a) Low laser power and (b) increased laser power. (Reproduced with permission from ref. 94)

for direct desorption/ionization. In general, neutral organic molecules are desorbed at power densities up to an order of magnitude lower than those necessary for the desorption of ions [87,90].

The separation of the desorption and ionization steps also opens up the possibility of selective ionization with a second laser based on resonance enhanced multiphoton ionization (REMPI). The multiphoton ionization method that has found most extensive application in analytical chemistry is resonant two-photon ionization (R2PI) [91]. In this process, one photon promotes the molecule to an excited electronic state and a second ionizes the molecule. The sum of these two photon energies must therefore be greater than the ionization potential of the molecule. To selectively photoionize a given molecule, however, not only does the laser have to be tuned to an intermediate electronic state for that molecule but also the absorption coefficients of other molecules that may be present must be considerably smaller. This requirement is more likely to be met if the desorbed neutrals are cooled by entrainment in a supersonic beam (section 5.3.4). Only under such conditions are the otherwise broad absorption bands of organic molecules resolved into separate, narrow vibrational transitions. With REMPI, only low laser intensities are needed for high ionization efficiencies and the technique provides very soft ionization [92]. However, an increase in laser power results in further absorption by the ions and can be used to provide controlled fragmentation of the molecule [93,94] (Fig. 5.6).

A disadvantage of separate desorption and ionization processes interfaced by a molecular beam system is the relatively low overall yield due to losses in the transport process. Some attempts have been made to use a desorption laser and a second multiphoton post-ionization laser, focused just above the sample surface, in the same vacuum chamber [95]. Unfortunately, attempts to post-ionize a wide range of laser-desorbed compounds with a molecular weight much above 500 dalton without excessive fragmentation using this configuration have so far not been successful [96].

A similar experimental configuration has been used by Becker and co-workers to demonstrate single photon ionization of lower molecular weight compounds desorbed from the solid state [97] or from frozen aqueous solutions [98]. In this case post-ionization uses a vacuum ultraviolet source that operates at an ionizing wavelength (118 nm) equivalent to a photon energy of 10.5 eV which is greater than the ionization potential of most organic compounds. Thus, unlike REMPI, single-photon ionization can be considered as a universal detector.

5.2.4.2 Desorption/ionization from a matrix

Surprisingly, single-step laser desorption/ionization is generally a softer technique which leads to less fragmentation than the two-step laser desorption/laser photoionization process. The reason for that may be that the molecule is able to dissipate excess energy by interaction with the surrounding molecules in the condensed phase, whereas this is not possible in the gas phase [89]. This observation led to the idea of dissolving an analyte in a matrix that is strongly absorbing at the laser wavelength used.

The most important function of such a matrix is as an intermediate in energy transfer from the laser beam to the sample [99,100]. Thus, an absorbing matrix can transform the laser energy into excitation energy for the solid system which leads to the sputtering of some surface molecular layers of matrix and sample. Another function of the matrix, which is present in large excess, is to separate, and therefore minimize the binding between, analyte molecules. Ionization of analyte molecules may also be assisted by the formation of photoionized matrix molecules which can act as intermediates in this process [101].

In their original work, Hillenkamp and co-workers [102] found that nicotinic acid was a suitable matrix for the desorption/ionization of intact proteins with a laser wavelength of 266 nm. By this means, ionized proteins with molecular masses in excess of 100 kDa were readily observed—considerably greater than anything previously achieved. Since the inception of this technique, a number of other, generally preferable, matrices have been developed, sometimes in conjunction with the use of different laser systems. These alternatives are discussed in more detail in section 5.3.4.1.

It seems likely that, just as for FAB/LSIMS (section 5.2.2.1), energy deposition in the matrix leads to the explosive sputtering of small, charged aggregates rather then of individual analyte ions. Volatile matrix molecules will be readily lost from these aggregates, leaving a cooled analyte or cluster ion for analysis. Sample preparation is obviously important in this respect since the use of a liquid matrix (3-nitrobenzyl alcohol) encourages the formation of large analyte clusters to a much greater extent than the commonly used solid matrices [103].

When viewed under a microscope, sample preparations that use solid matrices appear inhomogeneous and the mass spectrum recorded can vary from spot to spot in a single preparation. Factors such as the intensity of the analyte signal, suppression of the matrix signal, and shot-to-shot reproducibility all seem to be affected by the distribution of matrix and analyte within the sample preparation [103–106]. Spectra of improved quality have recently been obtained from single 2,5-dihydroxybenzoic acid (DHB) crystals grown from DHB matrix/protein mixtures [107]. The protein is believed to be homogeneously distributed within these crystals. Preliminary experiments with liquid matrices suggest that they offer better reproducibility and promote surface renewal as in FAB [103].

5.3 Practice

5.3.1 DCI instrumentation

The wire tip that receives the sample for DCI analysis is typically part of a probe assembly which is introduced into the source housing via a standard vacuum lock (Fig. 5.7). The ion source itself is a completely conventional CI source. The end of the probe bearing the tip seals on to the source block to maintain the higher pressure needed for CI operation and also picks up the current needed to rapidly heat the wire. The tip is located 1–2 mm behind the axis of the electron beam when the probe is in position. In addition to being easy to heat, the use of an electrically conducting tip avoids the loss of ion current that can result from the charging of an insulating tip.

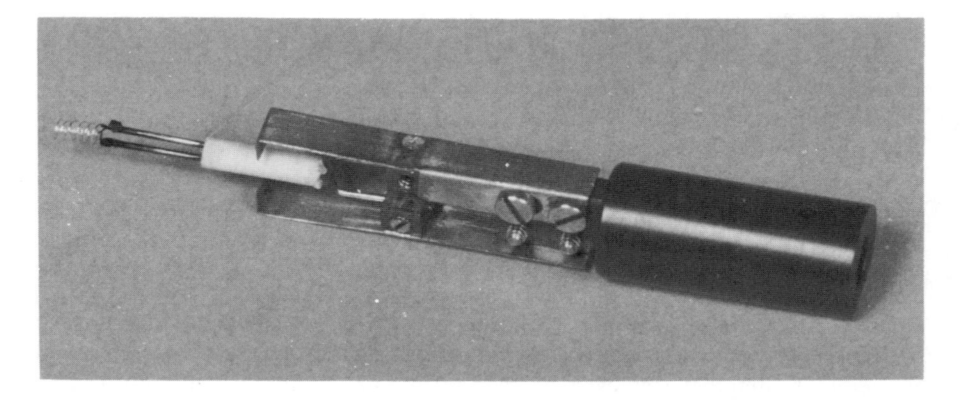

Fig. 5.7 DCI probe tip showing heated wire coil and contacts to pick up heating supplies. (Courtesy Kratos Analytical Instruments)

The tip is constructed from a few centimetres of wire that is more robust than that used for FD work, e.g. 50–200 μm diameter, coiled to a suitable shape to enter the ion source readily. The wire can be resistively heated to a maximum temperature of 500–1000°C at high rates and can conveniently act as its own temperature sensor. Desorption of many relatively involatile organic compounds takes place at lower temperatures than these, but the higher temperature affords a means of cleaning the wire between analyses or of carrying out pyrolysis experiments.

As was pointed out in the discussion of section 5.1.1, evaporation is favoured compared with decomposition at high temperatures. Ideally the sample is heated instantaneously to a suitable temperature and maintained at that temperature while spectra are recorded. If, however, too high a temperature is maintained then the evaporation process is rapid compared with the spectrometer scan time and the evaporation and decomposition processes become more difficult to resolve [3,8]. In the first instance, therefore, a programming rate of approximately 10°C s^{-1} is employed in the analysis of an unknown sample.

Just as in FD, there is a temperature at which the production of molecular ions is optimal for each compound [17,108]. For example, in experiments designed to demonstrate the vapour phase cationization of intact sucrose molecules evaporated from a heated wire, Kistemaker and colleagues [80] showed that a temperature of approximately 200°C was optimal for evaporation with minimum decomposition.

If a best emitter temperature can be determined from a run programmed at 10°C s^{-1} then better quality spectra may sometimes be obtained by setting the wire temperature to rise as rapidly as possible to this value. It is a general feature of DCI operation that the best spectra are obtained over a relatively short time period after this optimum temperature is reached. Thus, fast repetitive scanning is essential for the recording of representative spectra.

Other parameters may affect the quality of DCI spectra. For example, most investigators have found that the level of fragmentation is strongly dependent on the position of the tip carrying the sample in the ion source (e.g. refs. 9 and 15) and that the best results are obtained when evaporation takes place 1–2 mm from the axis of the electron beam.

5.3.1.1 Effect of probe conditions on DCI performance

The surface properties of the wire tip can have a considerable effect on the quality of the DCI spectrum obtained. Thus, various materials have been used including tungsten [109,110], rhenium [19,111], platinum [112], and platinum–iridium with similar results, so long as the surface remains clean. If, however, the wire is not cleaned by heating between each analysis a deterioration in the quality of the DCI spectra is observed.

An enhanced absolute yield of the quasi-molecular ion and a decrease in the relative abundance of fragment ions is seen for a number of compounds when a bare metal wire is replaced with one which has been electrophoretically coated with a polyimide material [113] (Figs. 3.8 and 5.8). The polyimide-coated wires, which are relatively easy to prepare, may be used for up to ten successive samples. Other reports of the advantageous use of polymer-coated tips have appeared in the literature. For example, Bencsath and Field [114] used a ceramic tube carrying an internal heating wire and coated externally with 'Dexsil' polymer which can be cleaned by flaming. Indirectly heated probe tips made of 'Vespel' [115] or coated with silicone [13] have also been used.

Unlike the use of an activated FD emitter, loading a sample on to a bare wire from solution can be very difficult since the wire is not adequately wetted and the solution and sample tend to accumulate at one point on the wire rather than providing an even sample deposit. Reproducible spectra of good quality are more readily obtained if the sample is deposited as an even film, and in this instance the electrospray loading technique [116,117] can be used to good effect (Fig. 5.9).

Electrospraying occurs automatically as soon as a positive voltage is applied to the needle of the syringe containing the sample in solution. The fine spray of charged droplets gives a very even deposit on the wire tip. Many solvents such as acetone, methanol, and ethanol spray satisfactorily, although solutions with a high water content cannot be accommodated (cf. section 6.3.2.1). The use of a hard

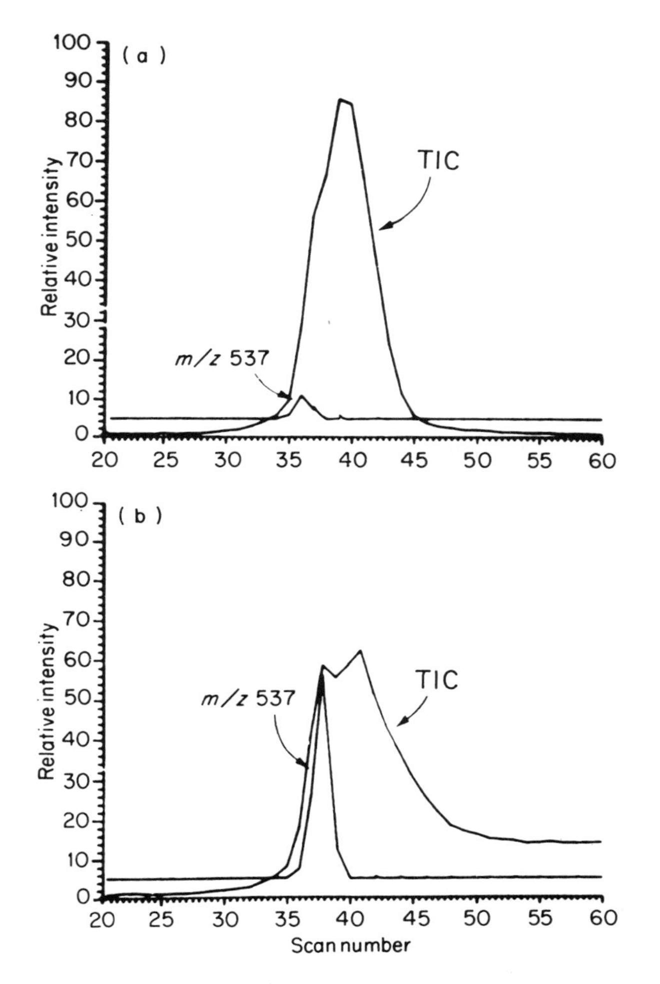

Fig. 5.8 Total ion and selected ion (m/z 537) profiles obtained from equal amounts of the pentapeptide (Phe–Asp–Ala–Ser–Val) desorbed from (a) bare wire and (b) polyimide coated wire. Ammonia reagent gas in each case (m/z 537 corresponds to $(M-H_2O + NH_4)^+$). (Reprinted with permission from V.N. Reinhold and S.A. Carr, *Anal. Chem.*, **54**, 499[113]. Copyright (1982) American Chemical Society)

ceramic probe tip as reported by Bencsath and Field [114] facilitates the loading of powdered solid samples that cannot be handled in solution.

An alternative to the use of wires is a return to the use of FD emitters pioneered by Hunt (section 5.2.1). Although emitters are more difficult to prepare, they are easier to load and need not be of the high quality used for field desorption [118]. In addition, the threshold sample size of approximately 100 ng for useful data found

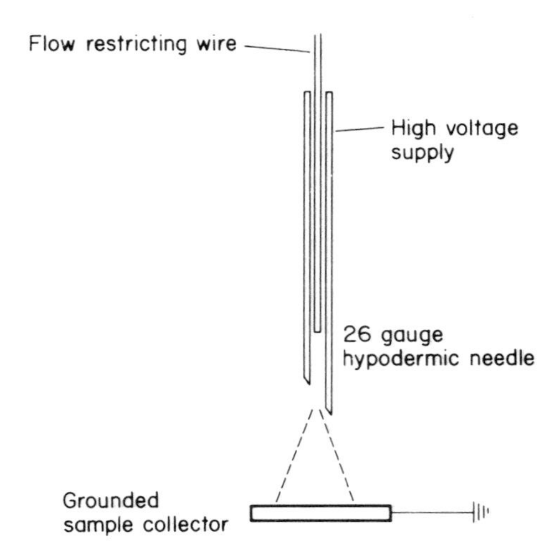

Fig. 5.9 Schematic diagram of apparatus for electrospraying dissolved samples

in some cases using uncoated wires [112] has not been observed using activated emitters.

5.3.1.2 Effect of CI conditions on DCI spectrum quality

The use of chemical ionization rather than electron impact conditions plays a fundamental part in the success of DCI. Even in 'in-beam' EI work, the consistent appearance of $(M+H)^+$ rather than M^+ ions suggests a chemical ionization process involving proton transfer from sample molecules. As might be expected, the choice or reagent gas is also important in determining the quality of DCI spectra. For example, Arpino and Devant [109] compared the DCI spectra of ascorbic acid using methane and ammonia as reagent gases and found that the protonated ion formed using methane has a much greater tendency to fragment than does the $(M + NH_4)^+$ ion. Again still less fragmentation is seen in the Cl^- attachment DCI spectrum of glucose than in the ammonia spectrum of glucose. Thus, less energetic reagent ions are most frequently used in DCI work, e.g. NH_4^+ in the positive ion mode and Cl^- and OH^- in the negative ion mode.

A primary objective in DCI work is to emphasize molecular weight information where the relevant ion may often be a weakly bound addition complex. The use of a somewhat higher reagent gas pressure so that fragmentation is relatively reduced can therefore be advantageous, even though the absolute sensitivity may be lower (section 3.2.3). Much published DCI data uses a reagent gas pressure of approximately 0.5 torr when surprisingly complex molecules may be volatilized intact and subsequently ionized (e.g. Fig. 4.4). Optimization of the source pressure using a test compound such as glucose under ammonia CI conditions may be a useful exercise.

An alternative technique in which K^+ ions for cationization are generated by heating a thermionic glass coated on a DCI filament has been used to provide DCI-like spectra of oligomers in a conventional EI source [119]. The oligomer sample itself is deposited on a thin metal ribbon held close to and heated radiatively by the DCI filament. Solid sources of CI reagent ions have also been discussed in sections 3.2.1 and 4.3.1.

5.3.2 Instrumentation and techniques for FAB/LSIMS

The principle of fast atom bombardment has been described previously (section 5.2.2.1). Energetic argon or xenon neutrals are produced by charge exchange of an accelerated ion beam with un-ionized gas present in the atom gun (Fig. 5.4):

$$Ar \rightarrow Ar^+ + e^- \tag{5.2}$$

$$\underset{\rightarrow}{Ar^+} + Ar^0 \rightarrow \underset{\rightarrow}{Ar^0} + Ar^+ \tag{5.3}$$

(The subscript arrow denotes an accelerated particle.)

This charge exchange process occurs with little or no loss of forward momentum, so that an ion beam accelerated to 8 keV produces a neutral beam of similar energy. With a gun voltage of 6–8 keV and a gas flow rate of 0.5 $cm^3\ min^{-1}$, approximately 25–50 μA equivalent ion current is available in the neutral beam.

An alternative source of energetic particles is a Cs^+ ion gun in which ions are produced by evaporation from a suitable caesium salt [26,120]. Typical operating conditions for a Cs^+ ion gun are a gun voltage of 35 keV and an ion current of approximately 2 μA [29,121] which results in an improved sensitivity and an improved ability to analyse higher molecular weight samples compared with the neutral gun. Other advantages are that an ion beam may be focused to give a spot size commensurate with the sample target size and that the use of an ion gun does not require the use of a gas such as xenon so that operation at a lower source housing pressure is possible.

The energetic neutral or charged beam strikes the sample surface at an angle of 60–70° to the normal and ions produced by the sputtering process are then extracted and focused into the analyser system. The sample matrix is deposited over an area of approximately 0.1 cm^2 on a metal support mounted on the end of a probe and held at normal ion source potential. The probe is introduced into the source through a vacuum lock. A metal such as copper or stainless steel is used as the support material. Facilities for heating the probe tip, e.g. by conduction from a heated source block, may also be provided. Such heating is essential to volatilize the liquid flow in the continuous-flow FAB technique (section 2.4.4).

Setting up for FAB operation may be accomplished using the peak at m/z 393 $(Cs_2I)^+$ from caesium iodide. This material may be applied to the probe tip as a saturated aqueous solution and is still effective even when the solvent has completely evaporated. Because ion production occurs over an extended time period and because sample fractionation is not observed, much slower scanning speeds are suitable for FAB operation compared with DCI or FD.

5.3.2.1 Sample preparation for FAB/LSIMS

As with all of the ionization techniques described in this chapter, sample preparation is of paramount importance. A general method of sample preparation for FAB/LSIMS analysis is as follows. The metal tip is cleaned using either solvent or acid to provide a surface that is free from impurities and which is readily wetted. One microlitre of a solution of the sample made up in a suitable solvent, such as water, methanol, or acetonitrile (1–10 $\mu g\, \mu L^{-1}$), together with any appropriate additives, such as an acid, often trifluoroacetic acid, or a cationizing salt, is then added to and thoroughly mixed with glycerol (2–3 μL) already dispersed on the clean probe tip.

The object of this preparation is to present the sample dissolved in glycerol as a continuous thin film on the probe tip. Significantly weaker spectra are produced if the solution fails to wet the surface and is presented as discrete drops rather than as a continuous film. As mentioned in section 5.2.2.1, a solution of the sample in a mobile medium such as glycerol provides a means by which the surface concentration of sample may be rapidly replenished following damage on bombardment [25], so that sample beams with good intensity may be prolonged for periods of 20 minutes or more. Even after this period the sample may often be rejuvenated by the addition of further glycerol to replace that lost by evaporation.

5.3.2.2 Effect of matrix in FAB/LSIMS

If a matrix is to provide an intense and persistent analyte ion current, it must have a low vapour pressure to keep the sample preparation in a liquid state while under vacuum and it must be a good solvent for the analyte [122]. A suitable matrix will also have sufficient electrical conductivity to prevent charging of the sample preparation.

Apparently suitable matrices can, however, also produce adverse effects. First, the extensive chemical background produced by most liquid matrices can compromise the detection of analyte ions present at lower levels. Second, an analyte with a low surface activity in the chosen matrix may exhibit a reduced sensitivity in FAB/LSIMS [123,124]. Third, a number of chemical reactions, mainly initiated by the neutral or ion beam, can occur in the liquid phase [51,125–131].

The choice of a matrix has been considered in a number of reviews [58,125,132–134], a majority of which [58,132,134] deal with the analysis of peptides and proteins by FAB/LSIMS. The physical properties of matrix materials have also been comprehensively documented [135]. Glycerol [125,136] is still widely used for peptides but usually does not give the best signal-to-noise ratio, and artefact formation has been noted [137,138]. Thioglycerol [125,139,140] is more volatile than glycerol and generally leads to higher quality spectra, although the reduction of disulphide bonds is a side reaction. Mixtures of glycerol and thioglycerol may also be used [132].

Another thiol-containing matrix that is useful for peptides and for a wide variety of other compound types is a mixture of dithiothreitol (five parts) aud dithioerythritol (one part) ('magic bullet') [141]. For example, Rose [132] has reported that peptides modified with an iron-containing chelating group gave spectra that showed apparent loss of iron in glycerol, while the use of a 'magic bullet' matrix gave spectra showing protonated molecular ions with the appropriate metal retained with the chelating group. A number of other thiols have also proved useful as matrix materials [e.g. ref. 142].

Apart from thiol-containing compounds, the other most suitable matrix for peptides is *m*-nitrobenzyl alcohol (*m*-NBA) [58,143]. Extensive use of *m*-NBA by Barber *et al.* [144] confirmed that the matrix is suitable for a wide range of samples, including organometallics and higher mass samples, such as small proteins, as well as less polar materials, such as cholesterol and chlorophyll (cf. section 5.3.2.4), which had proved difficult to analyse using other matrices.

The matrices described so far have been used in the majority of analyses although others, such as lower molecular weight glycols and their ethers [125,145–147], triethanolamine [125], diethanolamine [125], 2,4-ditertamylphenol [125], tetramethylene-sulphone [125,148], and even concentrated sulphuric acid [254], may sometimes be necessary to obtain useful spectra. For analyses in the negative ion mode, Rose *et al.* [149] have demonstrated the advantage of 3-aminopropane-1,2-diol which provides long-lasting spectra and appears to reduce suppression phenomena. Triethanolamine [145,150–152] and diethanolamine [145,153] are also useful general matrices for negative ion operation.

De Angelis *et al.* [142] have suggested the use of thiodiethylene glycol (TDEG) or a 1:1 mixture of TDEG and glycerol as a replacement for glycerol and as a matrix suitable for the analysis of organometallic compounds. TDEG is more fluid and volatile than glycerol and therefore yields more intense ion currents for a shorter period of time. It is also reported that TDEG can reduce some suppression problems that occur because of differences in surface activity in glycerol. For example, the tryptic digest of glucagon analysed in glycerol shows the more hydrophobic residue 19–29 (MH^+ = 1352) in preference to the more hydrophilic peptide 1–12 (MH^+ = 1357) [154]. The use of the TDEG matrix is sufficient to allow the $(M + H)^+$ ion from the 1–12 peptide, suppressed in glycerol, to appear in the spectrum with an intensity comparable to the other fragment.

As mentioned in section 5.2.2.2, surface activity plays a vital role in determining the relative intensities of molecular ions when mixtures are analysed by FAB/LSIMS. Thus, several workers have reported a correlation between the hydrophobic character of a peptide and the abundance of the molecular ion [154–158], i.e. the less hydrophobic the peptide the weaker its $(M+H)^+$signal.

In mixture analysis, the signal from a hydrophilic peptide is further suppressed in the presence of more hydrophobic peptides [154,157,158], since they are unable to compete effectively for the surface layers of the matrix. A change of matrix, as suggested earlier in this section, may circumvent these problems in mixture analysis. Another approach, which also reduces suppression in mixture analysis, is the use of the continuous-flow FAB technique [159] [section 5.3.2.4].

The use of *m*-NBA as a matrix exemplifies problems caused by chemical reactivity, in this case oxidation to the aldehyde under ion bombardment and subsequent reaction with analyte amino groups. The presence of additional ions at $(M+133+H)^+$or $(M+133-H)^-$ from this reaction will cause errors in molecular mass determination if these are not resolved from the protonated molecular species [160]. A similar formation of formaldehyde from glycerol and subsequent condensation of the aldehyde with analyte amino groups to form a Schiff base has been reported [137].

Reduction is one of the most widely documented beam initiated reactions. Cerny and Gross [161] have suggested that the formation of $[M+2H]^{+\cdot}$ and $[M+3H]^+$ ions from nucleosides in glycerol involves multiple protonation by the matrix followed by one- and two-electron reductions. The reduction of many other classes of analyte such as dyestuffs [162] and peptides [163] in glycerol has also been documented. Meili and Seibl [143] have demonstrated that whereas a corrin of MW 1120 undergoes reduction in glycerol a change of matrix to the aprotic *o*-nitrophenol octyl ether eliminates this problem. Examples have also been documented where a halogen atom is substituted by hydrogen in a beam-initiated reaction [127]. Keough [164] has demonstrated that the radicals formed from glycerol can react with *N*-alkyl pyridinium salts to form adduct ions. The same reaction does not occur in thioglycerol where reactive radical formation is minimized. Dass and Desiderio [138] have also reported that adduct ion formation, prevalent with some peptides in glycerol, is absent when thioglycerol is used.

5.3.2.3 Effect of additives in FAB/LSIMS

In general terms, FAB/LSIMS is better suited to compounds that have relatively basic or acidic centres, i.e. that can act as proton acceptors or donors, or to compounds that already have charged centres, e.g. quaternary ammonium salts and sodium salts of carboxylic acids. Additives that can promote the formation of ions in the condensed phase will result in correspondingly more intense spectra [cf. ref. 165]. Thus, an additive such as trifluoroacetic acid [29,58], acetic acid [166], or *p*-toluenesulphonic acid [167] will provide more intense positive ion spectra with compounds possessing basic centres whereas ammonium

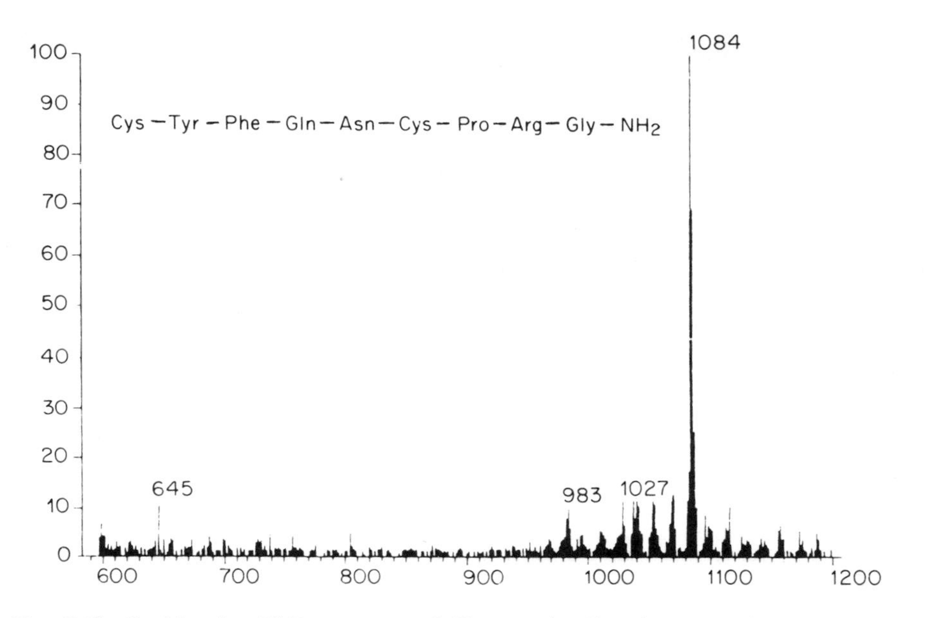

Fig. 5.10 Positive ion FAB spectrum of Vasopressin (Arg, human). (Courtesy Kratos Analytical Instruments)

hydroxide, sodium hydroxide [168] or tetrabutylammonium hydroxide [169] can be used to improve the yield of negative ions.

Peptide analysis provides a good illustration of this feature. Peptides containing an excess of basic amino acids will provide positive ion spectra with intense $(M+H)^+$ ions (Fig. 5.10) whereas peptides containing an excess of acidic amino acids provide $(M - H)^-$ ions rather than $(M+H)^+$ ions.

Addition of a salt such as sodium chloride to promote cationization is the preferred sample preparation method for oligosaccharide analysis. Cationization can also be a useful technique in many other cases. For example, although a peptide such as antiamoebin III does not give an intense $(M+H)^+$ ion, spectra containing intense $(M+Na)^+$ ions are obtained in the presence of a sodium salt (Fig. 5.11).

It has been reported that lithium, sodium, and potassium generate more abundant cationization products than rubidium and caesium with trehalose in glycerol [170]. These differences were explained on the basis of the relative affinities of the matrix and the saccharide for the available cations. Glycerol has a high affinity for alkali cations and can suppress cationization of the analyte. Cationization can be enhanced, however, by the use of a liquid matrix with lower cation affinity [171].

An excess of salts present as impurities can also suppress analyte ion formation. In this case, the use of a simple clean-up procedure based on a small-scale C_{18}

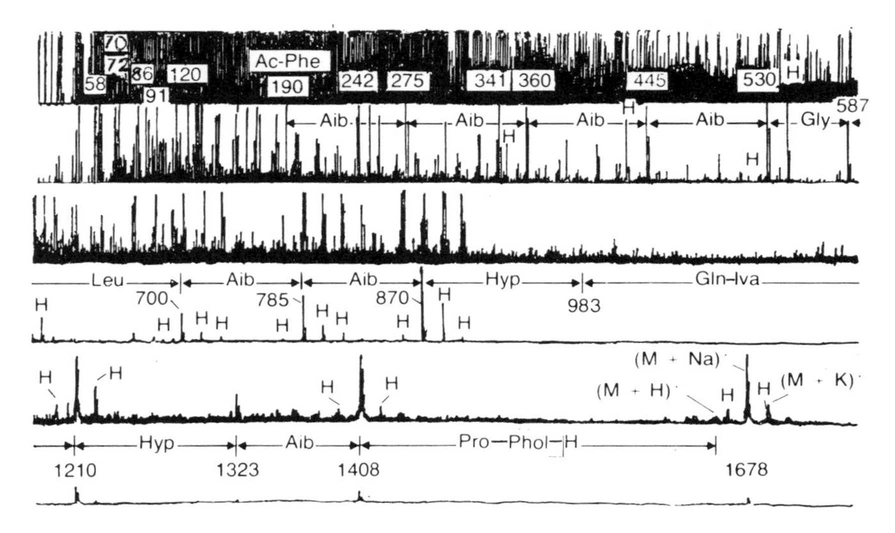

Fig. 5.11 Positive ion FAB spectrum of Antiamoebin III. (Reprinted with permission from M. Barber *et al.*, *Anal. Chem.*, **54**, 645A[25]. Copyright (1982) American Chemical Society)

reversed-phase LC column [172,173] is recommended, particularly with samples of biological origin. Samples containing high cation levels may be characterized by the appearance of cationized matrix peaks, e.g. $(92n + 23)^+$, from sodium salts in glycerol.

5.3.2.4 Effect of co-solvent in FAB/LSIMS: continuous-flow FAB/LSIMS

Efficient surface renewal leading to the maintenance of an intense ion beam requires that the sample be dissolved rather than slurried in the matrix material. For a sample such as chlorophyll A, which is totally insoluble in glycerol, no spectrum is obtained until approximately 1% of a solubilizing agent, Triton X-100 (**1**) has been added to the matrix [25].

$$ArO(CH_2CH_2O)_nH \qquad (\mathbf{1})$$

If a sample is insoluble or only poorly soluble in the matrix, it is usually dissolved in a co-solvent, which is miscible with the matrix, before the matrix is added. Typical co-solvents are methanol, ethanol, acetonitrile, water, dimethylsulphoxide [167,174,175], dimethylformamide [175], and tetrahydrofuran [176,177]. Röllgen and co-workers [178] have studied the dynamics of the supersaturated solutions that are formed as the more volatile co-solvent preferentially evaporates under vacuum.

A sample preparation technique which has been applied to samples insoluble in the matrix is the 'surface precipitation' method. In this method, the sample, dissolved in a suitable solvent, is carefully transferred to the surface of the matrix where the solvent is allowed to evaporate to leave an extensive surface layer of the analyte [179]. This method has also been applied to zwitterionic phospholipids [180] which are matrix soluble. In this case, chloroform and diethanolamine or glycerol were used as the solvent–matrix pair. The presence of a reduced level of background ions from diethanolamine in the spectra confirmed that the matrix surface was effectively covered with phospholipid molecules. In practice, the reduced background also assists in the ready identification of diagnostic sample ions. Satisfactory spectra were not achieved with these phospholipids by mixing the sample in the matrix.

Conventional FAB/LSIMS ionization usually produces a spectrum with some signal at every m/z value over the scanned mass range. These signals are mostly derived from ions produced by the matrix. Continuous-flow FAB has already been discussed as an LC–MS technique (section 2.4.4) but was originally developed to reduce such difficulties by using a water matrix with only a small concentration of an organic component such as glycerol [181]. Results have shown that this technique provides substantially better limits of detection. These improvements have been attributed to a decrease in the background chemical noise, due to the reduced organic content of the matrix (section 2.4.4), and also to an increase in the analyte ion yield from sample preparations which contain a higher proportion of water [182].

Continuous-flow FAB also shows a reduced ion suppression effect (cf. section 5.3.2.2), probably because of effective mixing of the thin liquid layer at the probe tip and the increased water content of the matrix. As an example, the continuous-flow FAB analysis of a mixture of seven peptides showed enhanced $(M+H)^+$ ions for four compounds and similar relative intensities for the other three compared with conventional FAB [183]. The $(M+H)^+$ ion intensity distribution was essentially the same as when the peptides were analysed individually by continuous-flow FAB, i.e. the presence of the other peptides had little effect on the relative intensity of the $(M+H)^+$ ions.

5.3.2.5 Sensitivity and signal-to-noise (S/N) in FAB/LSIMS

A definition of sensitivity in the FAB mode is somewhat more elusive than in some other ionization techniques, since it is not possible to define the amount of sample consumed during an analysis. At a certain bulk concentration an ideal sample will provide a constantly replenished monolayer at the surface of the glycerol. Thus, a 5×10^{-4} molar concentration of the surfactant cetylammonium bromide in glycerol achieves this condition and provides a spectrum showing only sample ions and no glycerol ions [25]. An increase in concentration above this value will not afford any increase in sample ion intensity for a given primary beam intensity. A decrease in sample concentration, due either to insufficient sample or to low solubility, will provide spectra containing an increased proportion of matrix ions.

A more practical measurement of sensitivity is therefore the amount of analyte required to provide a given signal intensity compared to the background signal due to matrix material [35]. For example, Caprioli and Fan [182] reported a signal-to-noise value of 3:1 at the $(M+H)^+$ ion from 5 pmol substance P (MW 1347) using a fast atom primary beam. Detection limits deteriorate rapidly with an increase in analyte molecular weight. Thus, van Breemen and Le [184] recorded the following detection limits [*S/N* = 2:1] using a probe tip modified by coating with nitrocellulose: dynorphin peptide (amino acids 1–13) (MW 1604), 0.6 pmol at RP 3000; human growth hormone releasing factor peptide (MW 3357), 75 pmol at RP 5000; α-amylase inhibitor (MW 7958), 6 nmol at RP 6000. A fast atom primary beam was also used for these experiments.

Continuous-flow FAB can give improved signal-to-noise figures, particularly because of the reduction in chemical background. For example, Caprioli and Fan [182] state that an *S/N* value of 10:1 may now be obtained at the $(M+H)^+$ ion from 100 fmol Substance P. A later experiment [185], in which the peptide angiotensin II (MW 1045) was introduced to the continuous-flow FAB probe from a capillary electrophoresis column, gave a very similar detection limit, i.e. an *S/N* value of 3:1 at the $(M+H)^+$ ion from 75 fmol angiotensin II.

If an array detector is available, an improved signal-to-noise ratio can be achieved for surface active analytes by taking advantage of the fact that the relative intensity of the analyte to matrix signal immediately after the the onset of energetic particle bombardment can be much enhanced compared with its subsequent value. Tyler *et al.* [186] have shown that pulsed FAB, in combination with an integrating array detector, offers a detection limit for the peptide eledoisin (MW 1188) in glycerol of 100 attomoles (1 amol = 10^{-18} mol). Sample preparation is an important factor in pulsed FAB. For example, although *m*-NBA is often a better matrix for conventional FAB, it does not generally offer the same *S/N* improvement with pulsed FAB exhibited by samples prepared in glycerol. The improvement seen with glycerol is attributed to the formation of surface layers of analyte that are removed in the first few hundred milliseconds of particle bombardment.

5.3.3 Instrumentation and techniques for PDMS

The flux of fission fragments from a ^{252}Cf source is relatively small so that the resulting secondary ion beam is weak and needs a more sensitive mass analyser, such as a time-of-flight mass spectrometer (TOFMS), which can record all the secondary ions produced. The general features of a PD mass spectrometer based on a time-of-flight analyser are shown in Fig. 5.12.

A 10-μCi ^{252}Cf source is placed just behind the foil that carries the sample. ^{252}Cf decays with a half-life of 2.65 years to give alpha particles and two multiply-charged fission fragments simultaneously emitted in opposite directions. A typical pair of fission fragments would be ^{106}Tc and ^{142}Ba with energies of 104 and 79 MeV respectively [187].

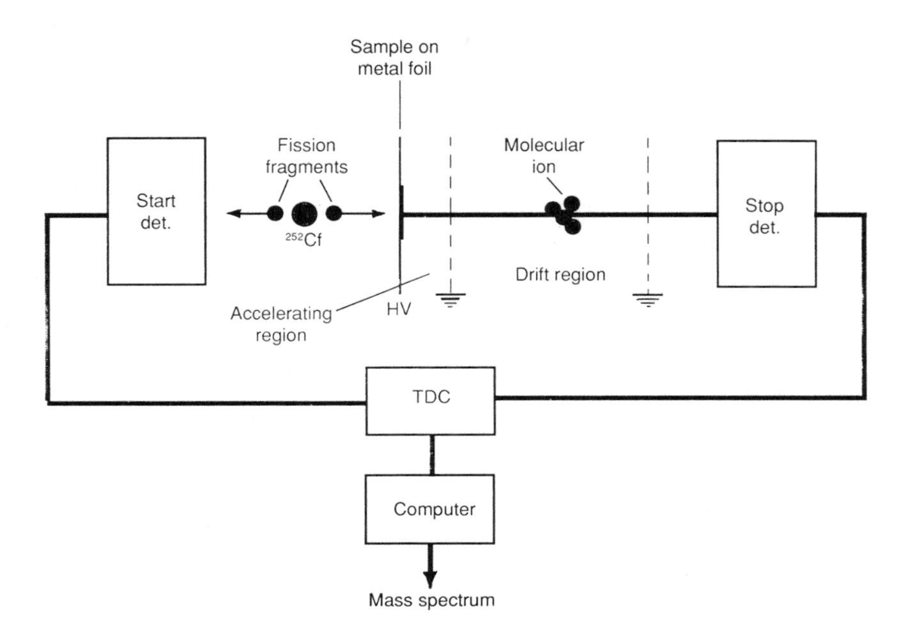

Fig. 5.12 Schematic diagram of plasma desorption instrument. (Adapted from ref. 67)

One of the pair of fission fragments hits the start detector and initiates the time-of-flight measurement, while the other penetrates the thin aluminium or aluminized polyester (Mylar) foil to the sample which is coated on the reverse side. Secondary ions desorbed by the fission fragment are then accelerated from the sample foil (held at 10–20 kV) towards a grounded grid which is located at the entrance to the TOF analyser.

Because of the low yield of secondary ions from each fission event, it is usual to accumulate a large number of spectra and for the total analysis time to be of the order of 5–10 minutes. The ion flight times which are recorded are then converted into mass-to-charge ratios by means of a suitable calibration. Practical methods of improving the accuracy of this calibration in PDMS have been discussed recently [188].

5.3.3.1 Sample handling and preparation for PDMS

Sample purity and presentation are of fundamental importance if good results are to be obtained by PDMS. Early work used the electrospray method (section 5.3.1.1) to deposit the sample as a thin layer on the foil base. By this means, molecular ion information could be recorded for a range of previously inaccessible compounds [187,189] including peptides and proteins with molecular weights up to about

14 kDa [190,191]. The electrospray method is, however, unsatisfactory for highly aqueous solvents [192] and spectrum quality is much poorer in the presence of alkali metal salts [67,193]. Similar problems have been noted with calcium ion contamination [194], but ammonium salts appear to have no adverse effect on spectrum quality [188].

A much more satisfactory procedure for peptides and proteins is to adsorb the sample on to a nitrocellulose layer that has previously been deposited on the sample foil. The nitrocellulose surface, which is negatively charged in aqueous media, offers a number of significant advantages. One of the most important of these is the ability to remove metal ions, by rinsing with deionized water or mild acid solutions, while adsorbed peptides remain bound to the surface.

The use of nitrocellulose, rather than electrospraying, also allows the deposition of a thinner and more even sample layer which is only weakly bound to the substrate (section 5.2.3). As a result, the yield of molecular ions is increased compared with the background signal. Another advantage is the improvement of peak shape which increases the effective mass resolution. The best nitrocellulose surfaces are thick homogeneous layers which are prepared by electrospraying 25–50 μL of a 2 μg μL^{-1} solution of nitrocellulose in acetone over a spot size approximately 7 mm in diameter [192].

Nitrocellulose is now the most commonly used sample preparation method for PDMS and has superceded an earlier suggestion [195] in which Nafion, a cation exchange polymer, was electrosprayed onto a Mylar foil. A number of authors have used positively charged surfaces that are complementary to nitrocellulose. For example, McNeal and Macfarlane [196] have shown that an immobilized cationic

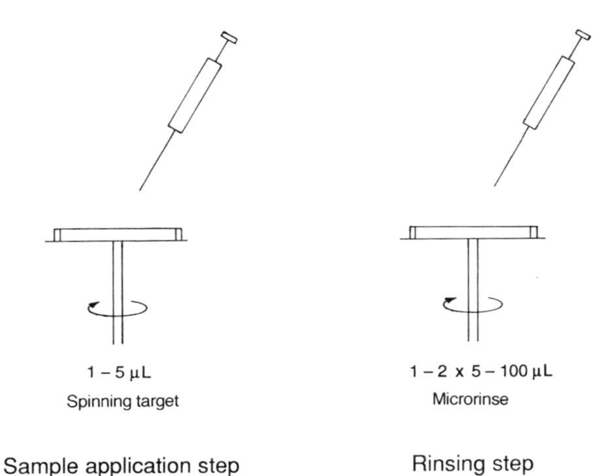

Fig. 5.13 Scheme for adsorption and washing of sample on a spinning nitrocellulose target. (Reproduced with permission from ref. 193)

surfactant, tridodecylmethyl ammonium chloride, has advantages for the analysis of polyanionic samples while Lacey and Keough [197] have investigated quaternized Nylon-66 electrosprayed on to Mylar.

The nitrocellulose technique has been improved and sensitivities in the low picomole range have been obtained by application of the sample solution to a spinning nitrocellulose target [192] (Fig. 5.13). By this means, a thin layer of the peptide or protein is deposited in the centre of the target whereas the more soluble inorganic impurities are carried with the solvent to the edge of the target. If an inspection of the spectrum suggests that the sample is still contaminated with salts, a washing procedure is applied to the same spin-deposited sample, which may then be analysed again since PDMS is virtually non-destructive.

With the nitrocellulose technique several workers have been able to observe molecular ions from small proteins with a molecular weight above 20 000 daltons [198–200]. Figure 5.14 shows a positive ion spectrum from porcine pepsin, with a molecular weight of 34 688 [198]. The response from a given protein is strongly dependent on the pH of the solution from which the sample is loaded on to the nitrocellulose surface [69,199]. Optimal PD mass spectra in terms of molecular ion intensity, minimum peak width, and highest relative intensity of more highly charged species are often obtained when analytes are loaded from solutions having pH values near their isoelectric points [69]. It is believed that the stable folded configuration achieved at the isoelectric point minimizes the interaction between protein and nitrocellulose.

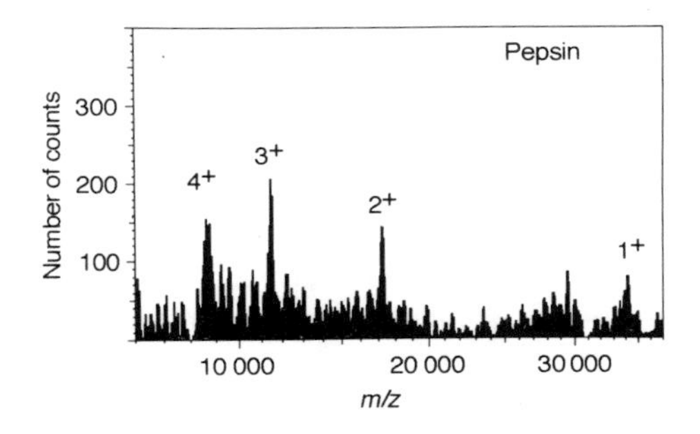

Fig. 5.14 Mass spectrum of positive ions from porcine pepsin (MW 34 688) adsorbed on a nitrocellulose backing. (Reproduced with permission from ref. 198)

Since PDMS analysis consumes only a small fraction of the sample presented, the remainder of the sample is then available for chemical or enzymatic reactions which may themselves be followed by a further PDMS analysis. These reactions may be carried out *in situ* on the nitrocellulose bound sample [192,201] or the sample may be recovered from the nitrocellulose by means of a suitable solvent system [193,202].

Suppression effects have been reported [203] when PDMS is used for the direct analysis of mixtures. Unlike the use of FAB for the analysis of peptide mixtures, no correlation is found between the relative hydrophobicity of a peptide and the abundance of its molecular ion in mixture analysis. On the other hand, there is a strong correlation between the relative net charge carried by the peptide and the molecular ion abundance. Thus, in positive ion PDMS, peptides with a more negative net charge under the conditions of the experiment can be suppressed in the presence of peptides with a more positive net charge, while the reverse is true in negative ion PDMS [203].

The relative abundances of analyte-related ions derived from mixtures of solid triglycerides and waxes in PDMS are found to vary with sample preparation and do not correspond to the bulk composition. In view of the packing characteristics of such lipids, these results support the idea that the ions are derived from the surface layers, rather than from the bulk of the solid. For example, it has been proposed that less efficiently packing, shorter chain compounds are forced to the surface where analysis takes place. On the other hand, mixtures that remain as liquids during analysis show ion abundances that are consistent with their bulk composition [204].

5.3.4 Instrumentation and techniques for LDMS

Although both continuous-wave and pulsed lasers were used by early experimenters, pulsed lasers are now almost exclusively used in both the UV and IR regimes. The short time scale of pulsed laser operation means that LDMS is not readily compatible with conventional scanning operation. On the other hand, the pulsed laser is an ideal ion source for a time-of-flight analyser which can record all the ions produced by each laser pulse. The virtually unlimited mass range of the TOF analyser is a particular advantage in matrix-assisted laser desorption/ionization analysis of high molecular weight materials (section 5.3.4.1). The Fourier transform mass spectrometer is also a very suitable instrument for laser desorption/ionization experiments and a number of authors have used this combination with both IR [205–208] and UV [77, 209, 210] lasers.

Early experiments on laser desorption/ionization, reported by Hillenkamp, were based on a commercial reflectron TOF instrument [89]. A similar instrument was later modified by Hillenkamp and co-workers by the addition of facilities for post-ionization [211]. In the latter case, the neutral molecules are desorbed by the first UV laser while a second UV laser is triggered with a variable time delay to photoionize them, in the gas phase, directly above the sample surface [89,96].

For molecular beam work, the sample can be volatilized by a pulsed CO_2 laser (Fig. 5.15). Volatilized neutrals are then entrained in a supersonic beam of argon atoms formed by expansion of argon gas at 1 atmosphere through a $100\,\mu m$ diameter orifice. The reduction in kinetic energy spread that results from entrainment is an important factor in the ability of the molecular beam technique

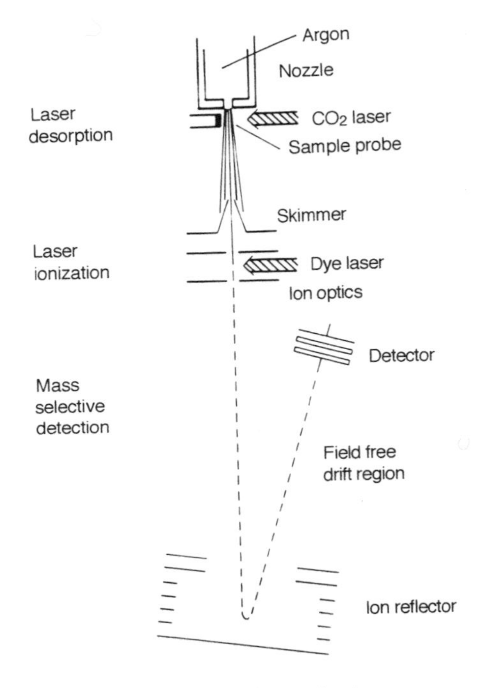

Fig. 5.15 Schematic diagram of laser desorption/resonance enhanced multiphoton ionization instrument based on a reflectron time-of-flight mass spectrometer. (Reproduced with permission from ref. 94)

to offer greatly improved mass resolution with a time-of-flight mass spectrometer [212]. The molecular beam is then interrogated and ionized by a transverse tunable dye laser beam in the source region. As previously mentioned (section 5.2.4.1), cooling of internal energy levels of analyte molecules in the expansion results in narrow absorption bands and therefore very selective ionization [88,94].

If laser desorption experiments are carried out in a conventional high pressure chemical ionization source, the lifetime of desorbed neutrals can be extended by collisions within the reagent gas plasma. In addition, the desorbed neutrals will also be ionized by the usual chemical ionization ion–molecule collision processes. By this means laser desorption can be made more nearly compatible with conventional scanning operation and provides a system by which chemical ionization spectra of relatively polar molecules may be recorded [213].

An alternative use of a laser in an electron impact source that is also compatible with a scanning instrument is to use the laser to volatilize the sample indirectly via the substrate. In a recent example [214], energy from a pulsed ruby laser is used to vaporize the sample by means of a fibre-optic coupling inserted in a hollow probe and reaching up to the sample holder (Fig. 5.16). Subsequent ionization of

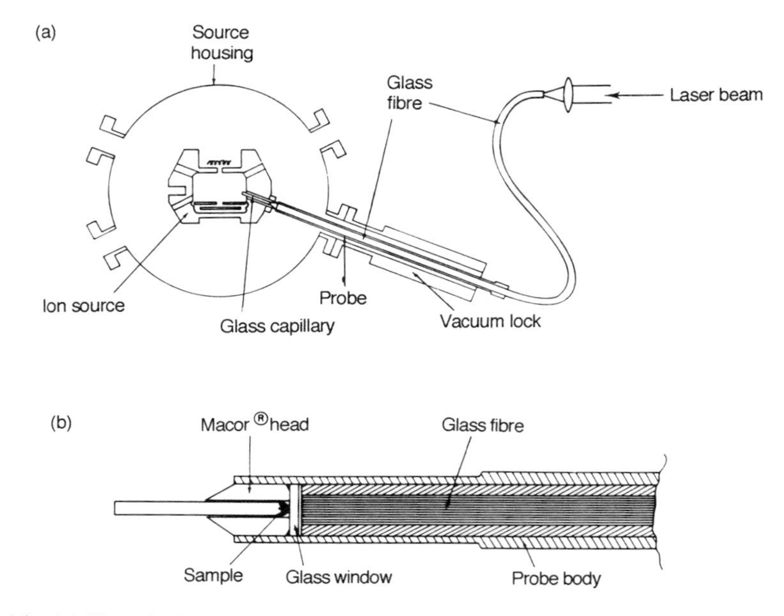

Fig. 5.16 (a) Use of a laser to vaporize sample into an electron impact ion source and (b) detail of sample probe. (Reproduced with permission from ref. 214)

the volatilized neutrals is by conventional electron impact. With this configuration, the relatively long duration of the signal suggests that sample evaporation lasts for a long time after the laser pulse. Decomposition is minimal and EI spectra of rifamycin macrocyclic antibiotics [215] show abundant molecular ions and fragments of diagnostic value. Interestingly, many important lower mass ions are easily observed owing to the lack of chemical noise often present in soft ionization methods.

5.3.4.1 Matrix assisted laser desorption/ionization (MALDI)

In their pioneering experiments on matrix-assisted laser desorption/ionization, Hillenkamp and co-workers [102, 216, 217] used the same basic instrument as for their earlier experiments, described in section 5.3.4 (Fig. 5.17). For MALDI of large biomolecules, with this instrument the laser beam is defocused to a diameter of 10–50 μm and is incident on the sample at an angle of 45° to the normal. Desorbed ions are then accelerated to an energy of 3 kV for mass analysis. At the detector the ions are post-accelerated to 20 kV for more efficient detection of high mass ions by the electron multiplier [218]. By this means, intact biomolecules, with molecular weights considerably in excess of anything previously attainable, were successfully ionized (Fig. 5.18).

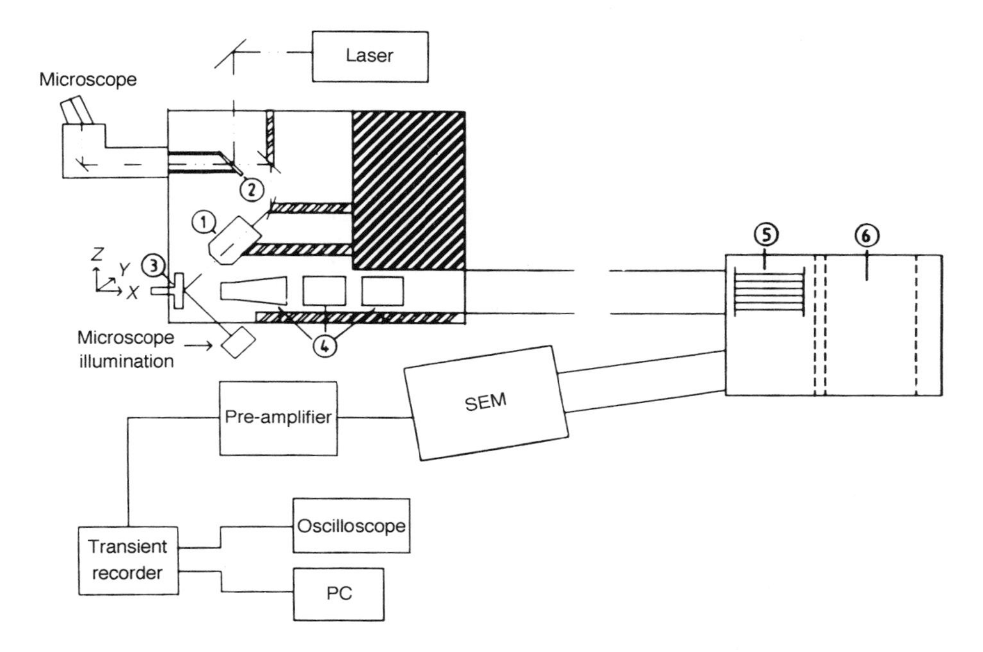

Fig. 5.17 Schematic diagram of LAMMA 1000 instrument used for laser desorption/ionization of proteins. (1) 10/0.2 microscopic objective, (2) dichroic mirror, (3) sample stage, (4) three element Einzel lens, (5) ion deflector, (6) ion reflector. (Reproduced with permission from ref. 218)

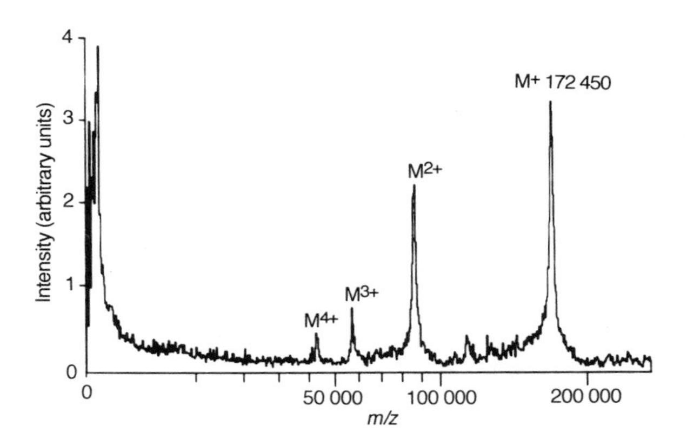

Fig. 5.18 Matrix-assisted laser desorption/ionization spectrum of glucose isomerase. Nicotinic acid matrix (λ=266 nm). (Reproduced with permission from ref. 222)

Other authors, e.g. Beavis and Chait [219], have used a different sample geometry. In this instrument, an area of $100 \times 300\,\mu m$ is irradiated with a laser beam that strikes the sample at an angle approximately 70° to the surface normal. In this case the ions produced are accelerated to an energy of approximately 20 kV for mass analysis in a linear, rather than a reflectron, time-of-flight instrument. A microchannel plate array without substantial post-acceleration is used for ion detection. Good results have also been reported by Sundqvist's group [220], again using a slightly different instrumental configuration.

The laser power density is a critical parameter for successful operation in the matrix-assisted mode. Differing values, e.g. approximately 10^6 W cm^{-2} [219] and 10^7 W cm^{-2} [218,220], have been reported by different authors. A recent publication [221] suggests that the power density threshold is approximately 10^6 W cm^{-2}, e.g. 10 mJ cm^{-2} with a 10 ns pulse width, and that the power density used should be no more than 20% above this threshold. The use of too high a power density can result in the loss of small neutral fragments from the molecular ion [222].

Initially, experiments reported by both Hillenkamp's group [102] and Beavis and Chait [219] used a Nd-YAG laser at 266 nm together with a matrix, such as nicotinic acid, which absorbs strongly at this wavelength. Subsequently, Beavis and Chait [223] reported that, with the correct choice of matrix, laser light of 355 nm from a Nd-YAG laser could also be used. This discovery also opened up the possibility of using other types of laser, such as a cheaper nitrogen laser (337 nm), as an alternative to the Nd:YAG laser used so far. Related experiments reported earlier by Tanaka *et al.* [224] had also demonstrated the desorption/ionization of intact proteins using a nitrogen laser. Their method used a preparation of the analyte together with an ultrafine cobalt powder and a liquid matrix, such as glycerol. It was suggested that the cobalt powder enhances absorption of laser energy while the glycerol matrix helps to replenish sample in the irradiated area.

Two of the most effective organic matrices for use at 337 or 355 nm are 2,5-dihydroxybenzoic acid (gentisic acid) [107] and 3,5-dimethoxy-4-hydroxycinnamic acid (sinapinic acid) [221]. Each of these new matrices has a reduced tendency to produce photochemically generated adduct ions and, because of the higher molecular weights of these matrix molecules, any adduct ion formed is easier to resolve from the molecular ion. The use of the longer wavelength also eliminates any possible absorption of laser light by the analyte itself, as most classes of biopolymers do not absorb light at this wavelength. It has also been reported that analyses using these matrices are relatively immune to the effects of inorganic and organic contaminants [225].

A typical sample preparation procedure [221,222] mixes 5 μL of the matrix solution (5–10 g L^{-1} in water or a water/organic mixture) with 0.5 μL of the analyte solution (10^{-5}–10^{-6} M in 0.1% trifluoroacetic acid or trifluoroacetic acid/organic mixture). An aliquot of this mixture (0.5–1 μL) which contains approximately 10^{-12}

mole of the analyte is loaded on to a sample probe and dried before introduction into the mass spectrometer.

In further experiments, Hillenkamp's group [226] demonstrated that an Er-YAG infra-red laser operating at a wavelength of 2.94 μm could be used with an appropriate matrix. In this case the best results were obtained at a power density of 10^6 W cm^{-2} for solid matrices and 5x10^6 W cm^{-2} for liquid matrices, although the power density is not as critical as in UV laser desorption/ ionization. A wide range of useful matrices was found, including glycerol, urea, carboxylic acids, and those already in use for UV laser desorption/ionization. The use of caffeic acid as a matrix promoted multiple charging in a number of cases. The mechanism of ionization and the role of the matrix in IR laser desorption/ionization remain unexplained, but the variety of useful matrices may be of advantage for the analysis of a wider range of biomolecules [227].

In recent experiments, Williams and co-workers [228,229] smeared aqueous solutions of oligonucleotide sodium salts or proteins on a cooled copper stage. By tuning a dye laser to resonant electronic transitions in either sodium or copper atoms, spectra that showed intense molecular ions and little fragmentation were recorded from the frozen solution. No other matrix material was added in these experiments so that samples were readily recovered.

5.4 Energy Sudden Methods in Use: A Comparison

DCI, PDMS, FAB/LSIMS, and MALDI are all techniques that can extend the range of organic mass spectrometry to include compounds that are too labile or involatile for analysis by conventional EI or CI techniques. For a newcomer to all four techniques, DCI represents the simplest instrumental investment since the use of a DCI probe may properly be seen as a means of making more effective use of an existing ion source and ionization technique (chemical ionization).

DCI has, however, a much more limited applicability than any of the other techniques, both in terms of molecular size and lability of samples that will produce molecular weight information, although this range may be extended by derivatization. For example, a series of substituted porphyrins readily gave molecular ions by FAB in every case whereas DCI gave molecular ions for only ten of the twelve samples [254]. With many simpler compounds, however, DCI has proved to be an excellent method for structural analysis, especially since significant fragment ions are found that may be absent or obscured by matrix ions in FAB [230,231]. Natural product classes that have proved particularly amenable to this technique include glycosides [230,232], dinucleotides [233], and oligosaccharides up to a molecular weight of about 3000 [231]. Further, earlier references to the use of DCI and DEI are given in Appendix 5.1.

Another limitation of the DCI technique is the rapid change of spectrum quality with more labile samples as the effects of thermal processes taking place on the surface of the heated wire become increasingly evident compared with the desorption of intact sample molecules. These changes, which necessitate the use of rapid scanning, can be moderated by the use of coated or other specially prepared wires [113]. Unlike some of the other techniques discussed in this chapter, DCI is relatively unaffected by the presence of inorganic impurities in the sample. Another reported use for a DCI probe is as the basis of a robotic probe-based sample inlet system [234].

For some years following its introduction, plasma desorption mass spectrometry provided a benchmark against which other ionization methods could be tested. The method scored a number of 'firsts' by the recording of molecular ions for compounds for which mass spectrometric analysis had previously been impracticable [235] and was always able to offer access to a somewhat higher molecular weight range than FAB/LSIMS. Subsequently, the method was used routinely for the analysis of a range of compounds of biological interest, principally peptides and small proteins, but also glyco- and phospholipids [236].

An important impetus to the use of the plasma desorption method came with the use of a nitrocellulose supporting matrix which provided an elegant means of handling and purifying very small amounts of peptides and proteins for analysis. Since plasma desorption is a virtually non-destructive technique, the sample may readily be modified after analysis, e.g. by enzymatic digestion or by chemical reaction, prior to further PDMS analysis. Such reactions can be carried out either *in situ* on the nitrocellulose surface [192,201], or after recovery from the surface by means of a suitable solvent [193,202] (Fig. 5.19). In addition, the PDMS method is relatively simple in practice with the possibility of either manual or automated operation.

A significant disadvantage of PDMS is that the technique is only compatible with time-of-flight instrumentation. Thus, unlike FAB/LSIMS, for example, PDMS

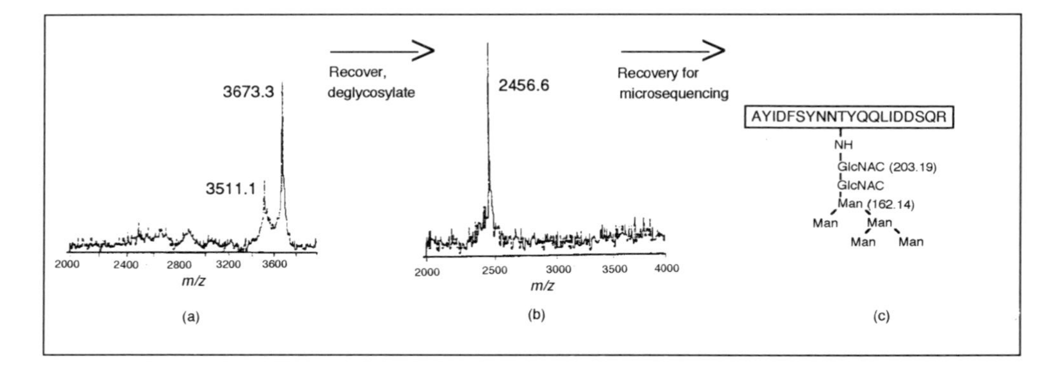

Fig. 5.19 (a) PDMS of glycopeptide, carboxypeptidase P (b) PDMS of deglycosylated peptide on recovered sample (c) sequence analysis on recovered sample (Courtesy Applied Biosystems, Inc.)

does not readily offer access to important ancillary techniques such as tandem mass spectrometry (Chapter 7). Again, the precision and accuracy of molecular weight assignment is poorer than with FAB/LSIMS for all but the highest molecular weight samples [29]. The subsequent introduction of electrospray ionization (Chapter 6), which is compatible with conventional quadrupole and sector instrumentation, meant that PDMS also lost its high mass advantage since electrospray ionization offers a greater useful mass range as well as better sensitivity, resolution, and accuracy of mass assignment [200,237].

From an instrumental viewpoint, FAB/LSIMS is also a simple technique. As a method of ionization, it provides stable, long-lasting ion beams of good intensity and is therefore entirely suitable for scanning instruments. The ease of operation, the ease with which either negative or positive ion spectra could be recorded, and the compatibility of the ion source with both magnetic sector and quadrupole instruments all contributed to the rapid acceptance of the technique. The introduction of FAB/LSIMS soon led to other important advances in instrumentation, such as the development of high field magnets for sector instruments to analyse the high mass ions formed.

A principal application of FAB/LSIMS has been to the structural analysis of biopolymers, i.e. peptides and proteins [238], oligosaccharides [239], and oligonucleotides [240]. The most publicized of these applications has been the determination of the amino acid sequence of peptides by means of MS–MS techniques, mainly using four-sector instruments (section 7.3.3). In addition to the analysis of higher molecular weight biopolymers, the use of FAB/LSIMS has also considerably extended the range of polar and labile materials, including many compounds of much lower molecular weight, that are amenable to analysis. Typical sample categories include glycosides [173], phospholipids [180], antibiotics [241], organometallics [242], drug and steroid conjugates [168], bile acids [243], dyestuffs [244], and quaternary ammonium salts [245]. FAB has, however, been less successful with non-polar samples so that FD (Chapter 6) is still the method of choice for the analysis of many synthetic polymers [246]. FAB/LSIMS is also not always an entirely suitable method for the analysis of unseparated mixtures since suppression effects (section 5.3.2.2) may lead to erroneous results [247,248].

A general problem with FAB/LSIMS, compared with a technique such as electrospray [Chapter 6], is the high background due to ions formed from the matrix material. One solution to this problem, and to the problems of mixture analysis mentioned above, is the use of continuous-flow FAB (section 5.3.2.4). Another solution is the use of tandem mass spectrometry techniques with FAB/LSIMS. For example, the use of collision-induced dissociation (CID) with tandem mass spectrometry techniques not only increases the amount of fragmentation, and therefore structural information, available [249,250] but also filters out matrix ion contributions when the appropriate analyte-related ion is selected (section 7.4.2). This reduction in background level means that FAB/LSIMS with tandem mass spectrometry can benefit from the integrating properties of an array detector (section 1.4.4) whereas conventional FAB/LSIMS operation, where the matrix-related background already determines the available sensitivity, generally does not. With conventional FAB/LSIMS, additional structural information may sometimes be obtained by re-recording a spectrum after chemical modification of the sample. For example, reactions such as acetylation, methylation, or reduction of disulphide bonds may be used with peptide samples [58].

The most recently introduced energy-sudden technique is matrix-assisted laser desorption ionization (MALDI). Just as with electrospray ionization (Chapter 6), MALDI shows a unique advantage compared with other methods for the analysis of biomolecules, since the yield of analyte-related ions does not appear to decrease with an increase in molecular weight. Thus, the measurement of protein molecular masses well in excess of 200 000 daltons has been demonstrated with MALDI [222].

Matrix-assisted laser desorption/ionization implemented on a time-of-flight instrument is an extremely sensitive technique. It has been estimated that no more than perhaps 10 attomoles (1 amol = 10^{-18} mol) are ablated during a single analysis [107]. In practice, however, about 1 pmol of analyte is required because of sample handling limitations. In comparison with electrospray (ES), MALDI is applicable to a wider range of biopolymer types, e.g. proteins, glycoproteins, oligonucleotides, and oligosaccharides [222]. With an appropriate choice of matrix, MALDI is also more tolerant than ES ionization towards

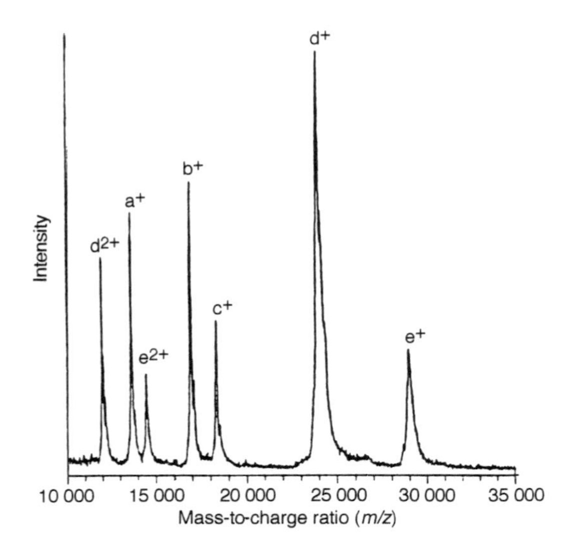

Fig. 5.20 Matrix-assisted laser desorption/ionization of a mixture of (a) bovine pancreatic ribonuclease A, (b) equine skeletal muscle myoglobin, (c) bovine milk β-lactoglobulin, (d) bovine pancreatic trypsinogen, (e) bovine erythrocyte carbonic anhydrase II (approximately 0.5 pmol of each component). (Reproduced with permission from ref. 251)

the presence of inorganic or organic contaminants [251]. The minimization of suppression effects means that MALDI is a suitable method for the direct analysis of unseparated protein mixtures (Fig. 5.20) and is the method of choice for a preliminary analysis of samples of unknown composition.

Mixture analysis can, however, also illustrate one of the problems of MALDI, i.e. the poor mass resolution available from laser desorption time-of-flight instruments. Adduct ions may not be resolved from the molecular ion with higher molecular weight analytes and sample heterogeneity, also unresolved, tends to increase with molecular weight. Under these circumstances, the accuracy of mass assignment deteriorates. Given adequate resolution, the best accuracy with a time-of-flight instrument has been obtained by the use of an internal mass calibrant [252]. Recently, improved resolution and accuracy of mass assignment has also been demonstrated using the MALDI technique with a sector instrument equipped with an array detector [253]. The advantages of MALDI compared with FAB/LSIMS on a sector instrument appear to be increased sensitivity, especially at higher mass, fewer suppression effects, and the absence of interfering matrix peaks. A disadvantage is that some kind of integrating detector, such as an array, must be available. A comparison of MALDI with electrospray ionization can be found in section 6.4.

Other advantages of the laser in mass spectrometry should not be forgotten, e.g. the ability to vaporize samples prior to ionization, in particular the ability to focus a laser beam to ablate selected areas of a sample, and the use of the laser as a method of post-ionization, in particular the exceptional selectivity in target compound analysis available with a tunable laser.

References

1. Fenn, J.B., Mann, M, Meng C.K., and Wong, S.F. (1990), *Mass Spectrom. Rev.*, **9**, 37.
2. Beuhler, R.J., Flanigan, E., Greene, L.J., and Friedman, L. (1974), *J. Am. Chem. Soc.*, **96**, 3990.
3. Cotter, R.J. (1980), *Anal. Chem.*, **52**, 1589A.
4. Doyle Daves, G.,Jr (1979), *Acc. Chem. Res.*, **12**, 359.
5. Reed, R.I. (1958), *J. Chem. Soc.*, 3432.
6. Dell, A., Williams D.H., Morris, H.R., Smith, G.A., Feeney, J., and Roberts, G.C.K. (1975), *J. Am. Chem. Soc.*, **97**, 2497.
7. Constantin, E., Nakatani, Y., Ourisson, G., Hueber, R., and Teller, G. (1980), *Tetrahedron Lett.*, **21**, 4745.
8. Cotter, R.J., and Fenselau, C. (1979), *Biomed. Mass Spectrom.*, **6**, 287.
9. Ohashi, M., Yamada, S., Kudo, H., and Nakayama, N. (1978), *Biomed. Mass Spectrom.*, **5**, 578.
10. Dessort, D., Van Dorsselaer, A., Tian, S.J., and Vincendon, G. (1982), *Tetrahedron Lett.*, **23**, 1395.
11. Baldwin, M.A., and McLafferty, F.W. (1973), *Org. Mass Spectrom.*, **7**, 1353.
12. Hansen, G., and Munson, B, (1980), *Anal. Chem.*, **52**, 245.
13. Carroll, D.I., Dzidic, I., Horning, M.G., Montgomery, F.E., Nowlin, J.G., Stillwell, R.N., Thenot, J.-P., and Horning, E.C. (1979), *Anal. Chem.*, **51**, 1858.

14. Beuhler, R.J., Flanigan, E., Greene, L.J., and Friedman, L. (1974), *J. Am. Chem. Soc.*, **96**, 3990.
15. Hansen, G., and Munson, B. (1978), *Anal. Chem.*, **50**, 1130.
16. Ohashi, M., Nakayama, N., Kudo, H., and Yamada, S. (1976), *Mass Spectrosc. (Japan)*, **24**, 265.
17. Hunt, D.A., Shabanowitz, J., Botz, F.K., and Brent, D.A. (1977), *Anal. Chem.*, **49**, 1160.
18. Soltmann, B., Sweeley, C.C., and Holland, J.F. (1977), *Anal. Chem.*, **49**, 1164.
19. Udseth, H.R., and Friedman, L. (1981), *Anal. Chem.*, **53**, 29.
20. Honig, R.E. (1972), in *Advances in Mass Spectrometry*, Vol. 2 (Ed. R.M.Elliott), p.25, Heyden & Son, London.
21. McHugh, J.A. (1975), *Methods of Surface Analysis*, Ch. 6, Elsevier, Amsterdam.
22. Benninghoven, A., Jaspers, D., and Sichtermann, W. (1976), *Appl. Phys.*, **11**, 35.
23. Grade, H., and Cooks, R.G., (1978), *J. Am. Chem. Soc.*, **100**, 5615.
24. Lafortune, F., Beavis, R., Tang, X., Standing, K.G., and Chait, B.T. (1987), *Rapid Commun. Mass Spectrom.*, **1**, 114.
25. Barber, M., Bordoli, R.S., Elliott, G.J., Sedgwick, R.D., and Tyler A.N. (1982), *Anal. Chem.*, **54**, 645A.
26. Aberth, W., Straub, K.M., and Burlingame, A.L. (1982), *Anal. Chem.*, **54**, 2029.
27. Barber, M., Bordoli, R.S., Garner, G.V., Gordon, D.B. Sedgwick, R.D., Tetler, L.W., and Tyler, A.N. (1981), *Biochem. J.*, **197**, 401.
28. Magee, C.W. (1983), *Int.J. Mass Spec. Ion Phys.*, **49**, 211.
29. Barber, M., and Green, B.N. (1987), *Rapid Commun. Mass Spectrom.*, **1**, 80.
30. Williams, P., and Gillen, G. (1989), in *Ion Formation from Organic Solids*, IFOS IV (Ed. A. Benninghoven), p. 15, John Wiley, Chichester.
31. Katz, R.N., and Field, F.H. (1989), *Int. J. Mass Spectrom. Ion Processes*, **87**, 95.
32. Hunt, D. F., Bone, W.M., Shabanowitz, J., and Ballard, J.M. (1981), *Anal. Chem.*, **53**, 1704.
33. Barber, M., Bordoli, R.S., Sedgwick, R.D., and Tyler, A.N. (1981), *Nature*, **293**, 270.
34. McNeal, C.J. (1982), *Anal. Chem.*, **54**, 33A.
35. Martin, S.A., Costello, C.E., and Biemann, K. (1982), *Anal. Chem.*, **54**, 2362.
36. Kambara, H. (1982), *Org. Mass Spectrom.*, **17**, 29.
37. Blain, M.G., Della-Negra, S., Joret, H., Le Beyec, Y., and Schweikert, E.A. (1989), *Phys. Rev. Lett.*, **63**, 1625..
38. Appelhans, A.D., and Delmore, J.E. (1989), *Anal. Chem.*, **61**, 1087.
39. Yen, T.Y., Barofsky, E., and Barofsky, D.F. (1991), in *Proceedings of the 39th ASMS Conference on Mass Spectrometry and Allied Topics*, Nashville, Tennessee, p. 142.
40. Williams, D.H., Bradley, C., Bojesen, G., Santikarn, S., and Taylor, L.C.E. (1981), *J. Am. Chem. Soc.*, **103**, 5700.
41. Caprioli, R.M. (1983), *Anal. Chem.*, **55**, 2387.
42. Sunner, J., Morales, A., and Kebarle, P. (1988), *Int. J. Mass Spectrom. Ion Processes*, **86**, 169.
43. Johnstone, R.A.W., Lewis, I.A.S., and Rose, M.E. (1983), *Tetrahedron*, **39**, 1597.
44. Caprioli, R.M. (1983), *Anal. Chem.*, **55**, 2387.
45. Sunner, J., Morales, A., and Kebarle, P. (1988), *Anal. Chem.*, **60**, 98.
46. Sunner, J.A., Kulatunga, R., and Kebarle, P. (1986), *Anal. Chem.*, **58**, 1312.
47. Sunner, J., Morales, A., and Kebarle, P. (1987), *Anal. Chem.*, **59**, 1373
48. Puzo, G., and Prome, J.-C. (1984), *Org. Mass Spectrom.*, **19**, 448.
49. Harada, K., Suzuki, M., and Kambara, H. (1982), *Org. Mass Spectrom.*, **17**, 386.
50. Lacey, M.P., and Keough, T. (1989), *Rapid Commun. Mass Spectrom.*, **3**, 46.

51. Detter, L.D., Hand, O.W., Cooks, R.G., and Walton, R.A. (1988), *Mass Spectrom. Rev.*, **7**, 465.
52. Kurlansik, L., Williams, T.J., Campana, J.E., and Green, B.N. (1983), *Biochem. Biophys. Res. Commun.*, **111**, 478.
53. Ligouri, A., Sindona, G., and Uccella, N. (1986), *J. Am. Chem. Soc.*, **108**, 7488
54. Ashton, P.R., and Rose, M.E. (1986), *Org. Mass Spectrom.*, 21, 388.
55. Barber, M., Bell, D.J., Morris, M., Tetler, L.W., Woods, M.D. Monaghan, J.J., and Morden, W.E. (1989), *Org. Mass Spectrom.*, **24**, 504.
56. Blais, J.-C., Cole, R.B., Viari, A., and Tabet, J.-C. (1990), in *Proceedings of the 38th ASMS Conference on Mass Spectrometry and Allied Topics*, Tucson, Arizona, p.193.
57. Takayama, M., Fukai, T., Nomura, T., and Nojima, K. (1989), *Rapid Commun. Mass Spectrom.*, **3**, 4.
58. Carr, S. (1990), *Advanced Drug Delivery Reviews*, **4**, 113.
59. Macfarlane, R.D. (1983), *Anal. Chem.*, **55**, 1247A.
60. Macfarlane, R.D., and Torgerson, D.F. (1976), *Science*, **191**, 920.
61. Macfarlane, R.D., and Torgerson, D.F. (1976), *Phys. Rev. Lett.*, **36**, 486.
62. Williams, P., and Sundqvist, B.U.R. (1987), *Phys. Rev. Lett.*, **58**, 1301.
63. Bitenski, I.S., and Parilis, E.S. (1987), *Nucl. Instrum. Meth.*, **B21**, 26.
64. Jonsson, G.A., Hedin, A.B., Håkansson, P.L., Sundqvist, B.U.R., Säve, G., Roepstorff, P., Nielsen, P.F., Johansson, K.E., Karnovsky, I., and Lindber, M.S.L. (1986), *Anal. Chem.*, **58**, 1084.
65. Chait, B.T. (1989), *Int. J. Mass Spectrom. Ion Processes*, **92**, 297.
66. Salepour, M., Fishel, D., and Hunt, J. (1988), *Rapid Commun. Mass Spectrom.*, **2**, 59.
67. Roepstorff, P. (1989), *Acc. Chem. Res.*, **22**, 421.
68. Jonsson, G., Hedin, A., Håkansson, P., and Sundqvist, B.U.R. (1988), *Rapid Commun. Mass Spectrom.*, **8**, 154.
69. Silly, L., and Cotter, R.J. (1989), *J. de Physique*, **50**, C2-37.
70. Håkansson, P., and Sundqvist, B.U.R. (1989), *Vacuum*, **39**, 397.
71. Craig, A.G., and Bennich, H. (1989), *Anal. Chem.*, **61**, 375.
72. Hedin, A., Håkansson, P., and Sunqvist, B.U.R. (1987), *Nuc. Instrum. Meth.*, **B22**, 491.
73. Macfarlane, R.D., McNeal, C.J., and Jacobs, D.L. (1989), *J. de Physique*, **50**, C2-21.
74. Zare, R.N., Hahn, J.H., and Zenobi, R. (1988), *Bull. Chem. Soc. Jpn.*, **61**, 87.
75. Stoll, R., and Röllgen, F.W. (1982), *Z. Naturforsch.*, **37A**, 9.
76. Van der Peyl, G.J.Q., Haverkamp, J., and Kistemaker, P.G. (1982), *Int. J. Mass Spectrom. Ion Processes*, **42**, 125.
77. Chiarelli, M.P., and Gross, M.L. (1987), *Int. J. Mass Spectrom. Ion Processes*, **78**, 37.
78. Van der Peyl, G.J.Q., van der Zande W.J., and Kistemaker, P.G. (1984), *Int. J. Mass Spectrom. Ion Processes*, **62**, 51.
79. Cotter, R.J., van Breemen, R.B., and Snow, M. (1983), *Int. J. Mass Spectrom. Ion Processes*, **49**, 35.
80. Van der Peyl, G.J.Q., Isa, K., Haverkamp, J., and Kistemaker, P.G. (1981), *Org. Mass Spectrom.*, **16**, 416.
81. Zare, R.N., and Levine R.D. (1987), *Chem. Phys. Lett.*, **136**, 593.
82. Van Vaeck, L., Bennet, J., Van Epsen P., Schweikert, E., Gijbels, R., Adams, F., and Lauwers, W. (1989), *Org. Mass Spectrom.*, **24**, 782.
83. Hillenkamp, F. (1983), in *Proceedings of Second International Conference on Ion Formation from Organic Solids*, Springer Series in Chemical Physics , Vol.25, p.190 (Ed. A.Benninghoven), Springer-Verlag, New York.
84. Linder, B., and Seydel, U. (1985) *Anal. Chem.*, **57**, 895.
85. Hillenkamp, F. (1989), *Microbeam Analysis—1989*, (Ed. P.E. Russell), p. 277, San Fransisco Press, San Fransisco, Calif.

86. Spengler, B., Bahr, U., Hillenkamp, F. (1989), *Institute of Physics Conference Series 94* (Section 2), p. 137.
87. Hillenkamp, F. (1989), in *Advances in Mass Spectrometry*, Vol. 11A (Ed. P. Longevialle), p. 354, Heyden & Son, London.
88. Grotemeyer, J., and Schlag, E.W. (1988), *Org. Mass Spectrom.*, **23**, 388.
89. Spengler, B., Karas, M., Bahr, U., and Hillenkamp, F. (1989), *Ber. Bunsen-Ges, Phys. Chem.*, **93**, 396.
90. Bahr, U., Spengler, B., Karas, M., and Hillenkamp, F., (1989), in *Ion Formation from Organic Solids*, IFOS IV (Ed. A. Benninghoven), p. 143, John Wiley, Chichester.
91. Lubman, D.M. (1988), *Mass Spectrom. Rev.*, **7**, 535.
92. Boesl, U., Neusser, H.J., Schlag, E.W., (1981), *Chem. Phys.*, **55**, 193.
93. Boesl, U. Neusser, H.J., and Schlag, E.W. (1982), *Chem. Phys.*, **87**, 1.
94. Grötemeyer, J., Boesl, U., Walter, K., and Schlag, E.W. (1986), *Org. Mass Spectrom.*, **21**, 645.
95. Engelke, F., Hahn, J.H., Henke, W., and Zare, R.N. (1987), *Anal. Chem.*, **59**, 909.
96. Spengler, B., Bahr, U., and Hillenkamp, F. (1989), *Institute of Physics Conference Series 94* (Section 9—RIS 88), p. 307.
97. Becker, C.H., (1989) in Ion Formation from Organic Solids, IFOS V, (Eds. A. Hedin, B.U.R.Sundquist, and A. Benninghoven), p. 167, John Wiley, Chichester.
98. Becker, C.H., Jusinski, L.E., and Moro, L. (1990), *Int. J. Mass Spectrom. Ion Processes*, **95**, R1.
99. Karas M., Bachmann, D., Bahr, U., and Hillenkamp, F. (1987), *Int. J. Mass Spectrom. Ion Processes*, **78**, 53.
100. Karas M., Bachmann, D., and Hillenkamp, F. (1985), *Anal. Chem.*, **57**, 2935.
101. Ehring, H., Karas, M., and Hillenkamp, F. (1992), *Org. Mass Spectrom.*, **27**, 472.
102. Karas M., Bahr, U., and Hillenkamp, F. (1989), *Int. J. Mass Spectrom. Ion Processes*, **92**, 231.
103. Chan, T.-W.D., Colburn, A.W., and Derrick, P.J. (1992), *Org. Mass Spectrom.*, **27**, 53.
104. Hillenkamp, F., Karas, M., Beavis, R.C., and Chait, B.T. (1991), *Anal. Chem.*, **63**, 1193A.
105. Chan, T.-W.D., Colburn, A.W., and Derrick, P.J. (1991), *Org. Mass Spectrom.*, **26**, 342.
106. Chan, T.-W.D., Colburn, A.W., Derrick, P.J., Gardiner D.J., and Bowden, M. (1992), *Org. Mass Spectrom.*, **27**, 188.
107. Strupat, K., Karas, M., and Hillenkamp, F. (1991), *Int. J. Mass Spectrom. Ion Processes*, **111**, 89.
108. Holland, J.F., Soltmann, B., and Sweeley, C.C. (1976), *Biomed. Mass Spectrom.*, **3**, 340.
109. Arpino, P.J., and Devant, G. (1979), *Analusis*, **7**, 348.
110. Beaugrand, C., and Devant, G. (1980), in *Advances in Mass Spectrometry*, Vol 8B (Ed. A. Quayle), p.1806, Heyden & Son, London.
111. Rapp, U., Dielmann, G., Games, D.E., Gower, J.L., and Lewis, E. (1980), in *Advances in Mass Spectrometry*, Vol 8B (Ed. A. Quayle), p.1660, Heyden & Son, London.
112. Bruins, A.P. (1980), *Anal. Chem.*, **52**, 605.
113. Reinhold, V.N., and Carr, S.A. (1982), *Anal. Chem.*, **54**, 499.
114. Bencsath, F.A., and Field, F.H. (1981), *29th Annual Conference on Mass Spectrometry and Allied Topics*, Minneapolis, p.587.
115. Cotter, R.J. (1979), *Anal. Chem.*, **51**, 317.
116. Murphy, R.C., Clay, K.L., and Mathews, W.R. (1982), *Anal. Chem.*, **54**, 336.
117. McNeal, C.J., Macfarlane, R.D., and Thurston, E.L. (1979), *Anal. Chem.*, **51**, 2036.
118. Roach, J.A.G., Malatesta, J.A., Sphon, J.A., Brumley, W.C., Andrzejewski, D., and Dreifuss, P.A. (1981), *Int. J. Mass Spectrom. Ion Phys.*, **39**, 151.

119. Simonsick, W.J., Jr (1989), *J. Appl. Polym. Sci.: Appl. Polym. Symp.*, **43**, 257.
120. McEwen, C.N. (1983), *Anal. Chem.*, **55**, 967.
121. Carr, S.A., Roberts, G.D., and Hemling, M.E. (1990), *Mass Spectrometry of Biological Materials* (Eds. C.N. McEwen and B.S. Larsen), p.87, Marcel Dekker, Inc.
122. Przybylski, M. (1983), *Z. Anal. Chem.*, **315**, 402.
123. Ligon, W.V., and Dorn, S.B. (1984), *Int. J. Mass Spectrom. Ion Processes* (1984), **57**, 75.
124. Ligon, W.V., Jr, and Dorn, S.B. (1984), *Int. J. Mass Spectrom. Ion Processes*, **61**, 113.
125. Gower, J.L. (1985), *Biomed. Mass Spectrom.*, **12**, 191.
126. DePauw, E. (1986), *Mass Spectrom. Rev.*, **5**, 191.
127. Schiebel, H.M., Schulze, P., Stohrer, W. D., Leibfritz, D., Jastorff, D., and Mauer, K.M. (1985), *Biomed. Mass Spectrom.*, **12**, 170.
128. Burlingame, A.L., Baillie, T.A., and Derrick, P. (1986), *Anal. Chem.*, **58**, 165R.
129. Busch, K.L., Czanderna, A.W., and Hercules, D.M. (Eds.), *Methods of Surface Characterisation*, Plenum Surface Analysis Series, Vol. 3, Plenum Press, New York.
130. Pachuta, S.J., and Cooks, R.G. (1987), *Chem. Rev.*, **87**, 647.
131. Cochran, R.L. (1986), *Appl. Spectrosc. Rev.*, **22**, 137.
132. Rose, K. (1989), *Spectrosc. Int. J.*, **7**, 39.
133. De Pauw, E., Agnello, A., and Derwa, F. (1991), *Mass Spectrom. Rev.*, **10**, 283.
134. Larsen, B.S. (1990) in *Mass Spectrometry of Biological Materials* (Eds. C.N. McEwan and B.S. Larsen), Marcel Dekker Inc.
135. Cook, K.D., Todd, P.J., and Friar, D.H. (1989), *Biomed. Environ. Mass Spectrom.*, **18**, 492.
136. Barber, M. Bordoli, R.S., Elliott, G.J., Sedgwick, R.D., Tyler, A.N., and Green, B.N. (1982), *J. Chem. Soc. Chem. Commun.*, **1982**, 936.
137. Lehmann, W.D., Kessler, M., and König, W.A. (1984), *Biomed. Mass Spectrom.*, **11**, 217.
138. Dass, C., and Desiderio, D.M. (1988), *Anal. Chem.*, **60**, 2723.
139. Cottrell, J.S., and Frank, B.H. (1985), *Biochem. Biophys. Res. Commun.*, **127**, 1032.
140. Schronk, L.R., and Cotter, R.J. (1986), *Biomed. Environ. Mass Spectrom.*, **13**, 395.
141. Whitten, J.L., Schaffer, M.H., O'Shea, M., Cook, J.C., Hemling, M.E., and Rinehart, K.L., Jr (1984), *Biochem. Biophys. Res. Commun.*, **124**, 350.
142. De Angelis, F., Nicoletti, R., and Santi, A. (1988), *Org. Mass Spectrom.*, **23**, 800.
143. Meili, J., and Seibl, J. (1984), *Org. Mass Spectrom.*, **19**, 581.
144. Barber, M., Bell, D., Eckersley, M., Morris, M., and Tetler, L. (1988), *Rapid Commun. Mass Spectrom.*, **2**, 18.
145. Takayama, M. (1991), *Org. Mass Spectrom.*, **26**, 1123.
146. Banoub, J.H., Shaw, J.H., Pang, H., Krepinsky, J.J., Nakhla, N.A., and Patel, T. (1990), *Biomed. Environ. Mass Spectrom.*, **19**, 787.
147. Aubagnac, J.-L., Claramunt, R.-M., Lopez, C., and Elguero, J. (1991), *Rapid Commun. Mass Spectrom.*, **5**, 113.
148. Vigato, P.A., Guerriero, P., Tamburini, S., Seraglia, R., and Traldi, P. (1990), *Org. Mass Spectrom.*, **25**, 420.
149. Rose, K., Savoy, L.-A., and Offord, R.E. (1989) in *Advances in Mass Spectrometry*, Vol.11A (Ed. P. Langevialle), p.484, Heyden & Sons, London.
150. Arita, M., Iwamori, M., Higuchi, T., and Nagai, Y. (1983), *J. Biochem.*, **94**, 249.
151. Arita, M., Iwamori, M., Higuchi, T., and Nagai, Y. (1984), *J. Biochem.*, **95**, 971.
152. Monaghan, J.J., and Morden, W.E. (1990), *Rapid Commun. Mass Spectrom.*, **4**, 436.
153. Tondeur, Y., Clifford, A.J., and De Luca L.M. (1985), *Org. Mass Spectrom.*, **20**, 157.
154. Naylor, S., Findeis, A.F., Gibson, B.W., and Williams, D.H. (1986), *J. Am. Chem. Soc.*, **108**, 6359.

155. Naylor, S., Skelton, N.J., and Williams, D.H. (1986), *J. Chem. Soc. Chem. Commun.*, **1986**, 1619.
156. Ligon, W.V. Jr (1986), *Anal. Chem.*, **58**, 485.
157. Clench, M.R., Garner, G.V., Gordon, D.B., and Barber, M. (1985), *Biomed. Mass Spectrom.*, **12**, 355.
158. Naylor, S., Moneti, G., and Guyan, S. (1988), *Biomed. Environ. Mass Spectrom.*, **17**, 393.
159. Caprioli, R.M. , Moore, W.T., and Fan, T. (1987), *Rapid Commun. Mass Spectrom.*, **1**, 15.
160. Barber, M. Bell, J.D. Morris, M., Tetler, L.W., Woods, M.D., Monaghan, J.J., and Morden, W.E. (1988), *Rapid Commun. Mass Spectrom.*, **2**, 181.
161. Cerny, R. L., and Gross, M.L. (1985), *Anal. Chem.*, **57**, 1160.
162. Gale, P.J., Bentz, B.L., Chait, B.T., Field, F.H., and Cotter, R. J. (1986), *Anal. Chem.*, **58**, 1070.
163. Musselman, B.D., and Watson, J.T. (1987), *Biomed. Environ. Mass Spectrom.*, **14**, 247.
164. Keough, T. (1988), *Int. J. Mass Spectrom. Ion Processes*, **86**, 155.
165. Busch, K.L., Unger, S.E., Vincze, A., Cooks, R.G., and Keough, T. (1982), *J. Am. Chem. Soc.*, **104**, 1507.
166. Martin, S.A., and Biemann, K. (1987), *Int. J. Mass Spectrom. Ion Processes*, **78**, 213.
167. Facino, R.M., Carini, M., Moneti, G., Arlandini, E., Pietta, P., and Mauri, P. (1991), *Org. Mass Spectrom.*, **26**, 989.
168. Veares, M.P., Evershed, R.P., Prescott, M.C., and Goad, L.J, (1990), *Biomed. Environ. Mass Spectrom.*, **19**, 583
169. Facino, R.M., Sparatore, A., Carini, M., Gioia, B., Arlandini, E. and Franzoi, L. (1991), *Org. Mass Spectrom.*, **26**, 951,
170. Puzo, G., and Promé, J.C. (1985), *Org. Mass Spectrom.*, **20**, 288.
171. Keough, T. (1985), *Anal. Chem.*, **57**, 2027.
172. Moon, D.-C., and Kelley, J.A. (1988), *Biomed. Environ. Mass Spectrom.*, **17**, 229.
173. Li, Q.M., Dillen, L., and Claeys, M. (1992), *Biol. Mass Spectrom.*, **21**, 408.
174. Li, Q.M., Van den Heuvel, H., Dillen, L., and Claeys, M. (1992), *Biol. Mass Spectrom.*, **21**, 213.
175. Giordano, G., Shyam, K., Sartorelli, A.C., and McMurray, W.J. (1992), *Org. Mass Spectrom.*, **27**, 633.
176. Edwards, D.M.F., Vékey, K., and Galimberti, M. (1991), *Biol. Mass Spectrom.*, **21**, 43.
177. Mallis, L.M., and Scott, W.J. (1990), *Org. Mass Spectrom.*, **25**, 415.
178. Junker, E., Wirth, K.P., and Röllgen, F.W. (1989), *J. de Physique*, **50**, C2-53.
179. Zhang, M.-Y., Liang, S.-Y., Chen, Y.-Y., and Liang, X.-G. (1984), *Anal. Chem.*, **56**, 2288.
180. Chen, S., Benfenati, E., Fanelli, R., Kirschner, G., and Pregnolato, F. (1989), *Biomed. Environ. Mass Spectrom.*, **18**, 1051.
181. Caprioli, R.M., Fan, T., and Cottrell, J.S. (1986), *Anal. Chem.*, **58**, 2949.
182. Caprioli, R.M., and Fan, T. (1986), *Biochem. Biophys. Res. Commun.*, **141**, 1058.
183. Caprioli, R.M., Moore W.T., Petrie, G., and Wilson, K. (1988), *Int. J. Mass Spectrom. Ion Processes*, **86**, 187.
184. van Breemen, R.B., and Le, J.C. (1989), *Rapid Commun. Mass Spectrom.*, **3**, 20.
185. Caprioli, R.M. (1990), *Anal. Chem.*, **62**, 477A.
186. Tyler, A.N., Romo, L.K., Frey, M.H., Musselman, B.D., Tamura, J., and Cody, R.B. (1992), *J. Am. Soc. Mass Spectrom.*, **3**, 637.
187. Sundqvist, B., and Macfarlane, R.D. (1985), *Mass Spectrom. Rev.*, **4**, 421.
188. Mann, M., Rahbek-Nielsen, H., and Roepstorff, P. (1989), in *Ion Formation from*

Organic Solids, IFOS V (Eds. A. Hedin, B.U.R. Sundquist, and A. Benninghoven), p. 47, John Wiley, Chichester.

189. Roepstorff, P., and Sundqvist, B. (1986), in *Biochemical Applications of Mass Spectrometry* (Ed. S. Gaskell), p. 269, John Wiley, Chichester.
190. Sundqvist, B., Håkansson, P., Kamensky, I., Kjellberg, J., Salehpour, M., Widdiyasekera, S., Fohlman, J., Peterson, P.A., and Roepstorff, P. (1984), *Biomed. Mass Spectrom.*, **11**, 242.
191. Sundqvist, B., Roepstorff, P., Fohlman, J., Medin, A., Håkansson, P., Kamensky, I., Lindberg, M., Salehpour, M., and Säve, G., (1984), *Science*, **226**, 696.
192. Roepstorff, P. (1990), in *Methods in Enzymology*, Vol. 193: *Mass Spectrometry* (Ed. J.A. McCloskey), Ch. 23, Academic Press, San Diego.
193. Roepstorff, P., Talbo, G., Klarskov, K., and Højrup, P. (1990), in *Biological Mass Spectrometry* (Eds. A. Burlingame and J.A. McCloskey), Ch.2, Elsevier, Amsterdam.
194. Deprun, C., and Szabó, L. (1989), *Rapid Commun. Mass Spectrom.*, **3**, 171.
195. Jordan, E.A., Macfarlane, R.D., Martin, C.R., and McNeal, C.J. (1983), *Int. J. Mass Spectrom. Ion Processes*, **53**, 345.
196. McNeal, C.J., and Macfarlane, R.D. (1989), in *Ion Formation from Organic Solids*, IFOS IV (Ed. A. Benninghoven), p. 63, John Wiley, Chichester.
197. Lacey, M.P., and Keough T. (1990), in *Proceedings of the 38th ASMS Conference on Mass Spectrometry and Allied Topics*, Tuscon, Arizona, p. 187.
198. Håkansson, P., and Sundqvist, B.U.R. (1989), *Vacuum*, **39**, 397.
199. Lacey, M.P., and Keough, T. (1989), *Rapid Commun. Mass Spectrom.*, **3**, 323.
200. Loo, J.A., Edmonds, C.G., Smith, R.D., Lacey, M.P., and Keough, T. (1990), *Biomed. Environ. Mass Spectrom.*, **19**, 286.
201. Chowdhury, S.K., and Chait, B.T. (1989), *Anal. Biochem.*, **180**, 387.
202. Chen, L., Kochersperger, M., Bozzini, M., and Yuan, P.-M. (1991), in *Proceedings of the 39th ASMS Conference on Mass Spectrometry and Allied Topics*, Nashville, Tennessee, p. 1420.
203. Nielsen P.F., and Roepstorff, P. (1989), *Biomed. Environ. Mass Spectrom.*, **18**, 131.
204. Showell, J.S., Fales, H.M., and Sokoloski, E.A. (1989), *Org. Mass Spectrom.*, **24**, 632.
205. Wilkins, C.L., Weil, D.A., Yang C.L.C., and Ijames, C.F. (1985), *Anal. Chem.*, **57**, 520
206. Brown, R.S., and Wilkins, C.L. (1986), *Anal. Chem.*, **58**, 3196.
207. Coates, M.L., and Wilkins, C.L. (1987), *Anal. Chem.*, **59**, 197.
208. Nuwaysir, L.M., Wilkins, C.L., and Simonsick, W.J.,Jr (1990), *J. Am. Soc. Mass Spectrom.*, **1**, 66.
209. Greenwood, P.F., Strachan, M.G., Willett, G.D., and Wilson, M.A., (1990), *Org. Mass Spectrom.*, **25**, 353.
210. Williams, E.R., Furlong, J.J.P., and McLafferty, F.W. (1990), *J. Am. Soc. Mass Spectrom.*, **1**, 288.
211. Spengler, B., Bahr, U., Karas, M., and Hillenkamp, F. (1988), *Analyt. Instrum.*, **17**, 173.
212. Grotemeyer, J., and Schlag, E.W., (1987), *Org. Mass Spectrom.*, **22**, 758.
213. Cotter, R.J. (1980), *Anal. Chem.*, **52**, 1767.
214. Cecchetti, W., Polloni, R., Maccioni, A.M., and Traldi, P. (1986), *Org. Mass Spectrom.*, **21**, 517.
215. Bravo, P., Cavalleri, B., Zerilli, L.F., Maccioni, A.M., Traldi, P., Cecchetti, W., and Polloni, R. (1989), *Biomed. Environ. Mass Spectrom.*, **18**, 301.
216. Karas, M., Bachmann, D., Bahr, U., and Hillenkamp, F. (1987), *Int. J. Mass Spectrom. Ion Processes*, **78**, 53.
217. Karas, M., and Hillenkamp, F. (1988), *Anal. Chem.*, **60**, 2299.

218. Hillenkamp, F., Karas, M., Ingendoh, A., and Stahl B. (1990), *Biological Mass Spectrometry* (Eds. A.Burlingame and J.A. McCloskey), Ch.3, Elsevier, Amsterdam.
219. Beavis, R.C., and Chait, B.T. (1989), *Rapid Commun. Mass Spectrom.*, **3**, 233.
220. Salehpour, M., Perera, I., Kjellberg, J., Hedin, A., Islamian, M.A., and Sundqvist, B.U.R. (1989), *Rapid Commun. Mass Spectrom.*, **3**, 259.
221. Hillenkamp, F., Karas, M., Beavis, R.C., and Chait, B.T. (1991), *Anal. Chem.*, **63**, 1193A.
222. Karas, M., Bahr, U., and Giessman, U. (1991), *Mass Spectrom. Rev.*, **10**, 335.
223. Beavis, R.C., and Chait, B.T. (1989), *Rapid Commun. Mass Spectrom.*, **3**, 436.
224. Tanaka, K., Waki, H., Ido, Y., Akita, S., Yoshida, Y., and Yoshida, T. (1988), *Rapid Commun. Mass Spectrom.*, **2**, 151.
225. Mock, K.K., Sutton, C.W., and Cottrell, J.S. (1992), *Rapid Commun. Mass Spectrom.*, **6**, 233.
226. Overberg, A., Karas, M., Bahr, U., Kaufmann, R., and Hillenkamp, F. (1990), *Rapid Commun. Mass Spectrom.*, **4**, 293.
227. Overberg, A., Karas, M., and Hillenkamp, F. (1991), *Rapid Commun. Mass Spectrom.*, **5**, 128.
228. Nelson, R.W., Thomas, R.M., and Williams, P. (1990), *Rapid Commun. Mass Spectrom.*, **4**, 348.
229. Schieltz, D.M., Chou, C.-W., Luo, C.-W., Thomas, R.M., and Williams, P. (1992), *Rapid Commun. Mass Spectrom.*, **6**, 631.
230. Sakushima, A., Nishibe, S., and Brandenberger, H. (1989), *Biomed. Environ. Mass Spectrom.*, **18**, 809
231. Reinhold V.N. (1987), in *Methods in Enzymology*, Vol. 138: *Complex Carbohydrates* Part E, (Ed. V. Ginsburg), Ch. 5, Academic Press, Orlando, Fla.
232. Bruins, A.P. (1980), *Anal. Chem.*, **52**, 605.
233. Isern-Flecha, I., Jiang, X.-Y., Cooks, R.G., Pfleiderer, W., Chae, W.-G., and Chang, C.-J. (1987), *Biomed. Mass Spectrom.*, **14**, 17.
234. Martin, D.J., and Bond, P.M. (1989), *Biomed. Environ. Mass Spectrom.*, **18**, 733.
235. Macfarlane, R.D., and Torgerson, D.F. (1976), *Science*, **191**, 920.
236. Cotter, R.J. (1988), *Anal. Chem.*, **60**, 781A.
237. Roepstorff, P., Klarskov, K., Andersen, J., Mann, M., Vorm, O., Etienne, G., and Parello, J. (1991), *Int. J. Mass Spectrom. Ion Processes*, **111**, 151.
238. Carr, S.A. (1990), *Advanced Drug Delivery Rev.*, **4**, 113.
239. Laine, R.A. (1990), in *Methods in Enzymology*, Vol. 193: *Mass Spectrometry* (Ed. J.A. McCloskey), Ch. 29, Academic Press, San Diego.
240. McCloskey, J.A. (1990), in *Methods in Enzymology*, Vol. 193: *Mass Spectrometry* (Ed. J.A. McCloskey), Ch. 41, Academic Press, San Diego.
241. De Angelis, F., Nicoletti, R., Bieber, L.W., and Botta, B. (1989), *Biomed. Environ. Mass Spectrom.*, **18**, 553.
242. Miller, J.M. (1989), *Mass Spectrom. Rev.*, **9**, 319.
243. Masui, T., Egestad, B., and Sjövall, J. (1988), *Biomed. Environ. Mass Spectrom.*, **16**, 71.
244. Harada, K.I., Masuda, K., Suzuki, M., and Oka, H. (1991), *Biol. Mass Spectrom.*, **20**, 522.
245. Simms, J.R., Keough, T., Ward, S.R., Moore, B.L., and Bandurraga, M.M. (1988), *Anal. Chem.*, **60**, 2613.
246. Rollins, K., Scrivens J.H., Taylor, M.J., and Major, H. (1990), *Rapid Commun. Mass Spectrom.*, **4**, 355.
247. Morris, H.R., Panico, M., Barber, M., Bordoli, R.S., Sedgwick, R.D., and Tyler, A. (1981), *Biochem. Biophys. Res. Commun.*, **101**, 623.

248. Morris, H.R., Panico, M., Judkins, M., Dell, A., and McDowell, R.A. (1981), *Proceedings of the ASMS 29th Conference on Mass Spectrometry and Allied Topics*, Minneapolis, p.375.
249. Tomer, K.B. (1989), *Mass Spectrom. Rev.*, **8**, 445.
250. Tomer, K.B. (1989), *Mass Spectrom. Rev.*, **8**, 483.
251. Beavis, R.C., and Chait, B.T. (1990), *Proc. Natl. Acad. Sci. USA*, **87**, 6873.
252. Beavis, R.C., and Chait, B.T. (1990), *Anal. Chem.*, **62**, 1836.
253. Annan, R.S., Kochling, H.J., Hill, J.A., and Biemann, K. (1992), *Rapid Commun. Mass Spectrom.*, **6**, 298.
254. Campana, J.E. (1989), *Org. Geochem.*, **14**, 171.

Appendix 5.1 Published applications of DCI and DEI

References cover the period up to the end of 1982

DCI

AMINO ACIDS AND DERIVATIVES. Arginine (22, 23, 27, 29). Creatine (23, 29). ANTIBIOTICS. Kanamycins (4). Anthracycline antibiotics (21). Aminoglycoside antibiotics (39). CARBOXYLIC ACIDS and DERIVATIVES. Potassium benzoate (23). Acyl hydroxamates (1). DRUGS and METABOLITES, MISCELLANEOUS. Fluphenazine oxide derivatives (31). DYESTUFFS (31). GLUCURONIDES (5, 16). Androsterone glucuronide (17, 25). Naphthyl glucuronides (2). GLYCOSIDES (10). Steroid glycosides (17, 46). G-strophanthin (22). Purpeaglycoside A (28). Digitoxigenin monodigitoxoside (28). Flavanoid glycosides (45). Naringin (24). ISOPRENOIDS. Phorbol ester (19). LIPIDS. Triglycerides (17, 44). Glycosphingolipids (3). 1-*O*-Stearoyl-2-hydroxy-sn-glyceryl-3-phosphorylcholine (19). Dimyristoyl phosphatidyl choline (37). *N*-Palmitoyl dihydrolactocerebroside (19). NITRO COMPOUNDS (12). NUCLEOTIDES and RELATED COMPOUNDS. Nucleosides (5, 11). Guanosine (23). Cyclic adenosine monophosphate (23). PENICILLINS and RELATED COMPOUNDS (15). PEPTIDES (19, 23, 35). Aspartyl-glycine (29). PHOSPHATE ESTERS. Dioxathon (23). PHOSPHONIUM SALTS (29). PROSTAGLANDINS and RELATED COMPOUNDS. Prostaglandin $F_{2\alpha}$ (17). Leucotriene E4 derivative (33). QUATERNARY AMINO COMPOUNDS (14). Choline chloride (23). SACCHARIDES. Sucrose (5, 17). Trehalose (24, 27). Raffinose (24). Melezitose (27). Glucose polymer permethylated (19). STEROIDS (17, 30). 3-Hydroxy steroids (38). Ergosterol peroxide (37). SYNTHETIC POLYMERS. Polystyrene (26). VITAMINS. Ascorbic acid (22, 27). Riboflavin (24).

DEI

AMINES. Ephedrine (42). AMINO ACIDS and DERIVATIVES. Leucine, cysteine, lysine, glutamic acid (36). Arginine (42). CARBOXYLIC ACIDS and

DERIVATIVES. Metal acetates (8, 18). Metal benzoates (8). CHLOROPHYLLS (20, 43). Phaeophytin, phaeophorbide (20). GLYCOSIDES. Flavanoid glycosides (45). LIPIDS. Phospholipids (9, 42). Glycolipids (7). PEPTIDES. Oligopeptides, *N*-carbobenzoxy (6). Leu-enkephalin (42). QUATERNARY AMINO COMPOUNDS (32). SACCHARIDES. Sucrose (40, 42). Glucose (40). Amino sugars (34). SULPHONIC ACIDS and DERIVATIVES. Sodium alkylsulphonates (41). VITAMINS. Ascorbic acid (42).

References for Appendix 5.1

1. Loo, Y. H., Miller, K. A., Nowlin, J., and Horning, M. G. (1980), *Life Sci.*, **26**, 657.
2. Krahn, M. M., Brown, D. W., Collier, T. K., Friedman, A. J., Jenkins, R. G., and Malins, D. C. (1980), *J. Biochem. Biophys. Methods*, **2**, 233.
3. Ariga, T., Murata, T., Oshima, M., Maezawa, M., and Miyatake, T. (1980), *J. Lipid Res.*, **21**, 879.
4. Takeda, N., Umemura, M., Harada, K., Suzuki, M., and Tatematsu, A. (1981), *J. Antibiot.*, **34**, 617.
5. Cotter, R. J., and Fenselau, C. (1979). *Biomed. Mass Spectrom.*, **6**, 287.
6. Ohashi, M., Tsujimoto, K., Tamura, S., Nakayama, N., Okumura, Y., and Sakurai, A. (1980). *Biomed. Mass Spectrom.*, **7**, 153.
7. Breimer, M. F., Hansson, G. C., Karlsson, K. A., Leffler, H., Pimlott, W., and Samuelsson, B. E. (1981), *FEBS Lett.*, **124**, 299.
8. Matsumoto, K., Kosugi, Y., Yanagisawa, M., and Fujishima, I. (1980), *Org. Mass Spectrom.*, **15**, 606.
9. Constantin, E., Nakatani, Y., Ourisson, G., Hueber, R., and Teller, G. (1980). *Tetrahedron Lett.*, **21**, 4745.
10. Hostettmann, K., Doumas, J., and Hardy, M. (1981), *Helv. Chim. Acta.*, **64**, 297.
11. Esmans, E. L., Freyne, E. J., Vanbroeckhoven, J. H., and Alderweireldt, F. C. (1980), *Biomed. Mass Spectrom.*, **7**, 377.
12. Yinon, J. (1980). *Org. Mass Spectrom.*, **15**, 637.
13. Roach, J. A. G., Malatesta, A. J., Sphon, J. A., Brumley, W. C., Andrzejewski, D., and Dreifuss, P. A. (1981), *Int. J. Mass Spec. Ion Phys.*, **39**, 151.
14. Hasiak, B., Ricart, G., Barbry, D., Couturier, D., and Hardy, M. (1981). *Org. Mass Spectrom.*, **16**, 17.
15. Gower, J. L., Beaugrand, C., and Sallot, C. (1981). *Biomed. Mass Spectrom.*, **8**, 36.
16. Bruins, A. P. (1981), *Biomed. Mass Spectrom.*, **8**, 31.
17. Carroll, D. I., Nowlin, J. G., Stillwell, R. N., and Horning, E. C. (1981). *Anal. Chem.*, **53**, 2007.
18. Matsumoto, K., and Kosugi, Y. (1981). *Org. Mass Spectrom.*, **16**, 249.
19. Reinhold, V. N., and Carr, S. A. (1982), *Anal. Chem.*, **54**, 499.
20. Constantin, E., Nakatani, Y., Teller, G., Hueber, R., and Ourisson, G. (1981), *Bull Soc. Chim. Fr. 2*, **1981**, 303.
21. Smith, R. G. (1982), *Anal. Chem.*, **54**, 2006.
22. Beaugrand, C., and Devant, G. (1980), *Adv. Mass Spectrom.*, **8B**, 1806.
23. Hunt, D. F., Shabanowitz, J., Botz, F. K., and Brent, D. A. (1977), *Anal. Chem.*, **49**, 1160.
24. Rapp, U., Dielmann, G., Games, D. E., Gower, J. L., and Lewis, E. (1980), *Adv. Mass Spectrom.*, **8B**, 1660.

25. Cotter, R. J. (1980), *Anal. Chem.*, **52**, 1767.
26. Udseth, H. R., and Friedman, L. (1981), *Anal. Chem.*, **53**, 29.
27. Arpino, P. J., and Devant, G. (1979). *Analusis*, **7**, 348.
28. Bruins, A. P. (1980), *Anal. Chem.*, **52**, 605.
29. Hansen, G., and Munson, B. (1978), *Anal. Chem.*, **50**, 1130.
30. Nowlin, J. G., Carroll, D. I., Dzidic, I., Horning, M. G., Stillwell, R. N., and Horning, E. C. (1979), *Anal. Lett.*, **12**, 573.
31. Brooks, P., and Heyes, W. F. (1982), *Biomed. Mass Spectrom.*, **9**, 522.
32. Lee, T. D., Anderson, W. R. Jr., and Doyle Daves, G., Jr (1981), *Anal. Chem.*, **53**, 304.
33. Murphy, R. C., Clay, K. L., and Mathews, W. R. (1982), *Anal. Chem.*, **54**, 336.
34. Ohashi, M., Yamanda, S., Kudo, H., and Nakayama, N. (1978), *Biomed, Mass Spectrom.*, **5**, 578.
35. Baldwin, M. A., and McLafferty, F. W. (1973), *Org. Mass Spectrom.*, **7**, 1353.
36. Traldi, P. (1982). *Org. Mass Spectrom.*, **17**, 245.
37. Sugnaux, F. R., and Djerassi, C. (1982), *J. Chromatogr.*, **251**, 189.
38. Tecon, P., Hirano, Y., and Djerassi, C. (1982), *Org. Mass Spectrom.*, **17**, 277.
39. Takeda, N., Harada, K., Suzuki, M., Tatematsu, A., and Kubodera, T. (1982), *Org. Mass Spectrom.*, **17**, 247.
40. Constantin, E. (1982), *Org. Mass Spectrom.*, **17**, 346.
41. Kosugi, Y., and Matsumoto, K. (1982), *Fresenius' Z. Anal. Chem.*, **312**, 317.
42. Dessort, D., Van Dorsselaer, A., Tian, S. J., and Vincendon, G. (1982), *Tetrahedron Lett.*, **23**, 1395.
43. Bazzaz, M. B., Bradley, C. V., and Brereton, R. G. (1982), *Tetrahedron Lett.*, **23**, 1211.
44. Merritt, C., Jr, Vajdi, M., Kayser, S. G., Halliday, J. W., and Bazinet, M. L. (1976), *J. Assoc. Off. Anal. Chem.*, **59**, 422.
45. Itokawa, H., Oshida, Y., Ikuta, A., and Shida, Y. (1982), *Chem. Lett.*, **1982**, 49.
46. Bruins, A. P. (1980), *Adv. Mass Spectrom.*, **8A**, 246.

Chapter 6

The Ionization of Labile Materials (Part II)

6.1 Field Desorption in Vacuum

Field desorption methods, which use very strong electrostatic fields to extract ions from a substrate in vacuum, provided one of the earliest means by which intact molecular ions could be formed from involatile or thermally unstable materials. The desorption of positive ions from the surface of a metallic anode (the emitter) under the influence of an intense electric field was first investigated by Müller in his development of the field-ion microscope [1]. The mass-spectrometric analysis of these ions was then introduced by Inghram and Gomer in 1954 [2]. The later development of field ionization as a soft ionization method in organic analysis and the subsequent introduction of field desorption as a method for the analysis of involatile molecules is principally due to Beckey [3].

Field ionization (FI) and field desorption (FD) can be distinguished by the way in which sample is supplied to the emitter. If the sample is introduced in the gas phase from another inlet system, this is field ionization. If the sample has been coated on to the emitter prior to analysis and is then ionized from this state, this is field desorption.

The intense electric fields (10^7–$10^8\,\mathrm{V\,cm^{-1}}$) used in field desorption and in field ionization are generally produced by the use of wire-based emitters (section 6.3.1.1) having at their surface multipoint microneedles (dendrites) of small radius of curvature (approximately 10^3 Å) (Fig. 6.1). The local field strength at the tip of a microneedle is given by

$$F = V\left\{(1-\alpha)\left[0.5r\ln\left(\frac{4d}{r}\right)\right]^{-1} + \alpha r^{-1}\right\} \tag{6.1}$$

where V is the applied voltage, r is the radius of curvature, d is the distance to the counter-electrode (Fig. 6.7) and α is a shape factor ($\approx$.0064). Thus, for an applied voltage of 8 kV and a counter-electrode spacing of 2 mm, and $F \approx 1.5 \times 10^8\,\mathrm{V\,cm^{-1}}$ [4].

Fig. 6.1 Scanning electron micrograph (×500) of high temperature activated FD emitter. (Reprinted with permission from W.D. Reynolds *Anal. Chem.*, **51**, 283A[4]: Copyright (1979) American Chemical Society)

The energy transferred in the field ionization process is equivalent to only a fraction of an electronvolt so that the excess internal energy in the ionized molecule is much lower than that resulting from EI or CI processes (section 3.1). Thus both FI and FD are basically extremely mild ionization methods which do not induce fragmentation, although there is naturally more thermal decomposition associated with the sample evaporation process in FI, as may be seen from the spectra in Fig. 6.2. The internal energy of the ion has to be distinguished from the translational energy spread of ions leaving the source (cf. section 1.3.2). The translational energy spread is high in field desorption since ions may be formed and accelerated over a range of potentials in the environment of the emitter surface. Three distinct mechanisms relevant to ion production may be recognized in field desorption mass spectrometry:

(a) *Electron tunnelling under the influence of the electric field.* In the presence of an intense local field, the potential energy profile for the interaction of an adsorbed sample with a metal surface is distorted so that there is a finite probability that an electron from the sample molecule can tunnel into the emitter [5]. Under these circumstances the energy required for the desorption of an ion is considerably less than that required for evaporation of the neutral molecule. This field ionization process leads to the preferential formation of $M^{+\cdot}$ and $(M+H)^+$ ions at the tip of field-enhancing microneedles. Sample molecules

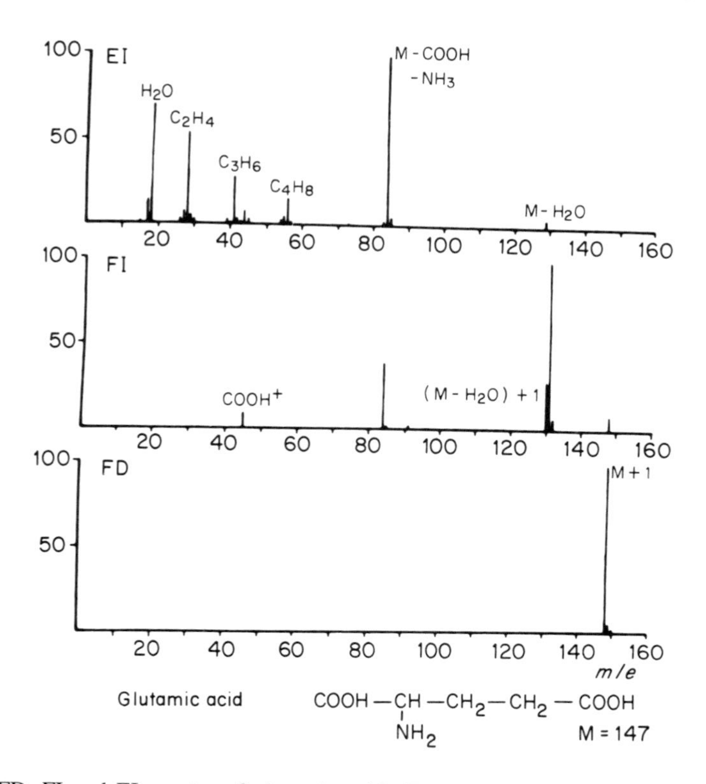

Fig. 6.2 FD, FI and EI spectra of glutamic acid. (Reproduced by permission of Elsevier Science Publishers BV [56])

may reach the tips either from the gas phase or via surface diffusion [6,7]. The observation of a $M^{+\cdot}$ ion is strong evidence for a gas phase process [7,8].

(b) *Proton or cation attachment.* If the sample is an electrolyte or contains salts or acids, ions are already present in the condensed phase. Such pre-formed ions may be extracted from the condensed phase by a somewhat weaker field than is required for electron tunnelling, typically 10^6 V cm^{-1} [7]. Thus, these processes may occur even when the sample is deposited on a smooth wire without microneedles (section 6.3.1.1). This mechanism, leading to the formation of $(M+H)^+$ and $(M+cation)^+$ ions, is the most important mechanism in field desorption. Ion formation by cationization can be very sensitive to experimental conditions, such as the relative amount of cation added to the sample and the composition of the sample matrix. A more detailed consideration of these effects is provided in section 6.3.1.2.

(c) *Thermal mechanisms.* With many samples, heating of the emitter is necessary to obtain an adequate FD spectrum. Thus, as the wire current is increased

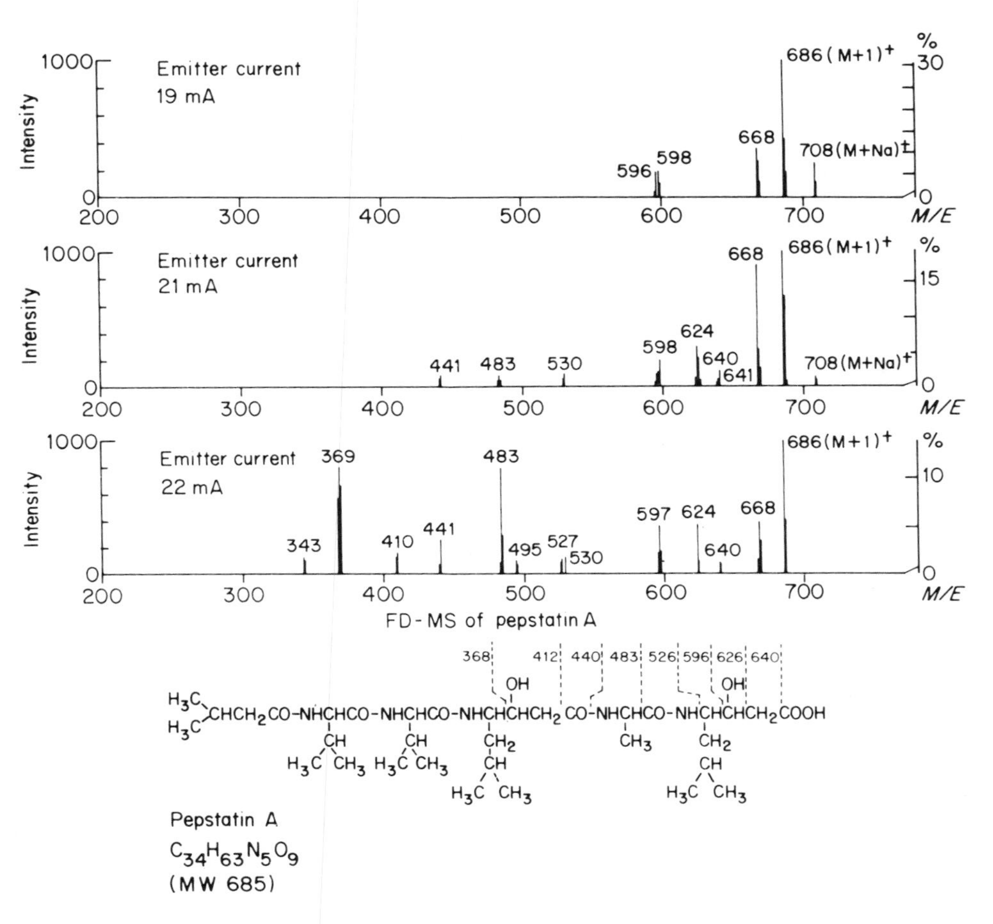

Fig. 6.3 FD spectra of pepstatin A recorded at different emitter currents. (Reproduced by permission of JEOL Ltd)

during an experiment, thermally induced fragmentation becomes increasingly evident. Structurally significant fragment ions are often conveniently obtained in this manner (Fig. 6.3). Thermal ionization under FD conditions also provides a highly sensitive analytical method for metal cations [9].

Desorption of analyte ions in vacuum can also be achieved without the use of an emitter or metal wire. For example, a high voltage applied to an involatile liquid introduced via a narrow capillary into a vacuum system causes the emerging liquid to assume a conical shape (cf. section 6.2). The field at the conical tip is then sufficient to desorb pre-formed ions present in the liquid phase. This technique, referred to as electrohydrodynamic ionization [10], has, however, gained little acceptance as an analytical procedure partly because of the difficulty of finding a suitable solvent to carry the analyte into the vacuum system and partly because of the complex nature of the spectra obtained [11]. On the other hand, much more practical techniques result when desorption is carried out in a higher pressure environment (section 6.2).

6.2 Desorption at Higher Pressures: Electrospray and Thermospray Ionization

Techniques for the desorption of ions from liquids in a higher pressure environment are based on the spraying of analyte solution to form a mist of small, charged droplets from which the ions are subsequently formed. Droplets can be generated in a variety of ways, all of which involve disruption of the liquid with an energy source. The source of energy can be thermal, aerodynamic, or electrical, or a combination of these. The gas that surrounds the droplets provides a source of thermal energy for the vaporization of solvent and acts to remove any excess translational energy from the desorbed ions. The two best known techniques which involve the formation of charged droplets are electrospray (ES) and thermospray (TS) ionization. The differences between ES and TS result mainly from the environment in which the droplets are created. ES is operated at atmospheric pressure and at a temperature that is close to ambient. TS is usually operated at a lower pressure, between 1 and 10 torr, and at an elevated temperature.

In electrospray, an electric field is generated by applying a high voltage directly to the spraying capillary, with a counter electrode located a few millimetres away

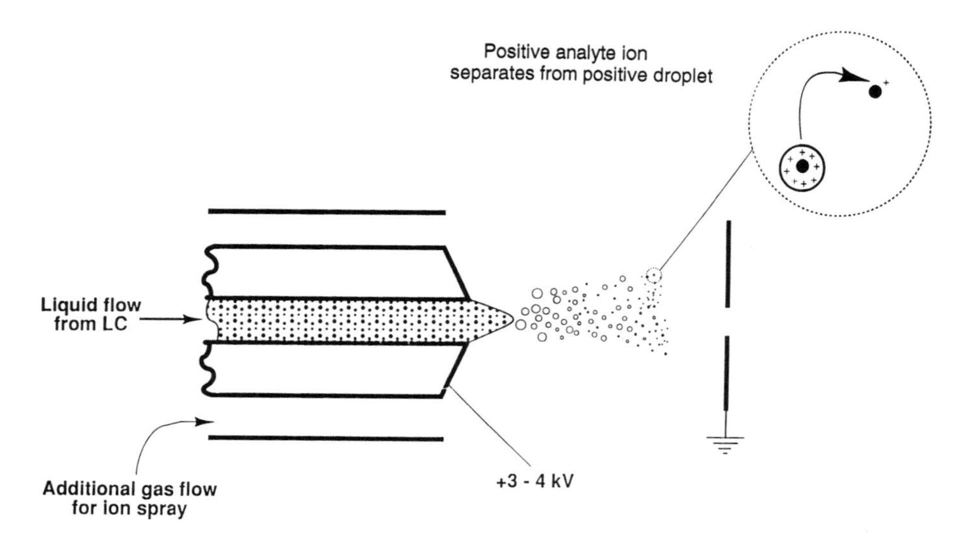

Fig. 6.4 Schematic representation of electrospray and ion spray processes

(Fig. 6.4). Under these conditions, the emerging liquid assumes an equilibrium conical shape ('Taylor cone') with a sharp tip from which a stream of droplets is ejected [13,14]. If the voltage is increased further, multiple spraying tips may appear [12]. At still higher voltages, a corona discharge is initiated in the surrounding gas. A partial separation of positive and negative ions present in the solution occurs near the capillary tip and the electrosprayed droplets therefore contain excess ions of one charge. For example, in a partly aqueous solution containing arginine, the excess ions in the positive ion mode will be H^+ and $(M+H)^+$ from arginine.

Thermospray (section 2.4.3) is another process that creates charged droplets. In this technique, a high percentage of the liquid flow is volatilized as it emerges from a heated metal capillary (Fig. 6.5). Under these conditions, the vapour that issues from the end of the capillary has sufficient energy to transform the remainder of the liquid flow into a mist of small droplets. In thermospray ionization, there is no externally applied electric field, since, at the somewhat lower pressure at which thermospray is normally operated, a glow discharge results, even at low field strengths. Instead, a relatively high concentration of a volatile electrolyte (0.1 M ammonium acetate) is added to the solution prior to volatilization. The violent disruption of the liquid containing electrolyte ions results in the formation of droplets each with a statistical distribution of excess charge [15]. The thermospray process therefore produces essentially equal populations of positively and negatively charged droplets, while electrospray produces only droplets of the same sign as the applied voltage.

A combination method of droplet formation which has some practical advantages is termed 'ion spray'. In this technique, the basic ES process is assisted by a coaxial nebulizing air flow. This technique has been widely used and is described separately in section 6.3.2.

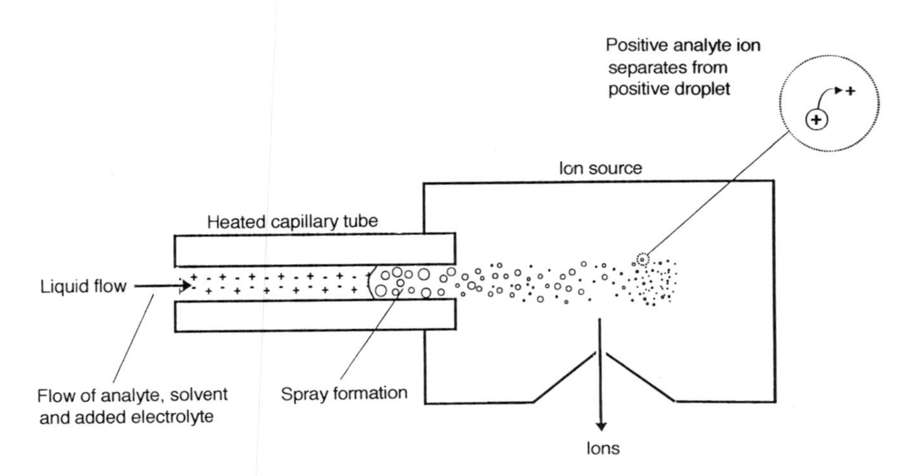

Fig. 6.5 Schematic representation of thermospray ionization process

There is general agreement that these charged droplets, however they are initially formed, evaporate until the increased surface charge density confers instability. At this stage (Rayleigh instability limit) the electrical forces due to the surface charge approach equality with those due to surface tension and the droplet disintegrates to produce several very much smaller charged droplets. The maximum permissible charge at the Rayleigh limit, q_r, is given by [16]

$$q_r = \left(\frac{8\pi}{e}\right)(\gamma\epsilon_0)^{1/2}r^{3/2} \tag{6.2}$$

where r is the droplet radius, γ is the surface tension, and ϵ_0 the permittivity of free space. It has been suggested that a succession of such fission processes, yielding smaller and smaller droplets, which are eventually the source of sample ions [section 6.2.1], occurs as evaporation continues [17].

Initial droplet diameters in pure electrospray are the smallest (1–2 μm) [18], while other techniques, such as thermospray, initially produce larger droplets which are perhaps some tens of micrometres in diameter. The small droplet size and high charge-to-mass ratio in ES are probably the principal reasons for a notable feature of ES spectra, i.e. the formation of multiply-charged analyte ions. The additional use of other forms of energy to create the droplets, e.g. the use of an air flow as in ion spray, is likely to decrease the charge density on the surface of the droplet and consequently reduce the extent of multiple charging [19]. Preliminary results suggest that the extent of multiple charging varies between electrospray ionization (very extensive) and conventional thermospray ionization (very little) with modified thermospray (section 6.3.3.2) and perhaps ion spray as intermediate cases [17,19].

6.2.1 Ion formation mechanisms

There are two general mechanisms by which gas-phase analyte ions may be formed from charged droplets. One possibility is for analyte ions already present in solution to be transferred to the gas phase. The other possibility is for neutral analyte molecules present in solution to be transferred to the gas phase and subsequently ionized by ion–molecule reactions.

Thermospray ionization can generate gas-phase analyte ions by either mechanism. Thus, under the conditions of most thermospray experiments, i.e. elevated temperature and relatively volatile analyte, gas-phase ion–molecule reactions based on ions derived from ammonium acetate and neutral analyte molecules are thought to be most important [20]. On the other hand, thermospray volatilization of solutions of ionic analytes in the absence of ammonium acetate (section 6.3.3.3) can involve the direct generation of gas-phase ions from the droplets.

In electrospray, analyte ions are only generated from the liquid phase. Even though gas-phase ammonium ions, for example, may be present, most compounds analysed by ES are insufficiently volatile at ambient temperature to generate neutral molecules, and hence are not observed as products of gas-phase ion–molecule reactions. Thus, ES relies on analyte ions being already present in solution either because the analyte is ionic or because it is associated with other ions present in solution, e.g. by protonation (see section 6.3.2.1).

While there is general agreement on the mechanism for the initial stages of charged droplet subdivision (section 6.2), different ideas have been used to explain the eventual formation of analyte ions directly from the liquid phase. According to one mechanism, the division of charged droplets at the Rayleigh limit would be repeated until finally the droplets contained a single analyte molecule. This analyte molecule would eventually be observed as a gas-phase ion provided that charge was retained by the analyte rather than being lost with evaporating solvent molecules. This simple, passive mechanism may apply under some conditions, but is unlikely to explain the distribution of charge states routinely observed in ES with higher molecular weight analytes (Fig. 6.6) or the change in charge state in ES spectra observed with changes in analyte chemistry [14,21].

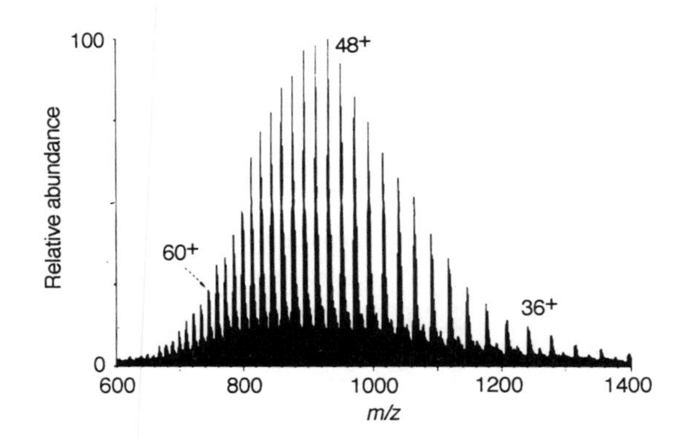

Fig. 6.6 Positive ion electrospray ionization spectrum of recombinant protein A (MW 45 000). The charge state of selected molecular ion peaks is indicated. (Reproduced with permission from ref. 125)

A more appealing mechanism for the formation of gas-phase ions is field desorption. Thus, as the solvent evaporates, the electric field at the surface of the charged droplet increases with the decreasing radius. An analyte molecule that has accumulated sufficient charge is then able to evaporate or desorb from the charged droplet, alone or in association with one or more molecules of solvent. The thermodynamics of this process have recently been discussed for organic cations and used, successfully, to predict trends in molecular ion intensities for amino acids analysed by electrospray [22].

The field, E, produced by the excess charge, q on a spherical droplet of radius r is given by

$$E = \frac{eq}{4\pi\epsilon_0 r^2} \tag{6.3}$$

where ϵ_0 is the permittivity of free space. Thus, if we denote the field needed for ion desorption by E_0, then the limiting charge, q_0, above which ion desorption occurs is given by

$$q_0 = \frac{4\pi\epsilon_0 E_0 r^2}{e} \tag{6.4}$$

whereas the maximum charge that can be retained at the Rayleigh limit is given by equation (6.2).

There has been considerable discussion of the relative sizes of these two limits. Iribarne and Thomson [23] suggested that for droplets of an appropriate size, the field required for ion desorption could be achieved without exceeding the Rayleigh stability limit, whereas Röllgen *et al.* [24] argued that this was not true, even for the smallest droplets. Recent work [15] has given a value of about $8\mathrm{x}10^8\ \mathrm{V\,m^{-1}}$ for E_0 and has demonstrated the closeness of the two limits for smaller droplets.

It seems, however, unlikely that any analysis of the stability of charged droplets solely in terms of the overall electrical field and surface tension will prove to be satisfactory. Changes in conditions near the surface of the droplet, perhaps due to the presence of analyte molecules, could give rise to a local disturbance of the surface and therefore a higher field which results in the expulsion of very small droplets by Rayleigh fission. It may, in fact, not be possible to distinguish such a fission process from the field desorption of highly solvated ions [15].

A particular feature of electrospray ionization is the fact that, certainly for molecular weights in excess of 1000, the molecular ions formed are multiply charged, e.g. by multiple protonation in the positive ion mode. The charge states observed form a continuous distribution and generally show an approximate proportionality to molecular weight so that, for many compounds, molecular ions appear in the m/z 500–2000 range (e.g. Fig. 6.6). Clearly, multiple charging should encourage the release of very high molecular weight ions from the droplet surface by a field desorption process [19].

The observation of ions that represent a given analyte in ES requires that the analyte is already ionized in the medium being sprayed and hence depends on the $\mathrm{p}K_\mathrm{a}$ of the analyte and the pH of the solution. For example, the ES sensitivity of organic nitrogen bases was found to be pH dependent, so that the protonated molecular ion signal was approximately proportional to the concentration of the protonated molecule in solution [25]. The rate of ion emission from droplets is also dependent on other properties such as, for example, the hydrophobicity of the analyte [21].

Fenn and co-workers have developed an electrostatic model which accounts for the number of charges on a single molecule in electrospray ionization [26]. For proteins the number of basic sites in the molecule is a guide to the expected charge distribution [17], but this may be modified by the difficulty of adding charges to a molecule that already has a number of charged centres [27,28]. Thus, the electrospray spectrum of a protein strongly reflects its solution ionization, although this may have been modified by subsequent gas-phase reactions. The charge states, or degrees of protonation in a positive ion spectrum, are therefore highly dependent on solution pH or on changes in solvent composition which may unfold the protein and leave more sites available for protonation.

There is ample evidence to suggest that gas-phase ion–molecule reactions play a dominant part in analyte ion formation under 'electrolyte' thermospray conditions, i.e. thermospray in the presence of a high concentration of an electrolyte such as ammonium acetate. Thus, a heated aqueous ammonium acetate solution is a source of gas-phase ions such as NH_4^+ and CH_3COO^-. These ions can ionize analytes by conventional CI processes, as confirmed by a decrease in such gas-phase reagent ion abundances when an analyte elutes in LC–MS [29]. These same authors [29] observed that solution pH had little effect on the abundance and nature of analyte-related ions in electrolyte thermospray. In another study, excellent thermospray spectra were obtained by the introduction of ammonium acetate and analyte solutions from separate heated capillaries [30]. These experiments demonstrated the absence of an electrical charging effect from the electrolyte and confirmed its role as a means of generating gas-phase CI reagent ions. A good correlation, as predicted by the Debye–Hückel theory, between gas-phase ion abundances and the ammonium acetate ionic activity in various solvents (section 6.3.3.3) has been reported by Liberato and Yergey [31].

In the absence of a volatile electrolyte, the abundance and type of ions in the thermospray spectra of model nucleosides acquired at various pH values followed solution-phase chemistry. Thus, the analysis of adenosine at low pH favoured the formation of $(M + H)^+$ ions whereas analysis at higher pH resulted in lower sensitivity and an increase in cationization by traces of sodium in the solution. When ammonium acetate was added, however, predominantly gas-phase reactions were observed and the spectra were unaffected by a change of pH [32].

6.3 Practice

6.3.1 Instrumentation and techniques for FD

A schematic representation of a field desorption source is shown in Fig. 6.7. The emitter is mounted on an insulated probe which is introduced into the source region via a vacuum lock. For positive ion FD or FI operation the probe tip and emitter are maintained at 8000–12 000 V relative to the counter-electrode which is approximately 2 mm distant. A field desorption source may also offer facilities for FI operation, e.g. for the direct analysis of crude oil fractions. In this case, the open

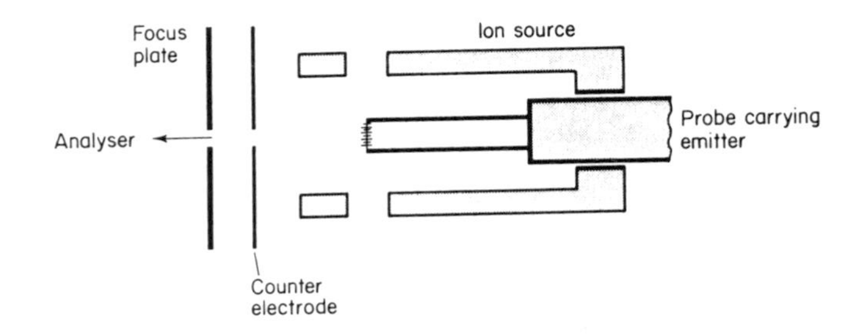

Fig. 6.7 Schematic diagram of field desorption source

construction of an FD source can be of advantage in allowing the close coupling of the FI sample inlet system and the emitter wire.

A major practical difficulty in FD operation results from the fragility of the 10 μm diameter tungsten wire which forms the basis of the emitter. The wire may be broken not only by rough or careless handling but also by electrical discharges in the source. It is preferable to use a programmable emitter heating supply rather than a manually controlled supply so as to avoid the sudden vaporization of large amounts of ionizable materials which may result in an electrical discharge. Electrical discharges will also occur to broken emitter wires or other irregularities on the counter-electrode, so that the surface of the counter-electrode should be kept clean and smooth and broken wires removed immediately the breakage occurs.

Setting up for FD operation is most conveniently achieved by maximizing the field ionization beam obtained from a volatile sample such as acetone (m/z 58) introduced via a reservoir inlet system. It should be appreciated, however, that a reduced FI sensitivity does not necessarily indicate a poor FD performance, e.g. where emitters are being used to desorb ions already present in the condensed phase. Sensitivity in the FD mode may be measured by loading a known amount of the order of 1 μg cholesterol on to the emitter using the syringe technique (section 6.3.1.2) and measuring the integrated ion current at m/z 386 during desorption (cf. section 1.6.1).

With an unknown sample, if no ion current is observed initially, then the emitter is heated using a steadily increasing current until desorption begins. Further heating after the onset of desorption may increase the ion intensity but will also eventually cause thermal fragmentation of the molecule (e.g. Fig. 6.3). As can be seen from Fig. 6.3, the appearance of an FD spectrum is critically dependent on the emitter heating current used. It has been found that the production of reproducible FD spectra is difficult with manual control of emitter heating and therefore the use of an emitter current controller is preferred. Several devices for this purpose have been described [33–37]. The use of such a device may also limit the loss of emitters due to electrical breakdown, as mentioned previously.

With scanning instruments, the preferred procedure for recording spectra is to scan repetitively over an appropriate mass range since the ion current obtained in field desorption is weak and fluctuating. Successive spectra, recorded by the data system, can then be integrated to provide either more reproducible spectra with better signal-to-noise characteristics or a composite spectrum that is representative of the complete sample in mixture analysis [33,38,39].

6.3.1.1 FD emitter preparation

The complex and time-consuming art of emitter preparation has always been a major problem in field desorption. Despite efforts directed towards improvement of this situation, the carbonaceous emitters originally introduced by Beckey and co-workers [40,41] have so far proved to be the most mechanically stable and efficient.

The most widely used technique for emitter activation produces carbon microneedles on a 10 μm diameter tungsten wire from benzonitrile in a specially constructed vacuum chamber (emitter activation rig). The temperature of the electrically heated wire is first adjusted to a value sufficient to pyrolyse the benzonitrile at various points on the wire without the application of any high voltage. These points then act as nuclei for further polymerization during the subsequent high voltage operation. At a higher voltage, growth takes the form of thicker roots, whereas the lower voltage growth is of thinner needles. Thus, the voltage on the counter-electrode is gradually decreased during this phase of growth to provide high surface area emitters having a characteristic branching structure of fine needles. The emitter is maintained at approximately 1200°C by electrical heating throughout this process which takes several hours. Lehmann and Fischer [42] have given a precise description of the conditions employed during each stage of emitter preparation and describe how these conditions may be maintained by monitoring the electrical resistance of the emitter during growth.

The structure of a well-prepared carbonaceous emitter with an average needle length of 30 μm (Fig. 6.1) allows the applied solution to spread over the enhanced surface area so as to leave a well-distributed film of sample on evaporation. This property and their small bulk, which allows them to be heated extremely rapidly, contributed to the success of early DCI experiments which used FD emitters as well as forming a basis for successful FD operation.

The use of untreated tungsten wires without microneedles was first reported by Röllgen and Schulten [43] following the realization that pre-formed ions could be desorbed at lower than normal field strengths (section 6.1). This technique has since been applied to the analysis of pre-formed ions from salts [44], oligosaccharides [45,46], peptides [47], nucleotides [47], and acids [46]. In many cases, sensitivities one or two orders of magnitude higher than those obtained with activated emitters have been reported because of the larger area available for desorption [44].

The addition of a viscous polymer such as polyethylene oxide or polyvinyl alcohol to the sample matrix gives further improvements to the untreated wire technique, leading to more continuous emission of quasi-molecular ions over a broader temperature range and starting at lower emitter temperatures [46](cf. ref. 48]. The formation of thin fibres from whose tips ion emission may occur has recently been observed in studies of the behaviour of polyethylene oxide films under the influence of strong fields by Giessman and Röllgen [7]. The use of a polyethylene oxide matrix has also provided the basis of a practical method for negative ion field desorption [49,50].

Emitters with very short microneedles may be prepared by heating an unactivated wire installed in an FD/FI source for only a few seconds [51]. Such emitters are more easily used than bare wires for the field desorption of pre-formed ions since they produce a field ionization current that is useful for setting up the source whereas bare wires do not. The same emitters can be used for the field desorption of non-ionic analytes with a sensitivity comparable to that of emitters with longer microneedles and are also very suitable for thermolabile materials. In practice, the ease of preparation of this type of emitter has to be weighed against the increased difficulty of loading samples on to a reduced surface area. Other, less time consuming, methods more recently developed can be used to prepare emitters with a larger surface area [52–54].

6.3.1.2 FD sample preparation

The simplest technique employed for sample loading is to dip the emitter into a solution of the sample contained in a vessel shaped so as to use the minimum volume of solution. An alternative technique which uses very little sample is to apply the solution to the emitter by means of a microlitre syringe preferably held in a micromanipulator [55,56]. A stereomicroscope or other viewing arrangement is helpful in controlling the application of solution to the very fine wire. This method also avoids memory problems which can occur with dipping since all the applied solution is loaded on to the emitter rather than on to the support legs as well.

Most solvents may be loaded using a syringe unless they are very volatile, in which case the solvent tends to evaporate before the drop leaves the needle. Difficulties can also be encountered if the surface tension is too high. Thus, aqueous solutions are more easily handled if they can be diluted with another miscible solvent such as methanol. The electrospray loading technique described in section 5.3.1.1 may also be used with FD emitters.

Many samples submitted for analysis contain alkali salts as contaminants, with the result that the field desorption spectra contain intense $(M+Na)^+$ or $(M+K)^+$ ions. This effect may be particularly noticeable if a sample has been loaded from a polar solvent such as methanol. While cationization is easily recognized and is often useful for the determination of molecular weight, an excess of inorganic impurities can suppress ion formation from the sample altogether [57,58]. Therefore, the use of a simple chromatographic(section 5.3.2.3) or solvent extraction [59] procedure

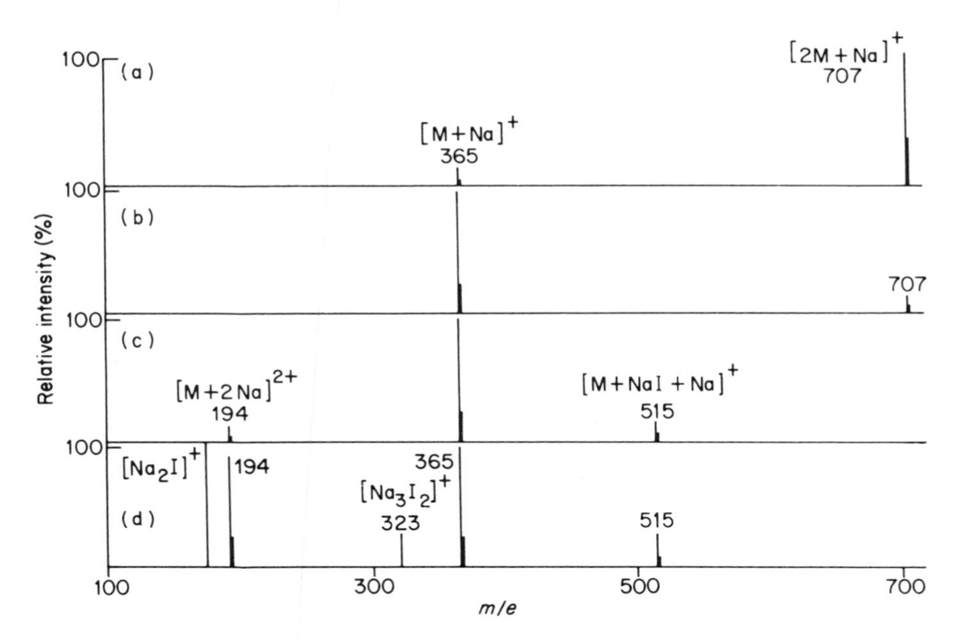

Fig. 6.8 Field desorption spectra of trehalose with different levels of added sodium iodide: (a) molar ratio of sugar/salt = 100; (b) molar ratio of sugar/salt = 10; (c) molar ratio of sugar/salt = 1; (d) molar ratio of sugar/salt = 0.1. (Reproduced with permission from ref. 60)

to clean up the sample just prior to analysis is recommended, particularly with samples of biological origin.

Cationization may also be deliberately promoted by the addition of an appropriate inorganic salt. For example, Prome and Puzo [60] found that the disaccharide trehalose analysed in bacterial extracts gave only a weak spectrum of the cationized material. After deliberately adding sodium iodide, a spectrum of the cationized material with a hundred- to a thousandfold increase in sensitivity was observed. In addition, the wire current required for desorption was lowered. A series of experiments in which different levels of sodium iodide were added to a sample of pure trehalose (Fig. 6.8) suggested that a molar ratio of added salt to analyte of between 1:10 and 1:1 gave the optimum intensity for the $(M+Na)^+$ ion (cf. ref. 57).

Alkali salts with large counter ions provide the best cation donors since a lower emitter temperature is needed to release the cation from these salts [61]. Iodides [60,62] or organic ions such as tetraphenyl borates [47] have been used in most cases. Smaller cations such as lithium are more readily solvated and therefore provide the greatest abundance of cationized sample ions [43,47] but require a much higher emitter temperature for desorption compared with a large ion such as caesium [63]. The choice of cation may therefore be determined by the thermal stability of the sample. Polyhydroxy compounds such as oligosaccharides are particularly amenable to cationization but the technique has also been applied to a number of other compound types.

Quasi-molecular ions formed by cationization are very stable and undergo little fragmentation. Thus, virtually all of the sample ion current in the lithium addition spectrum of raffinose is contained in the $(M+Li)^+$ ion. In practice, the favoured method for cationization is to dip the emitter in the inorganic salt solution, dry this by gentle heating in vacuum, and then add the sample [47]. Other methods give poorer sensitivity or less reproducible results.

Another approach to the prior formation of ions in the condensed phase is a simple acidification procedure which is applicable to basic and zwitterionic materials. For example, FD analysis of adenine using dilute hydrochloric acid as solvent gives an enhanced $(M+H)^+$ ion compared with the use of water [40]. Again, a wide range of zwitterionic materials desorb at reduced emitter currents to give very simple spectra (Fig. 6.9) following the addition of an equimolar quantity of a strong involatile acid such as *p*-toluene-sulphonic acid [64]. In the case illustrated in Fig. 6.9 the sulphonate group is so weakly basic that initially any water present is preferentially protonated. As a consequence, the use of hydrochloric acid which is removed together with water in the vacuum system is less successful as a method of cation formation.

6.3.2 Instrumentation for ES ionization

In an ES ion source, analyte solutions are typically sprayed from a stainless steel capillary of 0.1–0.15 mm i.d. in a region of atmospheric pressure. The outlet of the capillary is placed approximately 1 cm from a sampling orifice, which communicates with the mass spectrometer vacuum system (e.g. Fig. 2.17), and a positive or negative potential, depending on the ion polarity of interest, of 3-5 kV relative to the sampling orifice is applied to the capillary. The capillary tip may be flat, dome-shaped, or tapered. An appropriate flow of liquid (1–10 $\mu L\,min^{-1}$) is brought to the capillary, usually by means of a syringe pump. This liquid either already contains the analyte in solution or a solution of the analyte may be injected into the liquid flow. Alternatively, the liquid flow may originate from an LC system.

Electrostatic spraying of liquids takes place in several stages. As the voltage on the capillary is increased, the liquid begins to emerge as a succession of larger droplets. No ions can be observed at this stage. At a higher voltage, which depends on the spray geometry and on liquid properties (see equation 6.6), a continuous spray pattern is seen at the needle tip. The pattern observed is a liquid cone which emits an expanding fog of very small droplets. This mode produces stable analyte ion currents. Total ES currents for a typical water/methanol/5% acetic acid solution are in the range 0.1–0.5 μA. The electrospray process is limited to liquid flow rates of a few microlitres per minute since only a certain volume of liquid can be removed from the tip of the cone by the electrical shear force [14].

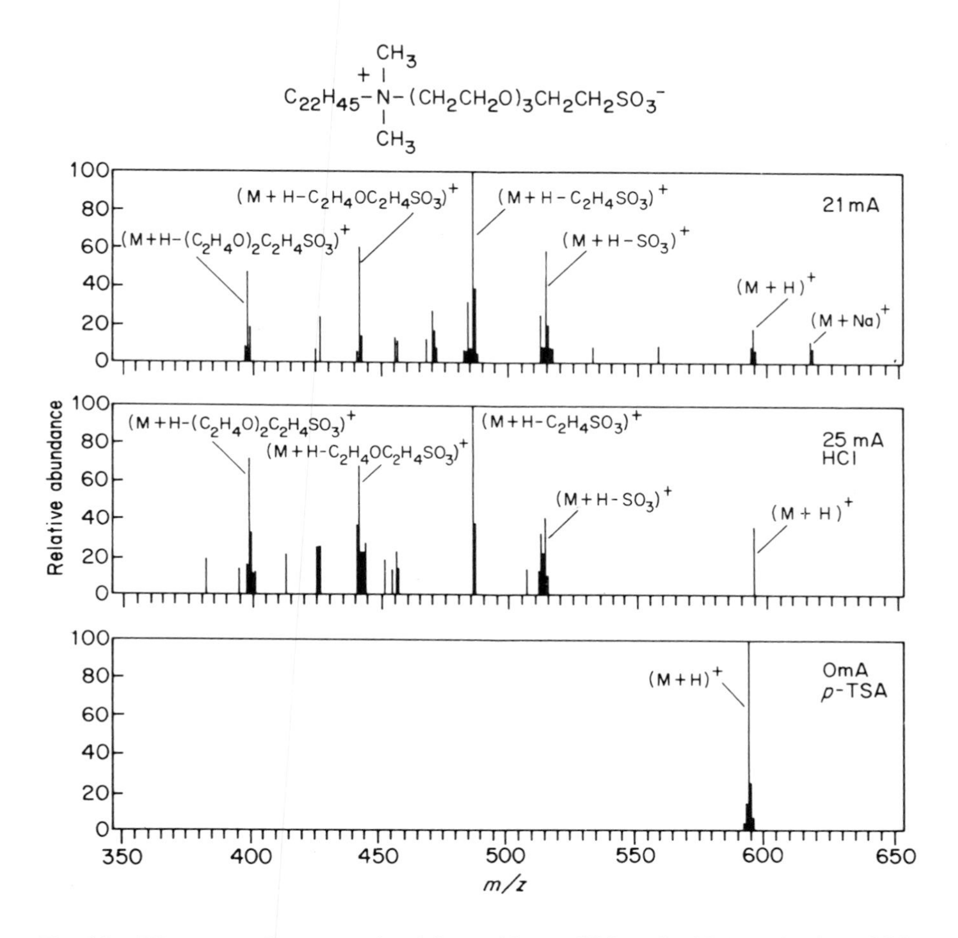

Fig. 6.9 FD spectra of an ammoniosulphate without addition of acid (upper), after addition of HCl (middle), and after addition of *p*-toluenesulphonic acid (lower). (Reprinted with permission from T. Keough and A.J. De Stefano, *Anal. Chem.*, **53**, 25[64]. Copyright (1981) American Chemical Society)

Typically, a 0.5–5 litre min^{-1} counter-current flow of warm (*ca.* 60°C) nitrogen gas is introduced between the spraying tip and the sampling orifice to aid desolvation of the highly charged electrospray droplets and to minimize the flow of neutral solvent molecules into the vacuum chamber. At higher capillary voltages, electrical breakdown of the surrounding gas may occur, creating a corona discharge that interferes with the observation of ions evaporated from the liquid. The same discharge will be observed at a lower needle voltage in the negative ion mode, presumably due to electron emission from the capillary tip. The onset of discharge conditions can be delayed to higher voltages by the introduction of an electron scavenging gas such as oxygen [14] or SF_6 [65] into the spraying area. Alternatively, a miscible liquid electron scavenger, e.g. CCl_4, may be added to the liquid flow [66].

A number of alternative source designs do not use a counter-flow of gas. Thus, droplets and solvated ions can be made to undergo desolvation as they pass through a long, directly heated metal capillary tube that communicates between atmosphere and the first vacuum stage of the mass spectrometer [67]. The residual solvent molecules are then removed from the ions by collisional activation which results from the voltages applied to the first vacuum stage (cf. section 6.3.2.2).

Desolvation by energetic collisions in a low pressure region has been observed with an ES source interfaced to an ion trap [68]. In this system, which also does not use a counter-current flow, or any heating, experiments suggest that negative ion electrospray of some compounds, notably sulphonates, results in the injection of highly solvated dianions that only desolvate within the ion trap. Cations derived from peptides, on the other hand, appear to be essentially desolvated prior to injection into the ion trap. A further alternative for desolvation is to incorporate a heater in the source block itself [69]. Operation of this heater at an elevated temperature (220–250°C) transfers heat to the electrospray chamber which is then maintained at approximately 50°C.

The use of a coaxial flow of nebulizing gas in conjunction with electrospray is referred to as 'ion spray' (Fig. 6.4) [70]. In this device, the coaxial gas jet supplies the majority of the energy required to disrupt the liquid stream while the electric field provides a net charge on each droplet. Unlike pure electrospray, the ion spray technique is able to produce stable analyte ion currents over a wide range of liquid flow rates, at least up to 200 $\mu L\ min^{-1}$, since the additional gas flow ensures effective nebulization. Another practical advantage, compared with electrospray, is that ion spray is able to handle highly aqueous solvents. Taken together, these two advantages emphasize the utility of ion spray as an LC–MS interfacing method.

6.3.2.1 Sample preparation for ES ionization

Analysis by electrospray requires the prior formation of analyte-related ions in solution. For compounds that are not ionic, analyte-related ions are usually prepared by solution chemistry methods. For example, peptides and proteins can be analysed in the positive ion mode after being dissolved in a mixture of equal proportions of methanol and water together with acetic acid (1–5%) to effect protonation of basic amino acids. Again, a compound such as polyethyleneglycol can be analysed after cationization, e.g. with sodium ions. An alternative method of ion formation uses a charge transfer reaction [71]. Thus, 2,3-benzanthracene, which itself gives no response in ES, provides an $M^{+\cdot}$ ion in the presence of an electron acceptor such as 2,3-dicyano-1,4-benzoquinone:

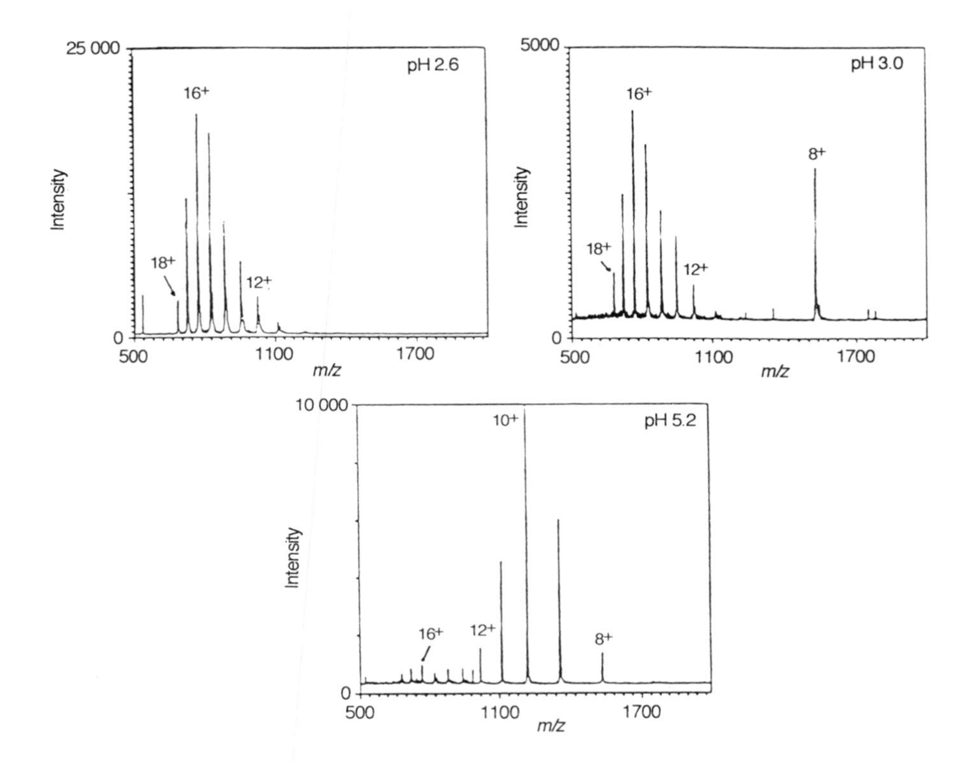

Fig. 6.10 ES spectra for cytochrome c, obtained from an aqueous solution with the lowering of pH by addition of acetic acid. (Reproduced with permission from ref. 125)

$$D + A \rightarrow (D^{+\cdot})_{solv} + (A^{-})_{solv} \qquad (6.5)$$

The solution chemistry of the analyte ions also determines the range of charge states that is observed for each analyte (eg. Fig. 6.10). For example, the charge state distribution observed for any protein is affected by the choice of solvent and pH since such changes can effect conformational changes in the protein which can make additional sites available for protonation [72–74].

The presence of certain impurities can lead to a more complex spectrum than is desirable. For example, Chait and co-workers [75] have shown that the presence of sulphate or phosphate anions in protein and peptide samples results in the formation of adduct ions such as $[M + nH + mH_2SO_4]^{n+}$ or $[M + nH + mH_3PO_4]^{n+}$ in addition to the expected protonated molecular ions. The presence of cations such as sodium or potassium can lead to similarly complex examples of adduct ion formation, as in Fig. 6.11. In each case, the formation of adduct ions not only results in a more complex spectrum but also reduces sensitivity by distributing the available

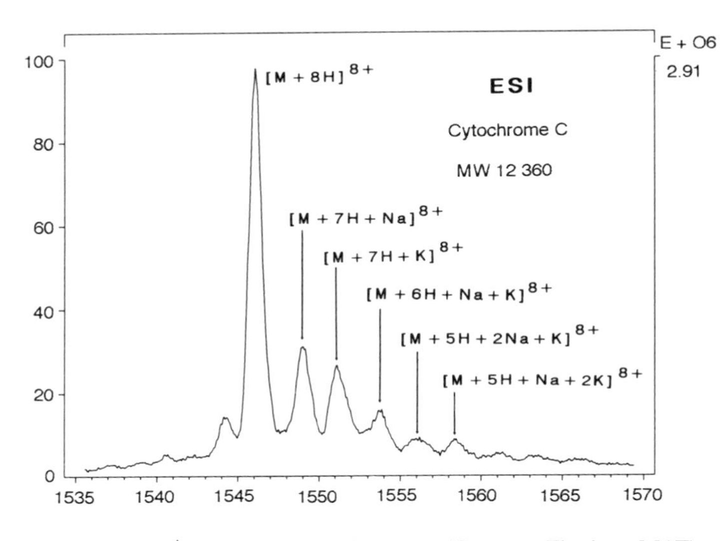

Fig. 6.11 8^+ peak from cytochrome c. (Courtesy Finnigan MAT)

ion current between a larger number of mass values. Desalting using a column chromatography method (section 5.3.2.3) or, in some cases, a precipitation method [75] is recommended prior to ES analysis.

The presence of any foreign electrolyte at a level in excess of 10^{-5} M appears [25] to reduce the analyte ion intensity in ES. Much higher levels, perhaps above 10^{-3} M [14,19], result in an unstable spray due to the increase in solution conductivity. Techniques that do not rely entirely on electrical forces for dispersion, e.g. ion spray and thermospray, can, however, provide a stable spray at much higher electrolyte concentrations.

The analyte ion current in electrospray is a function of analyte concentration. Thus, there is little change in the ion current from a given solution with change in liquid flow rate, despite the fact that the amount of analyte introduced into the source in unit time changes. On the other hand, there is a linear correlation between analyte concentration and ion current over a wide concentration range, at least up to 10^{-5} M. If the analyte concentration is increased much above the 10^{-5} M level, then the analyte ion current response becomes non-linear and increases much less than would be expected with the increase in concentration [67,76,77]. Experiments by a number of authors [76,77] have discounted the possibility that this diminished response is due to a lack of available charge in the electrospray process. Fenn and co-authors [76] have speculated that an analyte solubility limit in the evaporating droplets is a more likely cause.

In general, a solvent is chosen that sprays readily and is a good solvent for the analyte. Where a choice of solvent does exist, however, limited data [20,65] suggest that, unlike thermospray, sensitivity is highest in organic rather than aqueous liquid phases. Not all solvents are suitable for electrospray [66]. For example, less polar solvents, such as hexane or benzene, cannot be sprayed on their own [13,66]. Many other common organic solvents such as ethanol, methanol, isopropanol, acetonitrile, acetone, tetrahydrofuran, chloroform, and dichloromethane are suitable for spraying [66]. Other physical characteristics of the liquid are also important. For example, the threshold voltage (V_{th}) for spraying increases with increasing surface tension (γ) [13,78,79], as can be seen from the following equation:

$$V_{th} = A\left(2\gamma r_c \frac{cos\theta}{\epsilon_0}\right) ln\left(\frac{4h}{r_c}\right) \tag{6.6}$$

where r_c is the capillary radius, θ is the liquid cone half-angle (usually 49.3°), ϵ_0 is the permittivity of free space(8.8×10^{-12} farad metre^{-1}), h is the capillary–counter-electrode separation, and A is a constant with a value in the range 0.5–0.7 [65,79].

More highly conducting solvents can also present problems and generally require a reduced flow rate for stable spraying [78,79]. Distilled water exhibits a surface tension and a conductivity that both exceed suggested limits for these parameters [78]. However, electrospray of purely aqueous solutions has been demonstrated both by the use of a very sharp needle which reduces the onset voltage for spraying [78] and by the use in the surrounding atmosphere of SF_6, an efficient electron scavenger which allows the use of higher spraying voltages without discharge [65]. The heated source described by Allen and Vestal [69] has also been used to spray completely aqueous solutions, perhaps by reducing the surface tension and viscosity [17].

A practical alternative to the direct spraying of water is to dilute the aqueous solution, just prior to spraying, by means of a coaxial sheath flow of a miscible organic solvent (Fig. 6.12) [80]. The use of 2-methoxyethanol in the positive ion

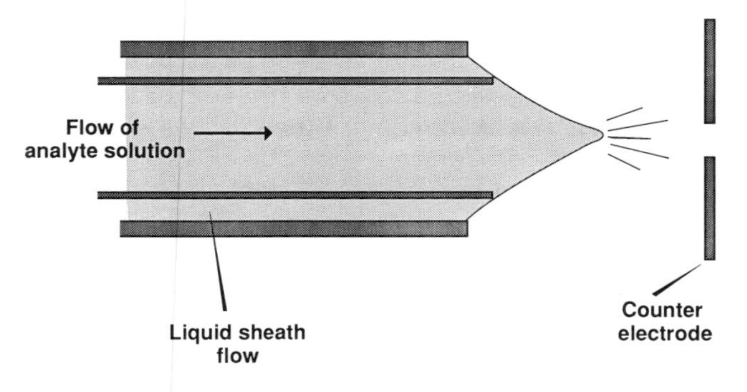

Fig. 6.12 Use of coaxial sheath flow in electrospray

mode and isopropanol in the negative ion mode had been recommended [81]. By this means, a range of solvent compositions, such as might be encountered in a gradient LC–MS experiment, may be accommodated. The ion spray technique (section 6.3.2) is also more suitable than electrospray for highly aqueous solutions.

6.3.2.2 Mass analysis of multiply charged ions

The analysis of multiply-charged ions formed by electrospray ionization has now been accomplished using most types of analyser, i.e. quadrupole, magnetic sector, ion trap, Fourier transform and time-of-flight. An advantageous feature of electrospray is that the m/z range of the mass analyser used need not be large because, as a result of multiple charging, ions above m/z 3000 are only rarely observed. Thus, a wide range of massive molecules, even those with a molecular weight in excess of 100 000, can be analysed using a mass spectrometer with a much more limited m/z range.

Multiple charging in ES creates a whole series of molecular ion peaks each of which represents a different charge state for the same compound (e.g. Fig. 6.6). The molecular weight may be determined from such a spectrum if it is assumed, as is invariably the case, that adjacent peaks in the series differ by only one charge. If we select two adjacent peaks from this series, then for an ion of measured mass-to-charge ratio, m_i, which originates from an analyte with molecular weight M_R and carries an unknown number of charges, i, supplied by proton attachment, m_i and M_R are related by

$$m_i = \frac{M_R + iM_H}{i} \tag{6.7}$$

where M_H is the mass of a proton. The mass-to-charge ratio, m_{i+1}, for an ion that carries one additional proton attached to the same analyte molecule is then given by

$$m_{i+1} = \frac{M_R + (i+1)M_H}{i+1} \tag{6.8}$$

so that i, the charge state, may be determined from the two m/z values using

$$i = \frac{m_{i+1} - M_H}{m_i - m_{i+1}} \approx \frac{m_{i+1} - 1}{m_i - m_{i+1}} \tag{6.9}$$

For charge-carrying adducts other than protons, e.g. Na^+ addition in the positive ion mode or proton abstraction in the negative ion mode, M_H may be replaced by M_A, where M_A is the adduct ion mass. Once the charge state has been calculated in this way, each peak in the series can be used to provide a separate estimate of the molecular weight, M_R.

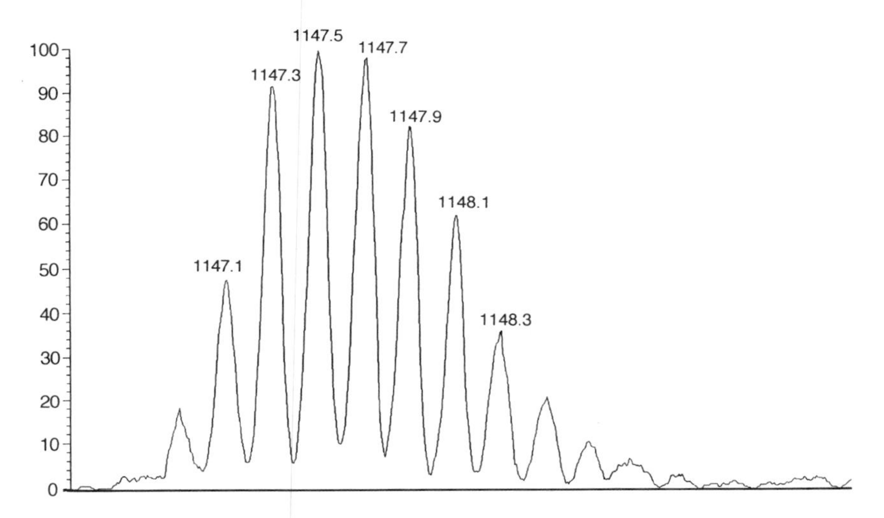

Fig. 6.13 Isotopically resolved 5^+ peak from bovine insulin (resolving power 7000). (Courtesy Kratos Analytical Instruments)

So-called deconvolution algorithms [82] offer another method of processing electrospray spectra. In these methods, all contributions from a charge state series are summed into a single peak (without adduct ion) which is plotted on a true mass rather than on an m/z scale. An alternative method of assigning charge state is provided by the use of higher resolution so long as any loss of instrumental sensitivity is not excessive. Thus, with resolution sufficient to resolve individual isotopic peaks, the number of these in an m/z unit unequivocally defines the number of charges (e.g. Fig. 6.13), without the need to identify other peaks in the same series.

Another apparent advantage of multiply-charged ions is in tandem mass spectrometry experiments. For example, the amount of energy available from collisional activation (E_{CM} in section 7.2.2) should show little variation with molecular weight since the increase in energy due to multiple charging (cf. equation 1.1) more or less compensates for the decrease associated with the increase in molecular weight (cf. equation 7.10). In addition, it would seem that bonds within multiply-charged ions may be more susceptible to cleavage because of coulombic repulsion [14,83].

Experiments have also shown that the effectiveness of collision induced decomposition (section 7.2) of electrospray ions is increased by internal energy accumulated from earlier collisions in the interface between atmospheric pressure and vacuum regions [84,85]. Collisional activation of ions in the interface is controlled by varying the potential drop across a region of intermediate pressure,

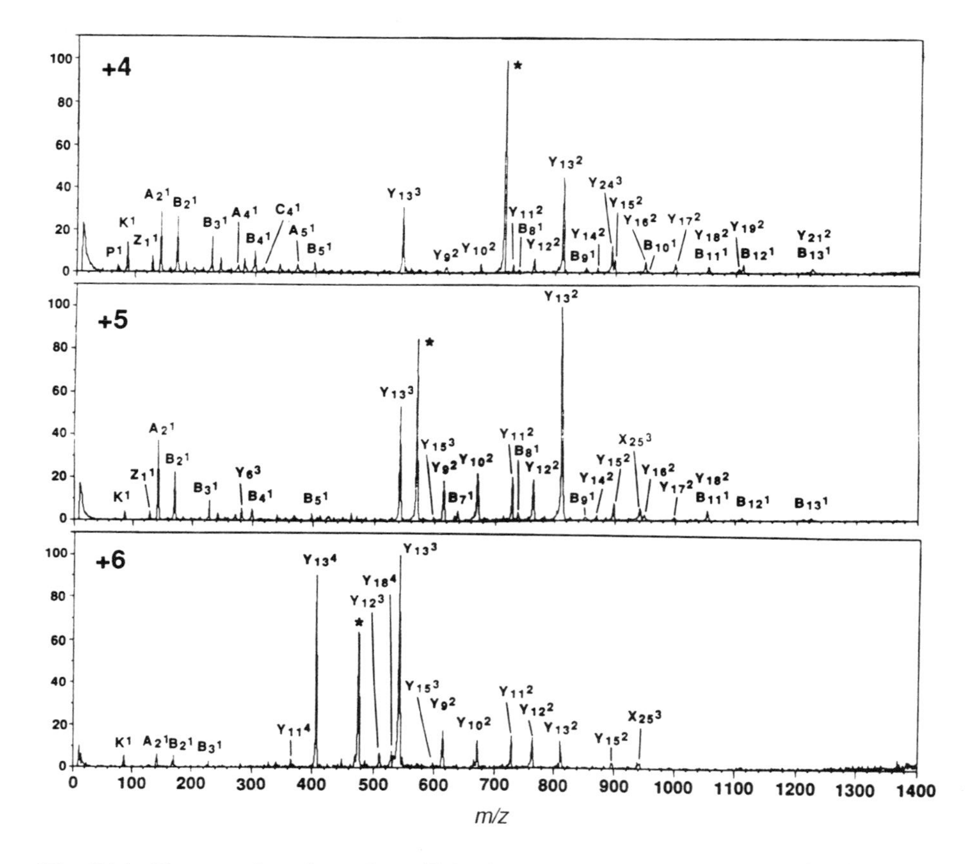

Fig. 6.14 Fragment ions from the collision-induced decomposition of the 4^+, 5^+ and 6^+ charge states of the mellitin molecular ion (molecular ion labelled *). The more intense rationalized sequence ions from the peptide are labelled using the nomenclature of ref. 133 with the charge state indicated by the superscript. (Reproduced with permission from ref. 134)

e.g. that bounded by the sampling orifice and skimmer in Fig. 2.17(b) [84]. The field strength and pressure in this region then determine the amount of energy acquired by an ion between collisions. In addition to assisting conventional CID, collision processes in the interface can also be used to induce dissociation of the multiply-charged ions prior to mass spectral analysis [84,86,87]. The fragmentation spectra produced by this method are often similar to conventional collisionally activated spectra [88]. In general, tandem mass spectra from multiply-charged ions (e.g. Fig. 6.14) present a difficult problem in interpretation since there is usually no independent measure of how the charge state of the fragment ions differs from that of the precursor ion.

6.3.3 Instrumentation for TS ionization

A detailed description of the thermospray source has already been given in section 2.4.3. It is sufficient at this stage to remind the reader of the two principal modes of operation with the thermospray source. The first is, true thermospray ionization, in which ions originate from a volatile electrolyte, usually ammonium acetate, added to the analyte solution. For the sake of clarity, this mode of operation will be referred to as 'electrolyte' thermospray ionization in the remainder of this discussion. The second is, discharge ionization, in which ion production is assisted by the operation of a filament or discharge electrode in the thermospray source.

6.3.3.1 Temperature effects in TS ionization

With electrolyte TS ionization, relatively stable analyte ion currents can only be obtained within a limited vaporizer temperature range. At lower temperatures the spray is too wet and the ion current is unstable. At higher temperatures the spray

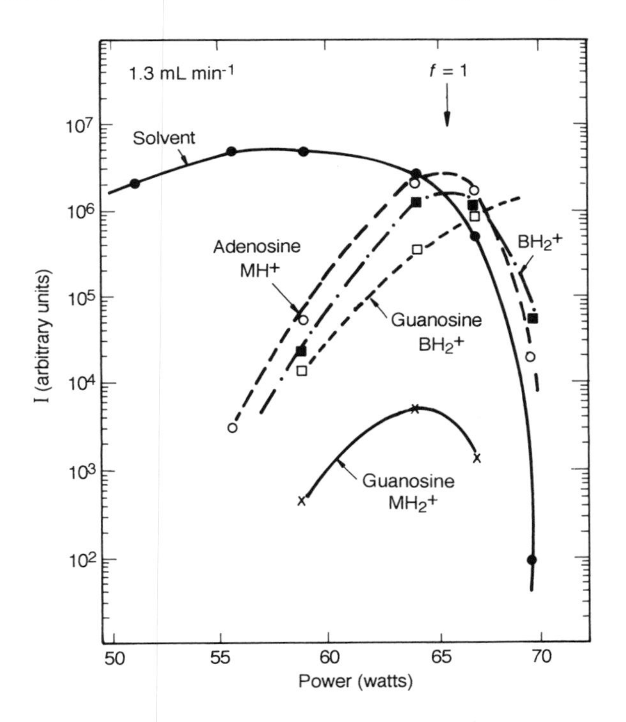

Fig. 6.15 Ion intensities as functions of power applied to a thermospray vaporizer without downstream heating. Test samples are adenosine and guanosine in 0.1 M aqueous ammonium acetate. f=1 indicates 100% solvent vaporization (Reprinted with permission from M.L. Vestal and G.J. Fergusson, *Anal. Chem.*, **57**, 2373[135]. Copyright (1985) American Chemical Society)

is completely dry and no ion current at all is recorded. A preliminary optimization of the temperature setting for a given mobile phase can be made using solvent ions, but a final optimization will ideally use analyte ions since ions that originate from the solvent do not optimize at the same temperature as analyte ions [89,90]. A compilation of typical TS operating temperatures is included in a recent review [90,91].

Spectrum quality can be critically dependent on temperature, especially that of the vaporizer (Fig. 6.15). In exceptional circumstances, a 10°C change in vaporizer temperature can represent the difference between being able to record a molecular ion and thermal decomposition of the analyte [92]. Temperature effects are more compound dependent and therefore less clear-cut than repeller effects (section 6.3.3.2), but higher temperatures, just like higher repeller voltages, discriminate against ammonium adducts compared with protonated molecular ions, and also tend to favour fragment ion production. Chemical reactions, e.g. solvolysis, can also occur at elevated temperatures [90,93,94].

The optimum vaporizer temperature also depends on the condition of the vaporizer capillary so that spray performance cannot always easily be reproduced [89]. The use of a short demountable capillary or sapphire orifice to terminate the main capillary at the probe tip has been suggested as a solution to this problem [90,95–97]. A recent general trend is to maintain vaporizer temperatures but to raise source block temperatures to prevent re-condensation of the sample after spraying [90]. Supplementary heating at the TS probe tip (Fig. 2.12) has also been implemented to minimize problems due to recondensation of less volatile samples [91,98].

6.3.3.2 Ion repeller and other instrumental effects in TS ionization

The need for more sensitive methods of detection and an interest in spectra that contain more structural information have led to investigations of the use of an ion repeller in the TS source (Fig. 6.16), either to enhance molecular ion sensitivity or to induce fragmentation in the recorded spectra.

Dramatic improvements in molecular ion sensitivity in electrolyte thermospray have been seen, principally for higher mass compounds, when a sharp 'needle' repeller is placed opposite to the ion sampling cone [95,96,99–101]. Most authors have coupled this use of a needle repeller with the use of a more restricted (e.g. 50–100 μm diameter) probe tip [95,96,99,100]. Reducing the diameter of the vaporizer orifice decreases the average droplet size in the spray and therefore increases the likelihood that analyte ions will be formed by ion evaporation.

The minimum diameter of the restriction is governed by practical considerations such as orifice plugging and the ability of the LC pump to operate at elevated pressures. Additional changes to normal thermospray conditions, including a reduction of the probe tip temperature and replacement of ammonium acetate by 1% acetic acid in the positive ion mode, have permitted the recording of multiply-

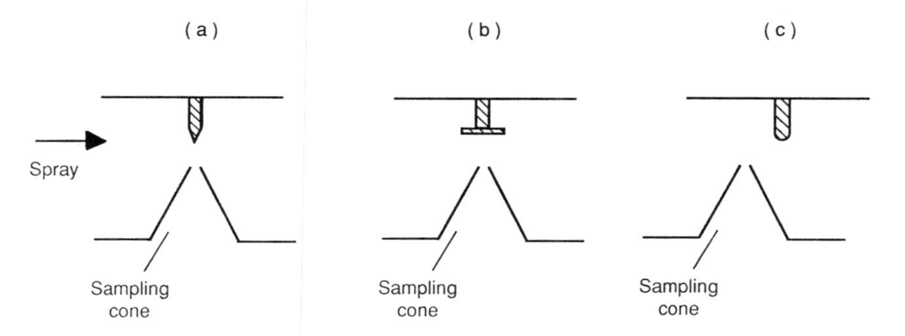

Fig. 6.16 Ion repeller configurations in a thermospray source (a) sharp repeller opposite sampling cone, (b) flat repeller opposite sampling cone, (c) repeller downstream from sampling cone

charged spectra of proteins such as myoglobin, although the level of charging and sensitivity are not as high as in electrospray [99,120].

A number of suggestions have been made to explain the function of the repeller, including an increase in ion mobility in the repeller field [96,101] and an electrospray type process from the tip of the needle [99]. Similar, but less pronounced, sensitivity enhancements have been observed for a flat plate repeller placed in an equivalent position in the ion source [98,100,101]. The magnitude of the effect with this type of repeller is very dependent on the internal geometry of the thermospray source [98].

When the TS interface is used in the discharge mode without added electrolyte, a high repeller potential applied to a flat repeller, positioned opposite to the ion sampling cone, can induce significant fragmentation of the analyte quasi- molecular ion [98,102,103]. Noise levels in such spectra are, however, generally high, especially at lower mass [103]. This fragmentation phenomenon has been attributed [98,102,103] to collisionally activated dissociation which occurs under the influence of the repeller field (cf. section 6.3.2.2). With electrolyte TS ionization, any such changes in the spectra with increased repeller potential are much less pronounced [89,103].

Baldwin and co-workers [104] have shown that fragmentation induced by a repeller is very dependent on its position in the ion source. Thus, in electrolyte thermospray, with a flat repeller 3 mm downstream from, rather than opposite, the cone, the height of the repeller had a tremendous influence on the outcome. For example, fragmentation of the diuretic drug furosemide (**1**) to give a furfuryl ion (m/z 81) barely occurs (Fig. 6.17) at any applied voltage with a height of 2 mm, while substantial fragment ion signals are seen at high voltages at 4 mm and at lower voltages at 6 mm. The intensity of the furosemide protonated molecular ion (m/z 331) can also be substantially increased by a change of the repeller voltage and maximizes at lower voltages compared with the furfuryl fragment

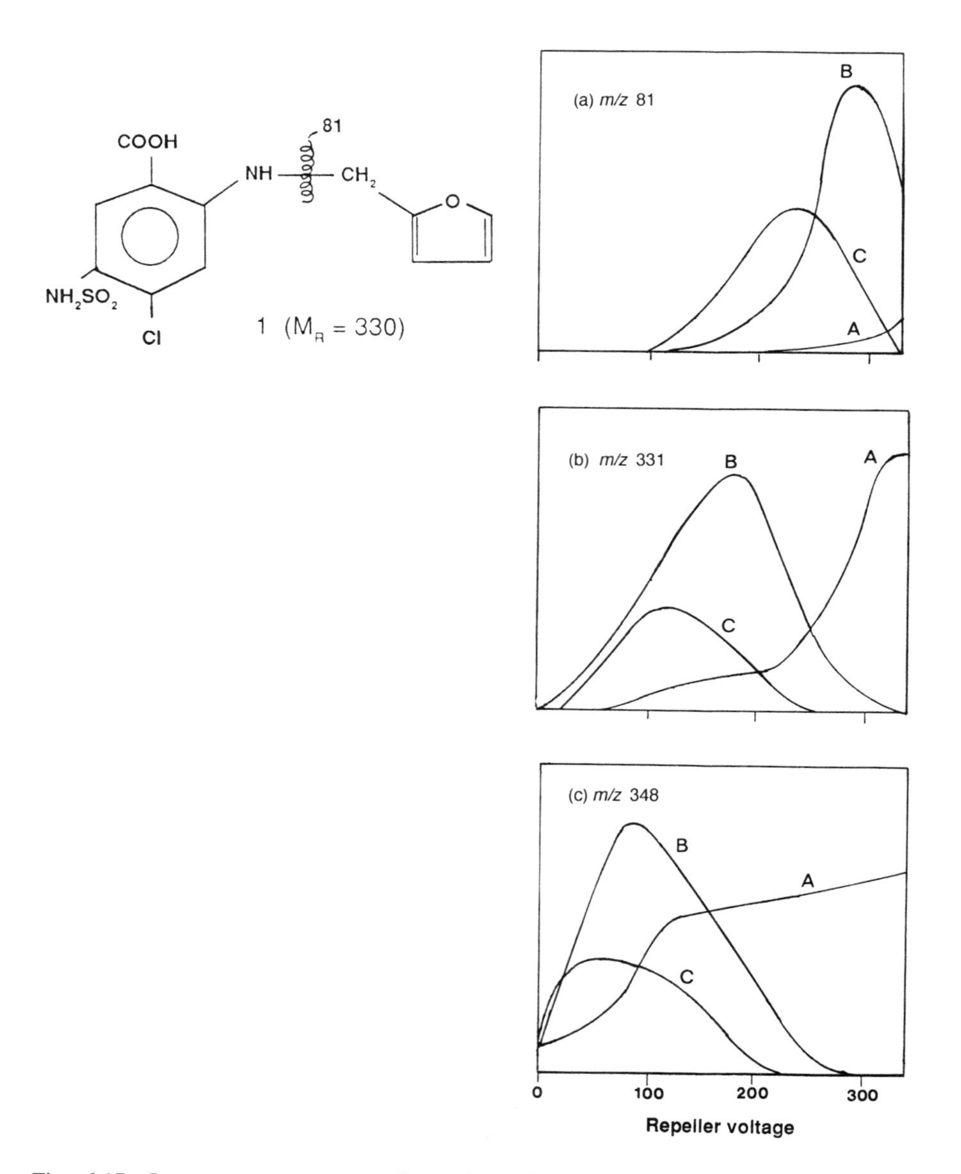

Fig. 6.17 Ion current versus repeller voltage for fragment ion (*m/z* 81), protonated molecular ion (*m/z* 321), and ammoniated molecular ion (*m/z* 348), from furosemide (structure 1) at repeller height settings of 2(a), 4(b) and 6(c) mm. (Reproduced with permission from ref. 104)

ion. Increased intensity of the ammoniated molecular ion is favoured by still lower voltages [104,105] consistent with the proposal that much of the protonated molecular ion current originates from dissociation of the ammonium adduct [106].

Bencsath and Field [107] have also reported that the use of a flat repeller (in this case parallel to the gas flow) downstream from the sampling orifice increases overall sensitivity and induces fragmentation in electrolyte thermospray operation. These, and other authors, have suggested that both of these effects are a result of the increased ion source residence time of ions slowed by a downstream repeller.

6.3.3.3 Sample preparation for TS ionization

Electrolyte thermospray ionization is limited to a few solvent and electrolyte combinations. Approximately 0.1 M aqueous ammonium acetate is the most common system. For neutral analytes, a large increase in the analyte signal at ammonium acetate concentrations up to 8×10^{-2} M was observed; no further increase was observed at higher ammonium acetate concentrations [108]. These same authors additionally evaluated a number of other electrolytes, including ammonium formate, and, in every case, found that ammonium acetate offered the greatest analyte sensitivity. Reference 89 also recommends the use of 50–100 mM ammonium acetate for the best sensitivity.

For moderately volatile analytes, ionization occurs by reactions between reagent ions (from ammonium acetate) and gaseous sample molecules [109,110]. $(M+H)^+$ and $(M+NH_4)^+$ ions (e.g. Fig. 6.18) are then the most common products from these gas-phase reactions. Cluster ions, formed by the addition of solvent molecules, are also sometimes observed and can vary in intensity with source design and operating conditions [111]. For example, with a mobile phase that contains acetonitrile, water, and ammonium acetate, the formation of an ion with the composition $(M+CH_3CN+H)^+$ is commonplace [111,112].

For ionic analytes, molecular ion intensity decreases with increasing electrolyte concentration [20], so that these samples are often vaporized in water, without the use of any electrolyte. For example, in the negative ion spectrum of a sulphonated azo dye, no significant analyte ion increase was observed for ammonium acetate concentrations between 10^{-5} and 10^{-2} M, and a suppression effect was seen at higher concentrations [113]. On the other hand, ionic samples dissolved in an insufficiently dissociating solvent require the addition of a volatile electrolyte to promote ionization [92].

The elimination of ammonium acetate can also result in the enhancement of the signal-to-noise ratio in cases where analytes are already partly ionized in solution. Thus, dopamine (MW 153) and noradrenaline (MW 169) at pH 4 show better signal-to-noise values, despite lower absolute responses, since the absence of ammonium acetate eliminates background due to ions from solvent clusters or from co-eluting materials that are no longer ionized [114].

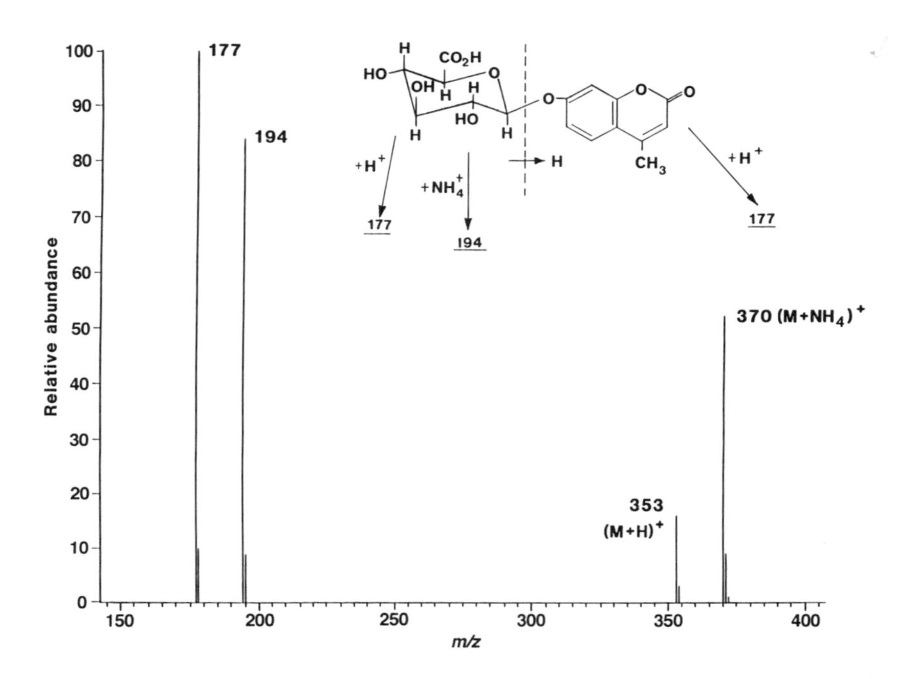

Fig. 6.18 Thermospray spectrum of 4-methyl-umbelliferyl glucuronide (MW 352). (Reprinted with permission from D.J. Liberato *et al.*, *Anal. Chem.*, **55**, 1741[136]. Copyright (1983) American Chemical Society)

In electrolyte TS, as the amount of organic solvent added to an aqueous ammonium acetate system is increased, the total ion current directly associated with either solvent or analyte decreases [115]. In many cases these ion currents can be related to the ammonium acetate ionic activity in the solvent as predicted by the Debye–Hückel theory. Methanol is a better organic modifier than acetonitrile in positive ion operation, while other polar solvents, such as ethanol or acetone, reduce the sensitivity still further. Recent experiments with a series of test compounds (adenosine, caffeine, and coniferic aldehyde) showed a slight decrease in sensitivity with an increase of methanol content from 0 to 40%. When the methanol content exceeded 40%, the sensitivity dropped significantly. The use of an external ionization source, filament or discharge, is recommended at these modifier levels [89].

Change of solvent composition can have a somewhat different outcome with more thermolabile analytes. For example, glutathione conjugates that give poor spectra in aqueous ammonium acetate yield relatively intense molecular ion signals when increased percentages of acetonitrile are added to the solution [94]. This is because the use of the modified solvent allows the use of a lower vaporizer probe temperature. Fragmentation is also affected by changes in solvent composition. Increased concentrations of organic modifier have similarly been shown to minimize certain solvolytic reactions [93].

The merits of various solvents used in normal phase liquid chromatography have been assessed for use in analyses by discharge mode TS [116]. These solvents were acetonitrile, acetone, methylene dichloride (MDC), and dichloroethane (DCE). The first two solvents gave an intense background and clusters with analyte ions and were not used further. The positive ion spectra of a range of aromatic compounds in MDC showed $(M+H)^+$ ions sometimes with less abundant solvent adduct ions. DCE was also found to be a very useful solvent for this type of LC–MS analysis.

6.4 Field Desorption Methods in Use: A Comparison

Although sharing common mechanistic features, the three ionization methods discussed in this chapter currently serve rather different application areas for the mass spectroscopist. Field desorption was, for some time, the only effective ionization technique for the analysis of labile and involatile compounds. Unfortunately, FD is an experimentally demanding technique. The preparation of emitters is time consuming and the analytical data obtained are strongly dependent on the emitter heating current. Negative ion FD operation is not readily available nor is FD suitable for instruments other than magnetic sector analysers because of the large energy spread amongst the ions formed.

FD was entirely supplanted by FAB/LSIMS (Chapter 5) for the routine analysis of polar materials and, in fact, in almost all areas other than the analysis of synthetic polymers. FD is still the preferred technique for the analysis of the latter materials [117], particularly since the lack of fragmentation allows the direct recording of a molecular weight profile for mixtures (Fig. 6.19). FD has been used to produce polymer molecular ions at masses in excess of 10 000 daltons [118].

Thermospray, as a combination of ionization and sample introduction methods, was introduced to fulfil a specific need, i.e. the effective interfacing of conventional (4.6 mm i.d.) LC columns to a mass spectrometer. It has succeeded admirably in this intention and is now in widespread use as an LC–MS technique. Current thermospray sources can accept the full flow from conventional columns with normal or reversed phase chromatographic systems under isocratic or gradient elution regimes. In addition, thermospray ionization is a soft ionization technique which offers molecular weight information from a wide range of thermally labile and polar analytes with acceptable sensitivity. A small selection of compound types, taken from ref. 119—azo dyestuffs, chlorinated herbicides, conjugates such as glucuronides, drugs and their metabolites, mycotoxins, nucleosides, small peptides, pesticides, phospholipids, surfactants, and triglycerides—indicates the applicability of the method.

Sensitivity in thermospray ionization is, however, very dependent on experimental conditions and on the nature of the analyte. In particular, sensitivity for

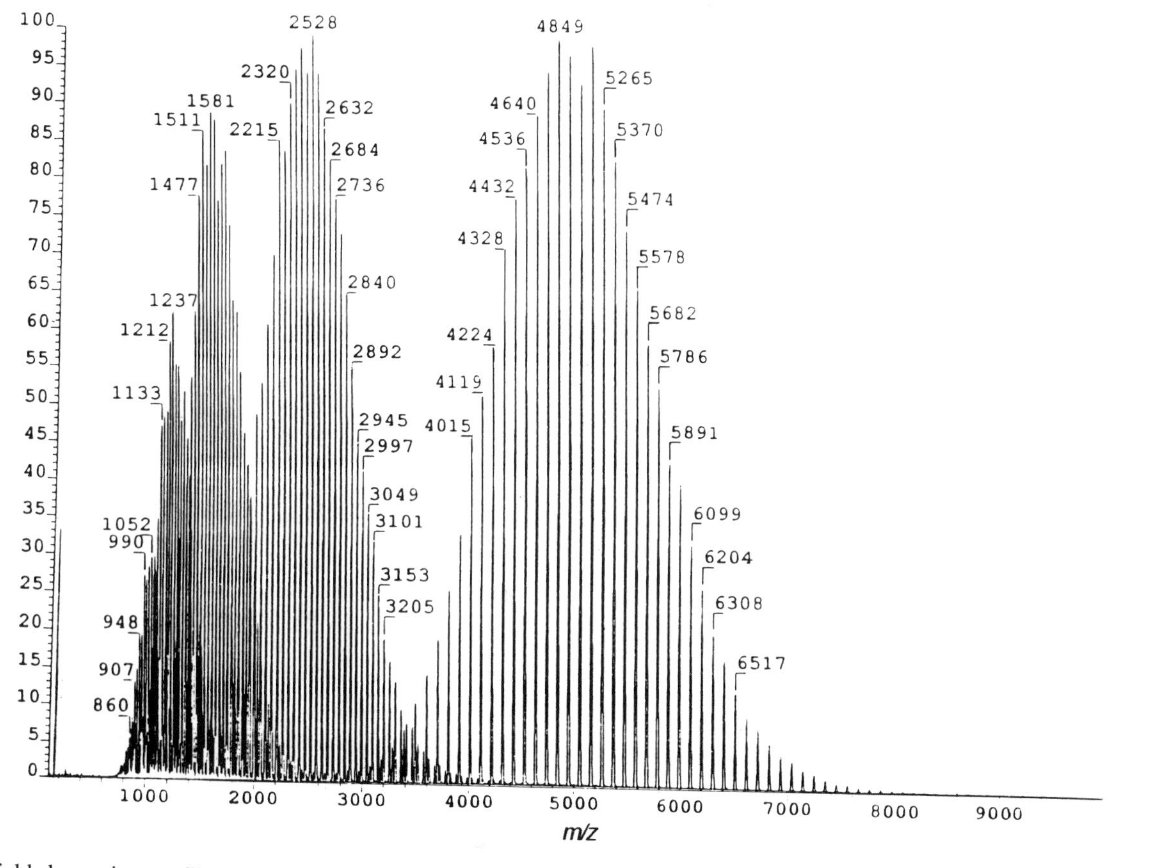

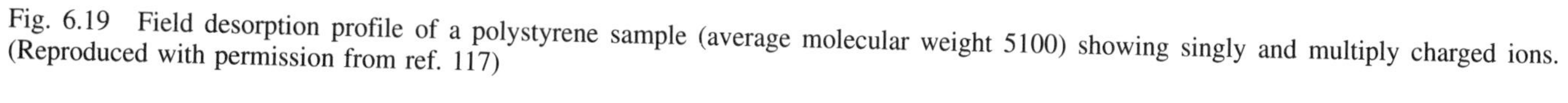
Fig. 6.19 Field desorption profile of a polystyrene sample (average molecular weight 5100) showing singly and multiply charged ions. (Reproduced with permission from ref. 117)

higher molecular weight involatile compounds such as peptides [120], nucleotides, oligosaccharides, and complex lipids is much poorer than for lower molecular weight materials. For this reason, thermospray has not been used routinely for the LC–MS analysis of these compounds, for which techniques such as continuous-flow FAB and, more recently, ion spray are preferred. Thermospray ionization often offers very little in the way of structural information, especially for lower molecular weight compounds, but these difficulties may be overcome by the use of tandem mass spectrometry techniques [Chapter 7] (Fig. 6.20).

The introduction of electrospray ionization provided the first practical means for the mass spectrometric analysis of very large biomolecules up to perhaps 100–200 kDa molecular weight. Electrospray ionization shows a number of unique advantages which led to its rapid acceptance by the mass spectrometry community. First, unlike 'energy-sudden' techniques, which were being applied to the analysis of biomolecules at that time, the yield of ions in electrospray shows little tendency to decrease with increasing molecular weight [14]. Second, the formation of multiply-charged ions made the analysis of high molecular weight analytes immediately accessible to instruments, such as quadrupole analysers, that only offer a restricted m/z range. In addition, multiply-charged ions are more readily detected than slower moving singly-charged ions (section 1.4.3). Third, the use of solvents such as water or methanol as a matrix leads to spectra that are relatively free from matrix-associated peaks (Fig. 6.21) with a consequent improvement in detection limits.

With analytes of moderate molecular weight, fast atom bombardment and electrospray ionization appear to offer similar sensitivities. For example, in the LC–MS analysis of tryptic digests of proteins, continuous flow-FAB and ion spray require 10–50 pmol of sample introduced into the ion source in each case [121,122]. A disadvantage of electrospray—a disadvantage shared with field desorption—is its susceptibility to the presence of alkali metal cations and other contaminants. Here again, ion spray offers advantages over electrospray in practical analytical situations.

Compared with another very high molecular weight ionization technique, matrix-assisted laser desorption ionization (MALDI), a major advantage of electrospray is improved resolution and, as a consequence, improved accuracy (equal to or better than 0.01%) for routine mass measurements [123]. Higher resolution is a particular advantage of instruments other than the quadrupole. For example, sector instruments have been able to demonstrate the resolution of individual isotope peaks at least up to molecular weight 14 000–15 000, with the accompanying prospect of a much increased accuracy of mass assignment [124].

Another advantage, compared with MALDI, is the suitability of ES for direct LC coupling. Despite the merits of direct mixture analysis by MALDI (section 5.4), there is no doubt that a coupled chromatographic method is, in principle, the most satisfactory approach to the analysis of mixtures by mass spectrometry. Again, just as with MALDI, electrospray does not generally lead to the formation of fragment

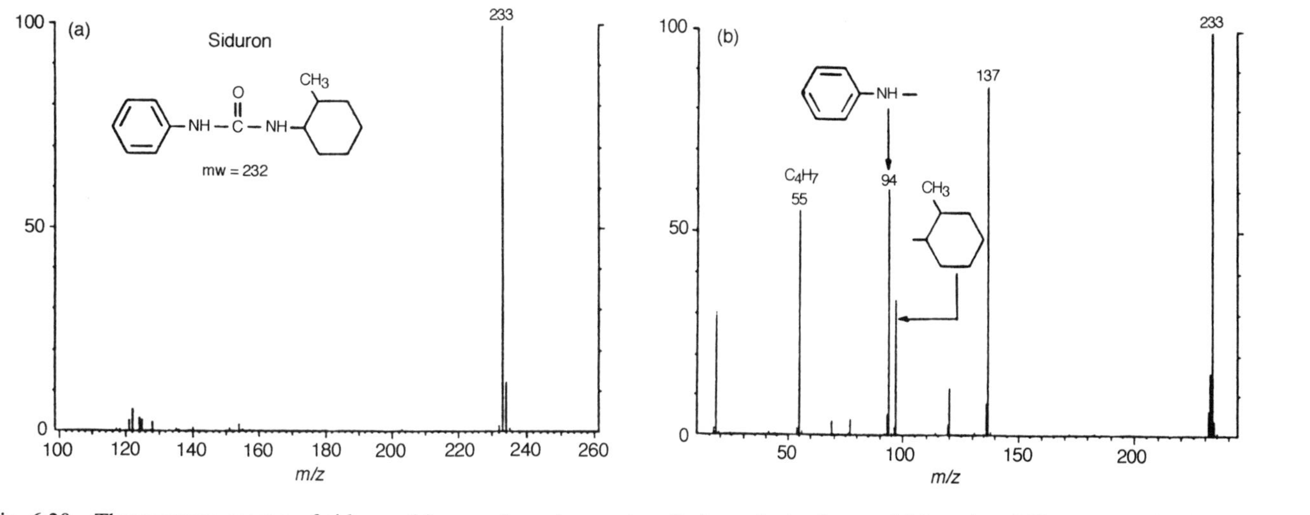

Fig. 6.20 Thermospray spectra of siduron: (a) normal spectrum using discharge ionization, and (b) tandem MS spectrum showing fragment ions from m/z 233 using electrolyte ionization. (Reproduced with permission from ref. 103)

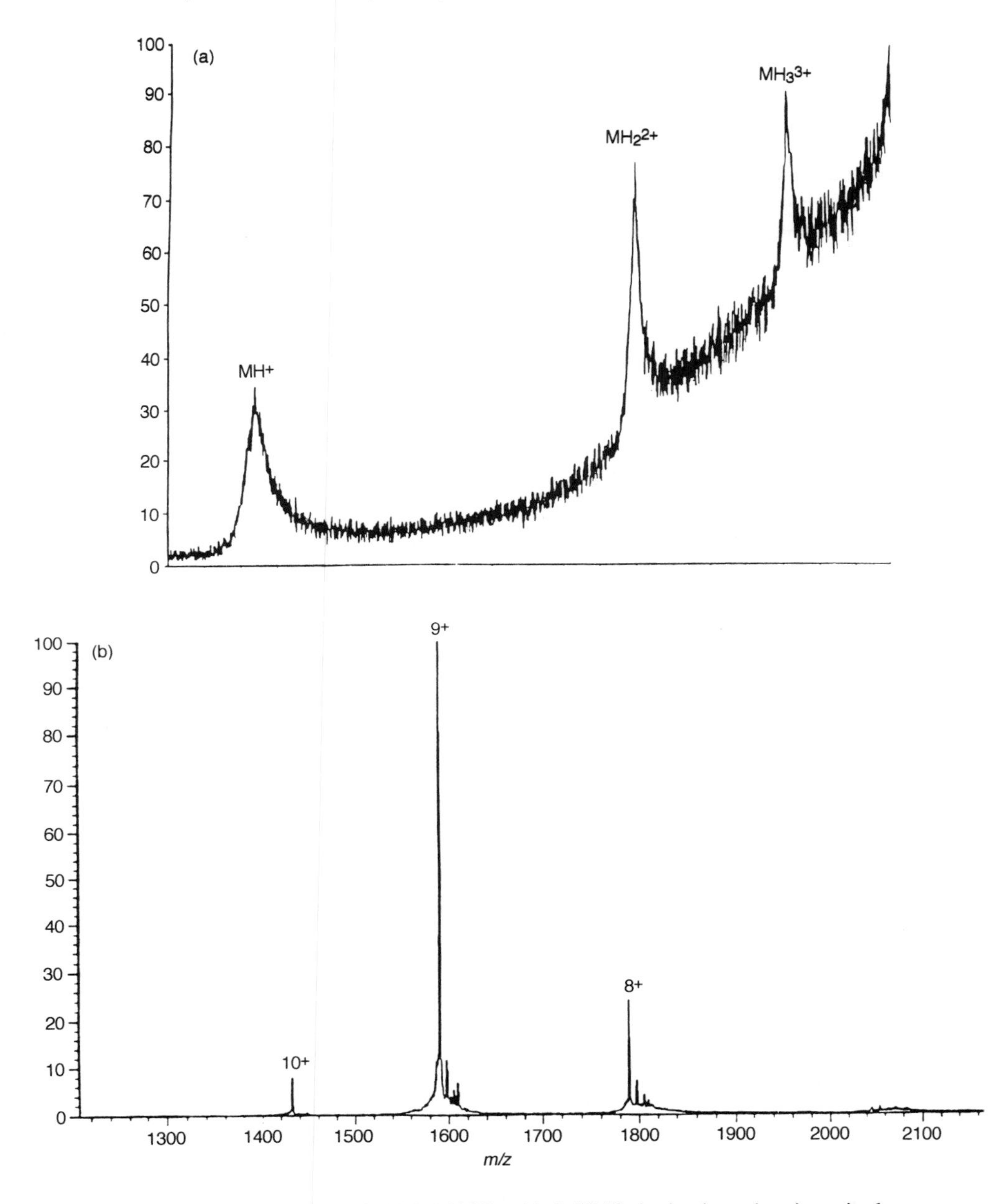

Fig. 6.21 Spectra of lysozyme (MW 14305): (a) LSIMS ionization showing singly-, doubly-, and triply-charged ions; (b) ES ionization showing 8^+, 9^+, and 10^+ ions and improved background levels. Both spectra were recorded using a magnetic sector instrument. (Courtesy Kratos Analytical Instruments)

ions. Thus, the ready availability of tandem mass spectrometry techniques with electrospray instrumentation can be a significant advantage.

Although its greatest impact has been in the analysis of biomolecules, such as peptides and proteins [125], and nucleotides [17], electrospray is also a useful technique for the ionization of a variety of smaller molecules. These molecules include highly ionic compounds, such as sulphonated azo dyes [126], a range of organometallic compounds [127,128], various toxins [129,130], phospholipids [131], and anthocyanins [132].

References

1. Müller, E.W. (1951), *Z. Phys.*, **131**, 136.
2. Inghram, M.G., and Gomer, R. (1954), *J. Chem. Phys.*, **22**, 1279.
3. Beckey, H.D. (1977), *Principles of Field Ionization and Field Desorption Mass Spectrometry*, Pergamon, London.
4. Reynolds, W.D. (1979), *Anal. Chem.*, **51**, 283A.
5. Beckey, H.D., and Schulten H.-R. (1975), *Angew. Chem. Int. Ed. Engl.*, **14**, 403.
6. Giessmann, U., Stoll, R., and Röllgen, F.W. (1980), in *Advances in Mass Spectrometry*, Vol. 8A (Ed. A. Quayle), p. 1047, Heyden & Son, London.
7. Giessman, U., and Röllgen, F.W. (1981), *Int. J. Mass Spectrom. Ion Processes*, **38**, 267.
8. Rogers, D.E.C., and Derrick, P.J. (1984), *Biomed. Environ. Mass Spectrom.*, **19**, 490.
9. Lehmann, W.D., and Kessler, H. (1982), *Stable Isotopes*, Elsevier, Amsterdam.
10. Cook, K.D. (1986), *Mass Spectrom. Rev.*, **5**, 467.
11. Stimpson, B.P., and Evans, C.A., Jr (1978), *Biomed. Mass Spectrom.*, **5**, 52.
12. Lüttgens, U., Röllgen, F.W., and Cook K.D. (1991), *NATO ASI Ser., Ser. B., 269 (Methods Mech. Prod. Ions Large Mol.)*, 185.
13. Michelson, D. (1990), *Electrostatic Atomization*, Adam Hilger, Bristol.
14. Mann, M. (1990), *Org. Mass Spectrom.*, **25**, 575.
15. Katta, V., Rockwood, A.L., and Vestal, M.L. (1991), *Int. J. Mass Spectrom. Ion Processes*, **103**, 129.
16. Lord Rayleigh (1882), *Philos. Mag.*, **14**, 184.
17. Smith, R.D., Loo, J.A., Edmonds, C.G., Barinaga, C.J., and Udseth, H.R. (1990), *Anal. Chem.*, **62**, 882.
18. Vestal, M.L., and Allen, M.H. (1991), in *Proceedings of the 39th ASMS Conference on Mass Spectrometry and Allied Topics*, Nashville, Tennessee, p. 445.
19. Fenn, J.B., Mann, M., Meng, C.K., and Wong, S.F. (1990), *Mass Spectrom. Rev.*, **9**, 37.
20. Covey, T.R., Bruins, A.P., and Henion, J.D. (1988), *Org. Mass Spectrom.*, **23**, 178.
21. Nohmi, T., Fenn, J.B., and Shen, S., (1991), in *Proceedings of the 39th ASMS Conference on Mass Spectrometry and Allied Topics*, Nashville, Tennessee, p. 443.
22. Sakairi, M., Yergey, A.L., Siu, K.W.M., Le Blanc, J.C.Y., Guevremont, R., and Berman, S.S. (1991), *Anal. Chem.*, **63**, 1488.
23. Iribarne, J.V., and Thomson, B.A. (1976), *J. Chem. Phys.*, **64**, 2287.
24. Röllgen, F.W., Nehring, H., and Giessmann, U. (1990) in *Ion Formation from Organic Solids*, IFOS V (Eds. A. Hedin, B.U.R. Sundqvist, and A. Benninghoven), p. 155, John Wiley, New York.
25. Ikonomou, M.G., Blades, A.T., and Kebarle, P. (1990), *Anal. Chem.*, **62**, 957.
26. Wong, S.F., Meng, C.K., and Fenn, J.B. (1988), *J. Phys. Chem.*, **92**, 546.

27. Guevremont, R., Siu, K.W.M., Le Blanc, J.C.Y., and Berman, S.S. (1991), in *Proceedings of the 39th ASMS Conference on Mass Spectrometry and Allied Topics*, Nashville, Tennessee, p. 1239.
28. Chait, B.T., Chowdhury, S.K., and Katta, V. (1991), in *Proceedings of the 39th ASMS Conference on Mass Spectrometry and Allied Topics*, Nashville, Tennessee, p. 447.
29. Smith, R.W., Parker, C.E., Johnson, D.M., and Bursey, M.M. (1987), *J. Chromatogr.*, **394**, 261.
30. Bütfering, L., Schmelzeisen-Redeker, G., and Röllgen, F.W. (1987), *J. Chromatogr.*, **394**, 109.
31. Liberato, D.J., and Yergey, A.L. (1986), *Anal. Chem.*, **58**, 6.
32. Voyksner, R.D. (1987), *Org. Mass Spectrom.*, **22**, 513.
33. Shiraishi, H., Otsuki, A., and Fuwa, K. (1979), *Bull. Chem. Soc. Jpn.*, **52**, 2903.
34. Smith, D.L., McCloskey, J.A., and Mitchell, J.K. (1981), *Anal. Chem.*, **53**, 1130.
35. Fraley, D.F., Woodward, W.S., and Bursey, M.M. (1980), *Anal. Chem.*, **52**, 2290.
36. Schulten, H.-R., and Nibbering, N.M.M. (1977), *Biomed. Mass Spectrom.*, **4**, 55.
37. Maine, J.W., Soltmann, B., Holland, J.F., Young, N.D., Gerber, J.N., and Sweeley, C.C. (1976), *Anal. Chem.*, **48**, 427.
38. Otsuki, A., and Shiraishi, H. (1979), *Anal. Chem.*, **51**, 2329.
39. Dougherty, R.C., Dreifuss, P.A., Sphon, J., and Katz, J.J. (1980), *J. Am. Chem. Soc.*, **102**, 416.
40. Schulten, H.-R., and Beckey, H.D. (1972), *Org. Mass Spectrom.*, **6**, 885.
41. Becky, H.D., Hilt, E., and Schulten, H.-R., (1973), *J. Phys. E. Sci. Instrum.*, **6**, 1043.
42. Lehmann, W.D., and Fischer, R. (1981), *Anal. Chem.*, **53**, 743.
43. Röllgen, F.W., and Schulten, H.-R. (1975), *Z. Naturforsch*, **30A**, 1685.
44. Röllgen, F.W., Giessmann, U., Heinen, H.J., and Reddy, S.J. (1977), *Int. J. Mass Spectrom. Ion Processes*, **24**, 235.
45. Röllgen, F.W., Heinen, H.J., and Giessman, U. (1977), *Naturwissenschaften*, **64**, 222.
46. Heinen, H.J., Giessman, U., and Röllgen, F.W. (1977), *Org. Mass Spectrom.*, **12**, 710.
47. Veith, A.J. (1977), *Tetrahedron*, **33**, 2825.
48. Wood, G.W., and Sun, W.F. (1982), *Biomed. Mass Spectrom.*, **9**, 72.
49. Ott, K., Röllgen, F.W., Zwinselman, J.J., Fokkens, R.H., and Nibbering, N.M.M. (1980), *Org. Mass Spectrom.*, **15**, 419.
50. Arpino, P.J., and Devant, G. (1979), *Analusis*, **7**, 348.
51. Giessman, U., Heinen, H.J., and Röllgen, F.W., (1979), *Org. Mass Spectrom.*, **14**, 177.
52. Rabrenovich, M., Ast, T., and Kramer, V. (1981), *Int. J. Mass Spectrom. Ion Processes*, **37**, 297.
53. Matsuo, T., Matsuda, H., and Katakuse, I. (1979), *Anal. Chem.*, **51**, 69.
54. Linden, H.B., Hilt, E., and Beckey, H.D. (1978), *J. Phys. E. Sci. Instrum.*, **11**, 1033.
55. Schulten, H.-R. (1979), *Int. J. Mass Spectrom. Ion Processes*, **32**, 97.
56. Beckey, H.J., Heindrichs, A., and Winkler, H.U. (1970), *Int. J. Mass Spectrom. Ion Phys.*, **3**, App. 9.
57. Lehmann, W.D. (1982), *Anal. Chem.*, **54**, 299.
58. Wood, G.W., Lau, P.Y., Morrow, G., Rao, G.N.S., Schidt, D.E., Jr, and Tuebner, J. (1977), *Chem. Phys.Lipids*, **18**, 316.
59. Wood, G.W., and Perkins, S.E. (1982), *Anal. Chem.*, **122**, 368.
60. Prome, J.-C., and Puzo, G. (1977), *Org. Mass Spectrom.*, **12**, 28.
61. Röllgen, F.W., and Schulten, H, -R. (1975), *Org. Mass Spectrom.*, **10**, 660.
62. Giessmann , U., and Röllgen, F.W. (1976), *Org. Mass Spectrom.*, **11**, 1094.
63. Prome, J.C., and Puzo, G. (1978), *Isr. J. Chem.*, **17**, 172.
64. Keough, T., and DeStefano, A.J. (1981), *Anal. Chem.*, **53**, 25.

65. Ikonomou, M.G., Blades, A.T., and Kebarle, P. (1991), *J. Am. Soc. Mass Spectrom.*, **2**, 497.
66. Hiraoka, K., and Kudaka, I. (1990), *Rapid Commun. Mass Spectrom.*, **4**, 519.
67. Chowdhury, S.K., Katta, V., and Chait, B.T. (1990), *Rapid Commun. Mass Spectrom.*, **4**, 81.
68. Van Berkel, G.J., Glish. G.L., and McLuckey, S.A. (1990), *Anal. Chem.*, **62**, 1284.
69. Allen, M.H., and Vestal, M.L. (1992), *J. Am. Soc. Mass Spectrom.*, **3**, 18.
70. Bruins, A.P., Covey, T.R., and Henion, J.D. (1987), *Anal. Chem.*, **59**, 2642.
71. Van Berkel, G.J., McLuckey, S.A., and Glish, G.L. (1991), in *Proceedings of the 39th ASMS Conference on Mass Spectrometry and Allied Topics*, Nashville, Tennessee, p. 1237.
72. Chowdhury, S.K., Katta, V., and Chait, B.T. (1990), *J. Am. Chem. Soc.*, **112**, 9012.
73. Chowdhury, S.K., and Chait, B.T. (1990), *Biochem. Biophys. Res. Commun.*, **173**, 927.
74. Katta, V., and Chait, B.T. (1991), *Rapid Commun. Mass Spectrom.*, **5**, 214.
75. Chowdhury, S.K., Katta, V., Beavis, R.C., and Chait, B.T. (1990), *J. Am. Soc. Mass Spectrom.*, **1**, 382.
76. Shen, S., Whitehouse, C.M., Wong, S.F., and Fenn, J.B. (1991), in *Proceedings of the 39th ASMS Conference on Mass Spectrometry and Allied Topics*, Nashville, Tennessee, p. 1235.
77. Raffaelli, A., and Bruins, A.P. (1990), in *Proceedings of the 38th ASMS Conference on Mass Spectrometry and Allied Topics*, Tucson, Arizona, p. 126.
78. Chowdhury, S.K., and Chait, B.T. (1991), *Anal. Chem.*, **63**, 1660.
79. Smith, D.P.H. (1986), *IEEE Trans. Ind. Appl.*, **IA-22**, 527.
80. Smith, R.D., Barinaga, C.J., and Udseth, H.R. (1988), *Anal. Chem.*, **60**, 1948.
81. Mylchrest, I., and Hail, M. (1991), in *Proceedings of the 39th ASMS Conference on Mass Spectrometry and Allied Topics*, Nashville, Tennessee, p. 316.
82. Mann, M., Meng, C.K., and Fenn, J.B. (1989), *Anal. Chem.*, **61**, 1702.
83. Rockwood, A.L., Busman, M., Udseth, H.R., and Smith, R.D., (1991), *Rapid Commun. Mass Spectrom.*, **5**, 582.
84. Smith, R.D., Loo, J.A., Barinaga, C.J., Edmonds, C.G., and Udseth H.R., (1990), *J. Am. Soc. Mass Spectrom.*, **1**, 53.
85. Smith, R.D., and Barinaga, C.J., (1990), *Rapid Commun. Mass Spectrom.*, **4**, 54.
86. Loo, J.A., Udseth H.R., and Smith, R.D. (1988), *Rapid Commun. Mass Spectrom.*, **2**, 207.
87. Smith, R.D., Loo, J.A., Barinaga, C.G., Edmonds, C.G., and Udseth, H.R. (1989), *J.Chromatogr.*, **480**, 211.
88. Smith, R.D., Loo, J.A., Edmonds, C.G., Barinaga, C.G., and Udseth, H.R. (1990), *Anal. Chem.*, **62**, 882.
89. Heeremans, C.E.M., van der Hoeven, R.A.M., Niessen, W.M.A., Tjaden, U.R., and van der Greef, J. (1989), *J. Chromatogr.*, **474**, 149.
90. Arpino, P. (1990), *Mass Spectrom. Rev.*, **9**, 1.
91. Vestal, M.L., (1990), in *Methods in Enzymology*, Vol. 193: *Mass Spectrometry* (Ed. J.A. McCloskey), Ch.5, Academic Press, San Diego.
92. Stout, S.J., Wilson, S.A., Kleiner, A.I. , DaCunha, A.R., and Francl, T.J. (1989), *Biomed. Environ. Mass Spectrom.*, **18**, 57.
93. Bean, M.F., Pallante, M.S.L., and Fenselau, C. (1989), *Biomed. Environ. Mass Spectrom.*, **18**, 219.
94. Bean, M.F., Pallante-Morell, S.L., Dulik, D.M., and Fenselau, C. (1990), *Anal. Chem.*, **62**, 121.
95. Fink, S.W., and Freas, R.B. (1989), *Anal. Chem.*, **61**, 2050.
96. McLean, M.A., and Freas, R.B. (1989), *Anal. Chem.*, **61**, 2054.

97. Davidson, W.C., Dinallo, R.M., and Hansen, G.E. (1991), *Biol. Mass Spectrom.*, **20**, 389.
98. Coutant, J.E., Ackermann, B.L., and Vestal, M.L. (1987), in *Proceedings of the 35th ASMS Conference on Mass Spectrometry and Allied Topics*, Denver, Colorado, p. 417.
99. McLean, M.A., Vestal, M.L., Vestal, C.H., Allen, M.H., and Field, F.A. (1990), in *Proceedings of the 38th ASMS Conference on Mass Spectrometry and Allied Topics*, Tucson, Arizona, p. 1138.
100. Robins, R.H., and Crow, F.W. (1988), *Rapid Commun. Mass Spectrom.*, **2**, 30.
101. Yinon, J., Jones, T.L., and Betowski, L.D. (1989), *Rapid Commun. Mass Spectrom.*, **3**, 38.
102. Niessen, W.M.A., Van der Hoeven, R.A.M., De Kraa, M.A.G., Heereman, C.E.M., Tjaden, U.R., and Van der Greef, J. (1989), *J. Chromatogr.*, **478**, 325.
103. McFadden, W.H., and Lammert, S.A. (1987), *J. Chromatogr.*, **385**, 201.
104. Harrison, M.E., Langley, G.J., and Baldwin, M.E. (1989), *J. Chromatogr.*, **474**, 139.
105. Lindberg, C., and Paulson J., (1987), *J. Chromatogr.*, **394**, 117.
106. Alexander, A.J., and Kebarle, P. (1986), *Anal. Chem.*, **58**, 471.
107. Bencsath, F.A., and Field, F.H. (1988), *Anal. Chem.*, **60**, 1323.
108. Voyksner, R.D., and Haney, C.A. (1985), *Anal. Chem.*, **57**, 991.
109. Bursey, M.M., Parker, C.E., Smith, R.W., and Gaskell, S.J. (1985), *Anal. Chem.*, **57**, 2597.
110. Parker, C.E., Smith, R.W., Gaskell, S.J., and Bursey, M.M., (1986), *Anal. Chem.*, **58**, 1661.
111. Barcelo, D. (1989), *Org. Mass Spectrom.*, **24**, 898.
112. Conchillo, A., Casas, J., Messeguer, A., and Abian, J. (1988), *Biomed. Environ. Mass Spectrom.*, **16**, 339.
113. Flory, D.A., McLean, M.M., Vestal, M.L., and Betowski, L.D. (1987), *Rapid Commun. Mass Spectrom.*, **1**, 48.
114. Gelpi, E., Abian, J., and Artigas, F. (1988), *Rapid Commun. Mass Spectrom.*, **2**, 232.
115. Liberato, D.J., and Yergey, A.L. (1986), *Anal. Chem.*, **58**, 6.
116. Thompson, M. (1988), in *Petroanalysis '87* (Ed. G.B.Crump), John Wiley, Chichester.
117. Rollins, K., Scrivens J.H., Taylor, M.J., and Major, H. (1990), *Rapid Commun. Mass Spectrom.*, **4**, 355.
118. Matsuo, T., Matsuda, H., and Katakuse, I. (1979), *Anal. Chem.*, **51**, 1329.
119. Arpino, P., (1992), *Mass Spectrom. Rev.*, **11**, 3.
120. Straub, K., and Chan, K. (1990), *Rapid Commun. Mass Spectrom.*, **4**, 267.
121. Hemling, M.E., Roberts, G.D., Johnson, W., Carr, S.A., and Covey, T.R. (1990), *Biomed. Environ. Mass Spectrom.*, **19**, 677.
122. Jones, D.S., Heerma, W., van Wassenaar, P.D., and Haverkamp, J. (1991), *Rapid Commun. Mass Spectrom.*, **5**, 192.
123. Chapman, J.R., Gallagher, R.T., Barton, E.C., Curtis, J.M., and Derrick, P.J. (1992), *Org. Mass Spectrom.*, **27**, 195.
124. Cody, R.B., Tamura, J., and Musselman, B.D. (1992), *Anal. Chem.*, **64**, 1561.
125. Smith, R.D., Loo, J.A., Ogorzalek Loo, R.R., Busman, M., and Udseth, H.R. (1991), *Mass Spectrom. Rev.*, **10**, 359.
126. Edmonds, C.G., Loo, J.A., Barinaga, C.J., Udseth, H.R., and Smith, R.D. (1989), *J.Chromatogr.*, **474**, 21.
127. Katta, K., Chowdhury, S.K., and Chait, B.T. (1990), *J. Am. Chem. Soc.*, **112**, 5348.
128. Curtis, J.M., Derrick, P.J., Schnell, A., Constantin, E., Gallagher, R.T., and Chapman, J.R. (1992), *Org. Mass Spectrom.*, **27**, 1176.
129. Pleasance, S., Quilliam, M.A., and Marr, J.A. (1992), *Rapid Commun. Mass Spectrom.*, **6**, 121.

130. Korfmacher, W.A., Chiarelli, M.P., Lay, J.O., Jr, Bloom, J., Holcomb, M., and McManus K.T. (1991), *Rapid Commun. Mass Spectrom.*, **5**, 463.
131. Weintraub, S.T., Pinckard, R.N., and Hail, M. (1991), *Rapid Commun. Mass Spectrom.*, **5**, 309.
132. Glässgen, W.E., Seitz, H.U., and Metzger, J.W. (1992), *Biol. Mass Spectrom.*, **21**, 271.
133. Roepstorff, P., and Fohlman, J. (1984), *Biomed. Mass Spectrom.*, **11**, 601.
134. Barinaga, C.J., Edmonds, C.G., Udseth, H.R., and Smith, R.D. (1989), *Rapid Commun. Mass Spectrom.*, **3**, 160.
135. Vestal, M.L., and Fergusson, G.J. (1985), *Anal. Chem.*, **57**, 2373.
136. Liberato, D.J., Fenselau, C.C., Vestal, M.L., and Yergey, A.L. (1983), *Anal. Chem.*, **55**, 1741.

Chapter 7

Tandem Mass Spectrometry: The Dissociation of Ions

7.1 Unimolecular Ion Dissociation: Metastable Ions in Magnetic Sector Instruments

The ionization of a neutral sample molecule can result in the formation of molecular ions with a wide range of internal energies. As a result, a number of molecular ions may be sufficiently energetic to undergo one or more dissociation processes. The extent of the competition between the available dissociation processes can be gauged from the value of the rate constant (k) for each process at a given internal energy (E). A simplified expression for the rate constant is given by

$$k = \nu \left(\frac{E - E_0}{E} \right)^{s-1} \tag{7.1}$$

where E_0 is the activation energy for the reaction in question, ν is a frequency factor, and s is the effective number of oscillators in the ion. Of the different reaction types, rearrangement reactions have low activation energies because they involve transition states with bond formation as well as dissociation and low frequency factors because of the need to achieve a particular geometrical configuration for the transition state. Simple bond cleavage reactions generally have higher activation energies and higher frequency factors which may be approximated by the bond vibrational frequency.

Whether the products of any particular dissociation process are actually recorded as part of the mass spectrum will be determined by a comparison of the appropriate rate constants with the instrumental time scale. For example, with an electron impact ion source and a magnetic sector analyser, reactions with a rate constant greater than 10^6 s^{-1} occur almost entirely within the ion source and mainly lead to fragment ions which are recorded as part of the normal spectrum. Those ions that react with rate constants less than 10^4 s^{-1} generally do not dissociate before they reach the detector and are therefore recorded as molecular ions. The spread of lifetimes associated with a given rate constant means that an intermediate value of, for example, 10^5 s^{-1} will correspond to the formation of significant numbers of both molecular and fragment ions. Alternatively, ions reacting with rate constants in the range 10^4–10^6 s^{-1} may dissociate in transit between the ion source and the detector and are then known as metastable ions [1].

Metastable ions are more likely to be associated with low activation energy processes, such as rearrangement reactions, than with higher energy processes, such as simple bond cleavage [1]. In practice the term 'metastable ion' often refers to the fragment ions which result from the unimolecular dissociation of metastable ions during transit. It is these fragment ions that are actually recorded by the detection system of the mass spectrometer.

A particular value of metastable ions in analytical chemistry is their specificity. For example, the observation of the dissociation

$$m_1{}^+ \rightarrow m_2{}^+ + m_3{}^0 \qquad (7.2)$$

in the mass spectrum of a particular compound would be more specific than the observation of possibly unrelated $m_1{}^+$ and $m_2{}^+$ ions in the same spectrum. Thus, by the use of metastable dissociations some details of the overall course of fragmentation for a given molecule can be put together. A prerequisite for this type of work with a magnetic sector instrument is a knowledge of how the observed mass of the metastable ion (m^*) is related to that of the precursor ion (m_1) and the fragment ion (m_2).

For the conservation of kinetic energy and momentum during the dissociation, the following equations apply:

$$m_1 v_1^2 = m_2 v_2^2 + m_3 v_3^2 \qquad (7.3)$$

$$m_1 v_1 = m_2 v_2 + m_3 v_3 \qquad (7.4)$$

where v_n is the velocity of an ion of mass m_n. Then, since $m_1 + m_2 = m_3$, the following result is obtained:

$$v_2 = v_3 = v_1 \qquad (7.5)$$

A magnetic sector acts as a momentum analyser (section 1.3.1) so that, if the product of the metastable dissociation of actual mass m_2 and velocity v_1 ($v_2 = v_3 = v_1$) is detected at a mass value corresponding to a normal ion of mass m^* and velocity v^*, then

$$m^* v^* = m_2 v_1 \qquad (7.6)$$

Since

$$\frac{1}{2} m^* v^{*2} = \frac{1}{2} m_1 v_1^2 = eV \qquad (7.7)$$

is true for any normal ion which has been accelerated through the full accelerating voltage V, the following equation results:

$$m^* = \frac{m_2^2}{m_1} \tag{7.8}$$

Thus, a metastable ion, which after acceleration decomposes in the field free region before the magnetic sector (Fig. 1.4), will be detected at a mass given by equation (7.8) in a normal scan (Fig. 7.1).

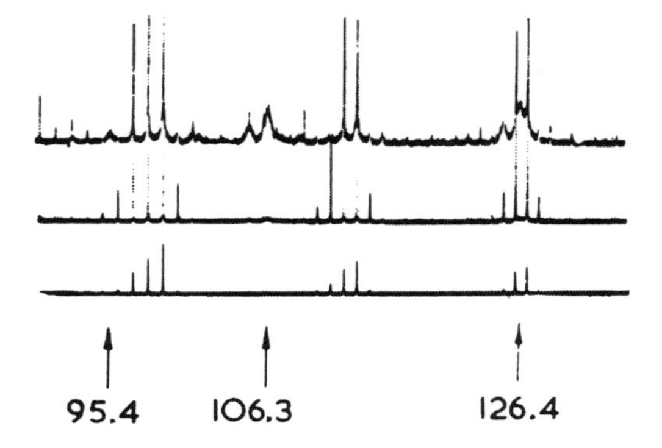

Fig. 7.1 Metastable ions from n-hexadecane recorded during a normal magnet scan

As can be seen from Fig. 7.1, metastable ions recorded during a normal magnet scan appear as relatively weak, broad peaks compared with those due to normal ions. The width of the peaks is a result of the energy released during dissociation, which can appear as a velocity component in any direction opposing or supplementing the original velocity of the ions. While the width of metastable peaks can give important information on energy release during dissociations (section 7.3.1), the diffuse nature and the low intensity of these peaks make them unfavourable for general analytical use. In addition, equation (7.8) cannot be solved uniquely for m_1 and m_2 so that the assignment of a metastable transition is not unambiguous.

These problems may be overcome by the use of special scanning techniques with a double focusing instrument or by the use of other appropriate instrumentation. Such procedures, in which, for example, an ion of a particular mass is deliberately selected and all fragmentations from that ion are recorded, are referred to as tandem mass spectrometry (MS–MS) experiments [2]. Individual MS–MS experiments are discussed in more detail together with the corresponding instrumentation in sections 7.3.1 to 7.3.7. Another important feature is that the analytical use of MS–MS is centred on collision-induced dissociation rather than unimolecular dissociation. This technique is discussed in the following section.

7.2 Collision-induced Dissociation (CID)

A wide peak in a normal scan in a magnetic sector instrument does not necessarily represent the unimolecular dissociation of a metastable ion. The dissociation may be collision induced. In this process, a fraction of the kinetic energy of a moving ion is converted to internal energy by collision with background gas molecules. Fragmentation follows this collisional activation as a second stage. Since collision-induced dissociations have a relatively high cross-section they may be much more numerous than unimolecular dissociations.

Collision-induced dissociations are more often caused deliberately. A collision region in a sector instrument is typically a short differentially pumped cell with narrow ion entrance and and exit slits and containing an inert gas [3,4]. The cell is ideally located at the focus point within a field free region. A collision

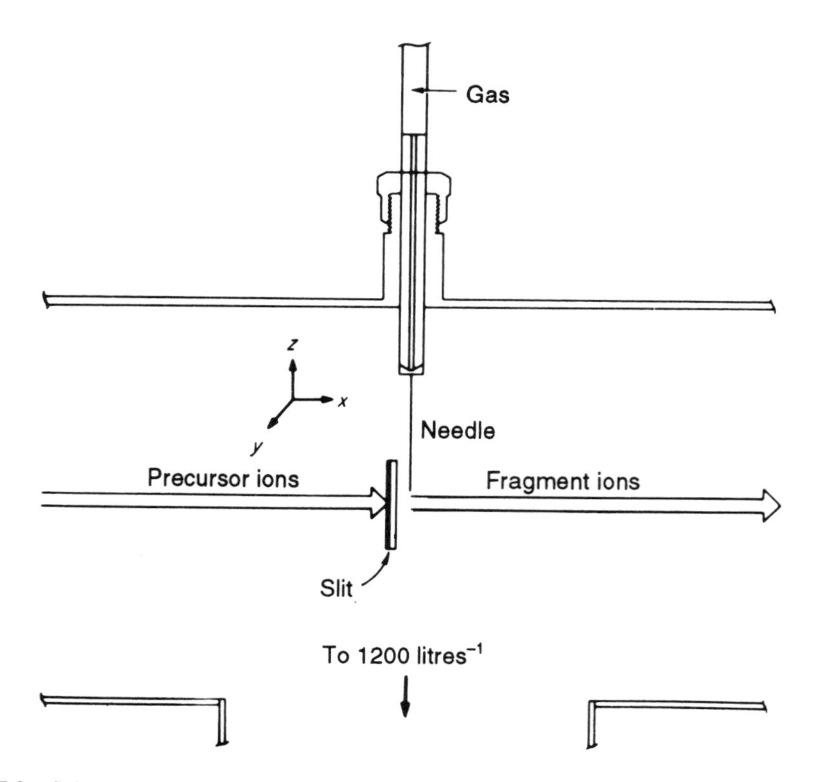

Fig. 7.2 Schematic diagram of collision region. (Reprinted with permission from G.L. Glish and P.J. Todd, *Anal. Chem.*, **54**, 842[5]. Copyright (1982) American Chemical Society)

cross-section of 10^{-14} cm^2 means that a path length of only 1 cm will ensure a sufficient probability of collision for most of the ions in a cell held at approximately $3x10^{-3}$ torr.

A particularly simple alternative construction for a collision cell [5] is illustrated in Fig. 7.2. In this design the molecular beam of gas formed by the hypodermic needle inlet offers a better approximation to the ideal point cell and a surprisingly high overall efficiency (Fig. 7.3). Efficiency can be evaluated using the fragmentation of the methane molecular ion [5,6]. The overall efficiency is then the product of the fragmentation efficiency and the collection efficiency where the fragmentation efficiency is the percentage of the ion current that comprises fragment ions, and the collection efficiency is the total ion current leaving the collision region as a percentage of the precursor ion current entering this region.

As can be seen from Fig. 7.3, the fragmentation efficiency in collision-induced processes increases with increasing collision gas pressure. Unimolecular processes do not show this pressure dependence. At higher pressures collisional scattering begins to predominate so that the overall efficiency falls again. The maximum overall efficiency for 8 kV precursor ions is achieved with a collision gas pressure that gives an approximately 70% attenuation of the precursor ion [3,5].

Another type of collision chamber is provided by a quadrupole mass filter containing an appropriate pressure of collision gas and operated in the r.f.-only (total ion) mode [7–9]. The r.f.-only quadrupole collision chamber provides focusing of scattered ions and shows a higher collection efficiency than a conventional cell. This

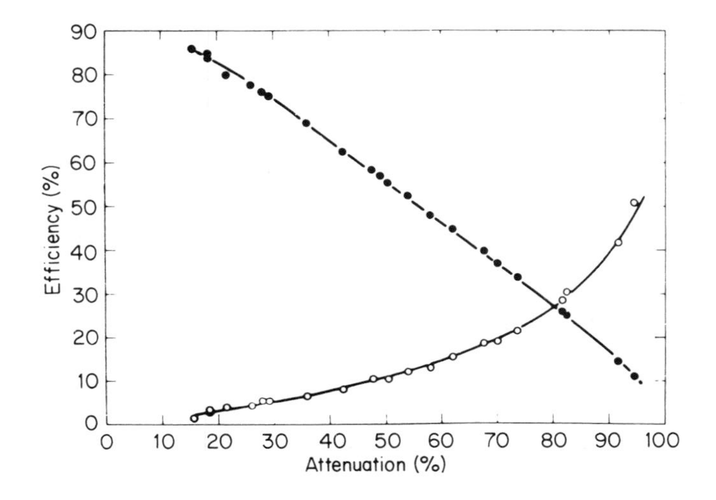

Fig. 7.3 Plots of collection efficiency (•) and fragmentation efficiency (○) versus precursor ion beam attenuation for the fragmentation of the methane molecular ion with helium collision gas. (Reprinted with permission from G.L. Glish and P.J. Todd, *Anal. Chem.*, **54**, 842[5]. Copyright (1982) American Chemical Society)

type of collision cell was originally introduced as a feature of the triple quadrupole instrument (section 7.3.2) but is also used in hybrid instruments (section 7.3.4).

An important difference results from the fact that the ions entering a quadrupole collision cell generally do so with low (generally $<100\,\text{eV}$) energies whereas those entering a collision cell on a sector instrument do so with kiloelectronvolt energies. This difference in energy and therefore in velocity has an direct effect on the interaction time of the ions and target molecules and therefore on the excitation mode available to the ions (section 7.2.1). Other differences follow from the physical differences in cell construction. Thus, the longer ion path in the quadrupole collision cell increases the possibility of multiple collisions and, in conjunction with the lower ion velocity, extends the time period over which fragmentations may be observed.

Collisional dissociation of an ion provides a fingerprint spectrum which is characteristic of its structure just as an EI spectrum is characteristic of molecular structure [10]. Collisional activation is also a valuable method of producing structural information when the primary ionization process, CI, FAB, ES, etc., does not introduce enough energy for fragmentation.

7.2.1 Mechanisms of collisional activation

The interaction time of a moving ion and target molecule can be discussed [11] in terms of the Massey parameter R_M [12]:

$$R_M = \frac{t_c}{\tau} \tag{7.9}$$

where t_c is the interaction time and τ is the characteristic period of the internal mode which is being excited. If $R_M \gg 1$, the collision period extends over many cycles of the internal motion, so that the probability of excitation of this mode is low since the system can continually adjust to the perturbation due to the collision. If $R_M < 1$, the collision period is short relative to the internal motion and now the process is referred to as an impulsive interaction [13] for which energy transfer is efficient. In the intermediate range, where $R_M \approx 1$, extremely high probabilities for energy transfer can be observed.

For an ion of mass 250 Da with a kinetic energy of 8 keV, the value of t_c for an interaction length of a few angstroms is approximately 6×10^{-15} s. Thus, collisions at kiloelectronvolt energies correspond to $R_M \approx 1$ for electronic modes in polyatomic ions, that have a similar period and electronic excitation is therefore expected to be efficient. Once the excitation has occurred, the energy can rapidly redistribute within the ion and the ion will then dissociate from excited vibrational states. Experimental evidence, e.g. the close parallel between collision-induced and electron impact spectra in a number of cases [14], seems to support this mechanism, although more recent studies [15] suggest that a vibrational contribution may also exist.

As the mass of an ion increases, its velocity at a given accelerating voltage decreases and direct vibrational excitation becomes more likely. Under these conditions, the impulse model is an important mechanism for vibrational excitation. In this mechanism, interaction is rapid enough for the effects of the collision to be localized to one part of the ion. The remainder of the ion is unaffected and can be regarded as a passive 'spectator'. The dissociation of higher mass precursor ions by collisional activation at kiloelectronvolt energies is also associated with a significant loss of translational energy. This loss can lead to errors in mass assignment of the resultant fragment ions (the so-called 'Derrick shift') in MIKE (mass-analysed ion kinetic energy) scan experiments and sensitivity losses in linked scan experiments (section 7.3.1.3).

For an ion of mass 250 Da with a kinetic energy of 20 eV, the value of t_c for an interaction length of a few angstroms is approximately 10^{-13} s. Thus, the possibility of direct excitation of electronic modes is precluded in low energy collisional activation. It is believed the formation of a relatively long-livéd collision complex [16], even with inert target gases, is a widespread and efficient mechanism for excitation of vibrational modes under low electronvolt conditions. It has been suggested that the lifetime of such a collision complex will decrease and the possibility of an impulsive collision mechanism will increase as the collision energy is increased [17].

7.2.2 Internal energy available from collisional activation

For collisional activation to result in dissociation of an ion during its transit through the collision region, the activation process must result in the creation of an appropriate level (equation 7.1) of internal energy. The internal energy available will be the sum of the internal energy present in the ion on formation and that imparted by collisional activation. Thus, preliminary collisional activation of ions in an electrospray source (section 6.3.2.2) has been shown to increase the effectiveness of subsequent collisionally induced dissociation of these ions [18]. Again, Alexander and Boyd [26] found that fragment ions of useful intensity could not be obtained from higher molecular weight peptides in a hybrid instrument unless a significant level of internal energy had already been deposited in the precursor ion on formation in the ion source.

The amount of kinetic energy converted into internal energy on collisional activation depends on a number of factors. In fact, the maximum amount of energy available is the relative kinetic energy in the centre-of-mass coordinate system (E_{CM}) which is given by

$$E_{\mathrm{CM}} = E_{\mathrm{LAB}} \left(\frac{m_{\mathrm{g}}}{m + m_{\mathrm{g}}} \right) \tag{7.10}$$

where m_g is the mass of the stationary target molecule, m is the mass of the ion, and E_{LAB} is the ion kinetic energy given by equation (1.1). Thus, if the relative kinetic energy after the collision is denoted by E'_{CM} and the internal energy taken up is denoted by Q we have

$$E_{CM} = E'_{CM} + Q \qquad (7.11)$$

so that Q achieves its maximum value when $E'_{CM} = 0$, i.e. when the colliding bodies stick together.

With low energy collisions and lower molecular weight precursor ions, a relatively large fraction of E_{CM} is converted into internal energy although, in general, this fraction decreases as the collision energy is increased [19]. If we now consider these same precursor ions colliding with kiloelectronvolt energies, the conversion to internal energy is relatively inefficient so that the mean internal energy of ions activated in low and high energy collisions may be quite similar. On the other hand, there will be a significant difference in the distribution of internal energies [20].

The distribution achieved by high energy collision has a high energy tail that is missing when ions are activated by low energy collisions. This high energy tail opens up the possibility of reactions that are not accessible in the low energy mode. For example, alkane elimination is commonly observed in 8 keV spectra of protonated alkyl amines but is much less evident in low electronvolt spectra because the reaction has a high activation energy [21]. For the same reason, charge remote fragmentation [22] is less often seen under low energy conditions. Highly endothermic reactions such as charge inversion ($M^- \rightarrow M^+$) [23] and charge stripping ($M^+ \rightarrow M^{2+}$) [24] are only seen under high energy conditions.

Consideration of equation (7.10) illustrates one of the potential difficulties in the collisional dissociation of higher mass ions. Thus, for a given value of the ion kinetic energy, E_{LAB}, and the target gas mass, m_g, increasing the mass of the precursor ion m in equation (7.10), reduces the value of E_{CM}. For example, the maximum amount of energy that can be deposited as internal energy as a result of a collision between an ion of m/z 3000 with 8 keV translational energy and a helium target molecule is approximately 10 eV. In addition, with large molecules, the average energy deposited per vibrational mode by collisional activation may be too small to induce fragmentation since it can be distributed among a large number of oscillators (equation 7.1).

7.2.3 Target gas effects in collisional activation

At low collision energies, since a large fraction of E_{CM} is converted to internal energy, the choice of target gas can have a significant effect on the amount of energy transferred into the ion and therefore on the appearance of the MS–MS spectrum. E_{CM} increases with target mass so that heavier targets result in increased fragmentation for a given value of E_{LAB} [25].

For example, in the fragmentation of leucine enkephalin (MW 556) in a hybrid tandem instrument, xenon was found to be more efficient than argon for the generation of fragment ions at a given value of E_{LAB}, under both single and multiple collision conditions [26]. On the other hand, an examination of the dissociation of simple ketones showed that the use of helium at low collision energy gave enough information to characterize and identify the samples. The use of a collision gas of higher molecular weight, e.g. nitrogen, produced more fragmentation but not necessarily more information [27]. The use of polyatomic targets can lead to less extensive fragmentation than the use of monatomic targets of similar mass, possibly because internal energy may be distributed among a large number of oscillators in the target molecule [28].

Chemical effects due to reaction with the target molecule can also be significant in low electronvolt CID spectra. These ion–molecule reactions can usually be avoided, so that a spectrum that contains only fragmentation products is recorded, by the use of an inert target gas. In general for low electronvolt CID a target gas, such as argon or nitrogen, which is massive enough to obtain sufficient energy transfer but which is essentially unreactive, is chosen.

A change of target gas has different effects in high energy CID. In practice, most high energy experiments have used helium as the target gas since it appears to be most efficient for energy transfer via direct electronic excitation and minimizes the loss of product ions due to scattering [29,30]. In addition, helium minimizes ion losses due to charge exchange with the target gas because of its high ionization potential. An interesting consequence of the minimization of scattering losses is the fact that fragment ions that show a much higher translational energy loss (section 7.2.1) can be recorded when helium is used as a target gas [31].

Uggerud and Derrick [32] have commented on the fact that helium is routinely used as a target gas in kiloelectronvolt activation of large organic and biological ions [33], even though equation (7.10) demonstrates that the combination of a low molecular weight target gas and a large precursor ion will reduce E_{CM} still further. They have proposed a method of calculating the efficiency with which E_{CM} can be transferred to internal energy in an impulsive process which suggests that this efficiency is greater when the target gas is of lower mass.

More recently, Jennings and co-workers [34,35] have demonstrated that a higher molecular weight target gas, such as argon, can be used for high energy activation of more massive ions when the collision cell is floated at a high voltage (section 7.3.3). Under these conditions, scattering processes are no longer a problem and argon can participate effectively in the vibrational processes that assume increased importance with higher mass ions. The use of argon rather than helium emphasizes some structurally significant fragmentation reactions and increases overall sensitivity for peptide ions with masses greater than 1000 Da.

At all collision energies, an increased target gas pressure increases both the number of ions that undergo collision and the possibility of multiple collisions for each ion. In practice, the energy deposited in an ion can be significantly increased by multiple collisions under low energy conditions. For example, the stable molecular ions of polycyclic aromatic hydrocarbons undergo only limited fragmentation under single collision conditions but give much more structural information with an increase in target gas pressure [24].

It has also been pointed out that when internal energy is added to an ion in small increments by means of a multiple collision experiment, the fragmentation can be completely different from that observed in single collision experiments. This is because the intermediate produced by stepwise activation can isomerize rapidly compared to the fragmentation process [36].

7.2.4 Alternative methods of ion dissociation

Cooks and his co-workers [37–39] have demonstrated that collisions with a metal surface provide an alternative means of inducing fragmentation of selected ions. The use of a surface as a collision target offers the possibility of introducing larger amounts of internal energy [37–39] into an ion and results in an energy distribution that is narrower than that observed in gas-phase collisional activation [40]. In a study of the fragmentation of isomeric furanocoumarins [41], surface-induced dissociation resulted in a level of fragmentation similar to that obtained with multiple-collision CID but much greater than that observed under single-collision conditions with the same collision energy.

In practice, surface-induced dissociation(SID) typically involves collisions of precursor ions on a metallic surface at an approximate 45° angle of incidence. The products of SID depend on the energy of the ion and on the nature of both the ion and the surface. In an investigation using polycyclic aromatic hydrocarbon molecular ions [42], collisions of ions with an energy between 50 and 100 eV resulted in the formation of ion–surface reaction products in competition with dissociation reactions. Above 100 eV the molecular ion was completely dissociated and ion–surface reaction products were not observed. If the collision energy was increased to several hundred electronvolts, sputtering of materials adsorbed on the collision surface was then found to dominate over dissociation of the impacting ion. A further comparison of the energetics of sputtering and dissociation processes has been provided by Ens and co-workers [43]. Recent dissociation experiments using glancing collisions with a surface in a four-sector instrument have demonstrated promising sensitivity for peptides with a molecular weight in excess of 2000 [44].

Photodissociation provides a further alternative to CID. The narrow, well-defined, energy distribution in photodissociation can be an advantage compared with the ill-defined energy distribution in collisional activation. A major disadvantage of photodissociation is the much smaller cross-section compared with that for CID. Laser photodissociation is much more effective with ions stored in FTMS [45] or ion trap [46] instruments than with fast ion beams [47], since the ion–photon interaction time can be much longer in trapping instruments.

Applications of photodissociation have utilized ultraviolet [45,48], visible [48,49], and infra-red [50] radiation. Of course, only those ions that absorb at the wavelength of the incident radiation can be photodissociated. The use of a high flux electron beam as an alternative means of inducing dissociation has been demonstrated more recently [51,52]. With this latter technique, there is no requirement for the ion of interest to contain an absorbing chromophore.

7.3 Instrumentation

7.3.1 Double-focusing magnetic sector instruments

In section 7.1 it was shown that metastable ions which fragment in the field free region between the ion source and the magnetic sector of a single-focusing instrument can be recorded during a normal magnet scan. While a double-focusing instrument of conventional geometry will allow metastable ions that fragment in the second field free region (Fig. 1.5) to be recorded in the same way, a number of solutions to the unambiguous assignment of masses to a fragmentation such as that illustrated in equation (7.2) are possible using double-focusing instruments.

Let the ion m_2 formed from ion m_1 subsequent to acceleration be recorded when the accelerating voltage, electrostatic sector voltage, and magnetic field have values V^*, E^*, and B^* respectively. Prior to this analysis let the instrument be tuned to a normal ion of mass m_0 using the corresponding values V_0, E_0 and B_0. Now the kinetic energy and momentum of the ion of mass m_0 are given by

$$\text{Kinetic energy} = \frac{1}{2}m_0 v_0^2 = eV_0 \tag{7.12}$$

$$\text{Momentum} = m_0 v_0 = (2m_0 eV_0)^{1/2} \tag{7.13}$$

respectively and for m_2 by respectively

$$\text{Kinetic energy} = \frac{1}{2}m_2 v_1^2 = \left(\frac{m_2}{m_1}\right) eV^* \tag{7.14}$$

$$\text{Momentum} = m_2 v_1 = \left(\frac{2m_2^2 eV^*}{m_1}\right)^{1/2} \tag{7.15}$$

If m_2 is formed from m_1 in the first field free region of an instrument with either conventional or reversed geometry then both the electrostatic and magnetic analysers must transmit m_2. In this case the following equations apply:

$$\frac{E^*}{E_0} = \left(\frac{m_2}{m_1}\right)\left(\frac{eV^*}{eV_0}\right) = \left(\frac{m_2}{m_1}\right)\left(\frac{V^*}{V_0}\right) \tag{7.16}$$

$$\frac{B^*}{B_0} = \frac{(2m_2^2 eV^*/m_1)^{1/2}}{(2m_0 eV_0)^{1/2}} = \left[\frac{m_2}{(m_1 m_0)^{1/2}}\right]\left(\frac{V^*}{V_0}\right)^{1/2} \tag{7.17}$$

If m_2 is formed from m_1 in the second field free region of a reversed geometry instrument then the electrostatic sector must again transmit m_2 but the magnetic sector must transmit the ion m_1. In this case equation (7.16) and the following equation apply:

$$\frac{B^*}{B_0} = \frac{(2m_1eV^*)^{1/2}}{(2m_0eV_0)^{1/2}} = \left(\frac{m_1}{m_0}\right)^{1/2}\left(\frac{V^*}{V_0}\right)^{1/2} \tag{7.18}$$

Similarly for an ion m_2 formed in the second field free region of a conventional geometry instrument, equation (7.17) and the following equation apply:

$$\frac{E^*}{E_0} = \frac{eV^*}{eV_0} = \frac{V^*}{V_0} \tag{7.19}$$

There are a number of different ways in which an analysis of this fragmentation can be effected, depending on the nature of the information required. Possible methods are:

(a) Tuning to a selected precursor ion (m_1) and recording all fragment ions (m_2) originating from m_1. This method provides a fingerprint spectrum for the ion m_1 in terms of its subsequent fragmentation ($m_0 = m_1$ = constant).
(b) Tuning to a selected fragment ion (m_2) and recording all precursor ions (m_1) corresponding to this ion. This method illustrates the mode of formation of the selected ion and may, for example, be used to detect the components of a mixture giving a fragment ion typical of a particular compound class ($m_0 = m_2$ = constant).
(c) Recording all fragmentations such that $(m_1 - m_2) = m_3$ = constant. This method may be used to detect the components of a mixture that fragment via a specific neutral loss characteristic of a particular compound class ($m_0 = km_3$ = constant).
(d) Monitoring a specific fragmentation ${m_1}^+ \rightarrow {m_2}^+$ characteristic of a particular compound or compound class. This method (selected reaction monitoring) is fully described in section 8.3.2.

The restrictions introduced by a choice of one of the scanning methods (a to c above) can be introduced into the appropriate equations selected from (7.16) to (7.19) together with a further restriction that one of the parameters V^*, E^*, or B^* remains constant (= V_0, E_0, or B_0 respectively). For example, if it is required to record all the fragment ions originating from a selected precursor ion in the first field free region and the accelerating voltage is to remain constant then equations (7.16) and (7.17) give the result $E^*/E_0 = B^*/B_0$. Thus, the scan has to maintain a constant B/E ratio.

The solutions to equations (7.16) to (7.19) listed in Table 7.1 sometimes describe a straightforward scan of one parameter (V, E, or B), but in many cases two

Table 7.1 Scan types required to record specific fragmentations. Scan types are defined so that a P, Q scan is a linked scan in which P and Q (both functions of the parameters V, E, and B) are varied so that the ratio of P to Q remains constant [193].

Fragmentations recorded	V = constant	B = constant	E = constant
(a) Fragmentations in 1FFR, either geometry			
Fragment ion spectrum (m_1 = constant)	B, E	$V^{1/2}, E$	B^2, V^{-1}
Precursor ion spectrum (m_2 = constant)	B^2, E	V	V
Constant neutral loss spectrum (m_3 = constant)	$B, [E^{-2} - (E_0E)^{-1}]^{-1/2}$	$V, [E + E^2(kE_0)^{-1}]$	$B^2, (V - V_0)^{-1}$
(b) Fragmentations in 2FFR, reversed geometry			
Fragment ion spectrum (m_1 = constant)	E	E	$V^{1/2}, B$
Precursor ion spectrum (m_2 = constant)	B^2, E^{-1}	V^2, E	V, B
Constant neutral loss spectrum (m_3 = constant)	$B^2, (E_0 - E)^{-1}$	$E, [V - V^2(kV_0)^{-1}]$	$B^2, (V^{-1} - V_0V^{-2})^{-1}$
(c) Fragmentations in 2FFR, forward geometry			
No *specific* scan types available. Both B and V, E scans record all fragmentations			

Table 7.2 Features of commonly used scan types

Scan type	Scan name	Resolution available	Energy release information	Field free region used	Other features	Literature reference
B^2, E	B^2/E linked scan	Fragment ion resolution ~ 500 Displayed (precursor ion) resolution < 100	Yes	1	Gives all ${m_1}^+$ from selected ${m_2}^+$. Alternative to accelerating voltage scan	57, 189
V	Accelerating voltage (V) scan	—	Yes	1	Gives all ${m_1}^+$ from selected ${m_2}^+$. Source conditions vary during scan therefore best restricted to narrow scans	190
B, E	B/E linked scan	Precursor ion resolution < 100 Displayed (fragment ion) resolution ≥ 500	No	1	Gives all ${m_2}^+$ from selected ${m_1}^+$	57, 189, 191
$B, [E^{-2} - (E_o E)^{-1}]^{-1/2}$	Constant neutral loss scan	Selected loss resolution < 100 Displayed resolution several hundred	No	1	Gives all ${m_2}^+$ from all ${m_1}^+$ by loss of constant ${m_3}^0$	141, 192, 193
E	MIKE scan	Precursor ion resolution 500–2500 Displayed (fragment ion) resolution < 100	Yes	2	Gives all ${m_2}^+$ from selected ${m_1}^+$. Sensitivity lower than with 1FFR scans. Reversed geometry only	5, 55
B	Normal magnetic scan	Displayed resolution low	Yes	2	Gives all ${m_1}^+$ from all ${m_2}^+$. Low sensitivity. Assignment of transitions ambiguous. Overlap with normal ions. Forward geometry only	54 (pp. 38–40)
B^2, E^{-1}	B^2E scan	Displayed resolution is better than that from B^2/E linked scan	No	2	Gives all ${m_1}^+$ from selected ${m_2}^+$. Reversed geometry only	64

of these parameters are scanned in a fixed relationship. These scans are known as linked scans (section 7.3.1.2). Commonly used scan types with some of their characteristics are listed in Table 7.2.

From a practical viewpoint, scans that involve a change of accelerating voltage are less satisfactory since the instrumental sensitivity changes as the accelerating voltage changes. These changes lead to a distortion of relative intensities during scanning and to an inability to use the technique at all over large mass ranges which require a large change in accelerating voltage. Probably the most useful application of accelerating voltage scans has been in the accurate measurement of energy release during a metastable transition [53]. The width of the metastable peaks may be measured very accurately and the energy release (T electronvolts) calculated from equation 7.20 where ΔV is the peak width in volts [54] (cf. equation 7.21).

$$T = \left(\frac{m_2^2 V}{16 m_1 m_3}\right)\left(\frac{\Delta V}{V}\right)^2 \tag{7.20}$$

7.3.1.1 MIKE scanning

Reversed geometry double-focusing sector instruments (Fig. 1.6) offer what is, at least conceptually, a very straightforward method of unambiguously assigning mass values to the dissociation recorded in equation (7.2). The initial magnetic sector is set to pass only precursor ions of mass m_1. Those precursor ions that fragment in the second field free region, whether by a unimolecular or by a collision-induced process, will produce fragment ions of reduced kinetic energy which can be recorded by decoupling the electrostatic analyser voltage from the accelerating voltage and then scanning it downwards from its normal value.

This scan of the electrostatic sector voltage while maintaining a constant magnetic field and a constant accelerating voltage is referred to as a MIKE (mass-analysed ion kinetic energy) scan [55] (Table 7.2). There is no change in source sensitivity during a MIKE scan and the scan mass range is unlimited since the electrostatic sector voltage may be scanned to zero if necessary. A typical MIKE scan of the unimolecular fragmentations of the molecular ion of 24-methylcholesterol generated under EI conditions is shown in Fig. 7.4.

During any fragmentation process, energy released is converted into translational energy so that the fragment ion may be given an additional velocity component in any direction, opposing or supplementing the original velocity of the ion. The resulting range of ion energies then requires a range of electrostatic analyser voltages for transmission so that a MIKE spectrum is a low resolution spectrum. The width of the MIKE peak (ΔV volts) is related to the energy release for singly charged ions (T electronvolts) by the following equation (cf. equation 7.20):

$$T = \left(\frac{m_1^2 eV}{16 m_2 m_3}\right)\left(\frac{\Delta E}{E}\right)^2 \tag{7.21}$$

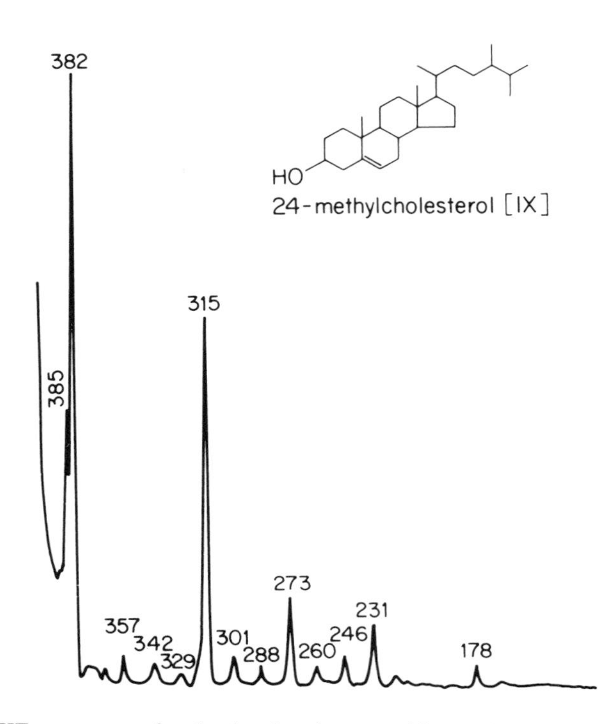

Fig. 7.4 MIKE spectrum of unimolecular decompositions of the molecular ion of 24-methylcholesterol ($m_1 = 400$, $m_2 = m_1(E^*/E_0)$). (Reproduced from ref. 146)

and can be regarded as being strongly characteristic of the structure of the product ion m_2. While this characteristic is certainly of value in studies of ion structure and of the energetics of ion formation, its analytical application has been limited to one or two examples of isomer differentiation (section 7.4.3).

The resolution available in selecting the precursor ion m_1 from the original spectrum is another practical consideration. A precursor ion resolution of approximately 500 at full sensitivity may be expected in MIKE scans with a higher resolution available at reduced sensitivity. For example, Stahl and Tabet [56] have demonstrated the MIKE analysis of the isobaric ions C_5H_9(69.0704) and C_4H_5O(69.0340) which requires a resolving power of 1900, with no contribution from the MIKE spectrum of one ion to that of the other.

7.3.1.2 Linked scanning

Linked scans in which the magnetic field strength B and the electrostatic sector voltage E are maintained in a fixed relationship throughout the scan can be used on both conventional and reversed geometry instruments and are applicable to the analysis of specific fragment or precursor ions and of specific neutral losses (section 7.4). Unlike scans involving a change of accelerating voltage, no refocusing of the ion beam is required to maintain sensitivity during the scan. Most linked scans record high intensity ions formed in the first field free region over a wide mass range and are compatible with the rapid scan speed needed for GC-MS operation [57–60]. First field free region dissociations are also used extensively in selected reaction monitoring experiments (section 8.3.2).

In practice there are two methods of effecting linked scans. First, direct measurement of the magnetic field using a Hall-effect probe, whose output is then used to control the scanning of the electric sector in the required manner during a normal magnet scan [61,62]. Second, provided that magnet scans are reproducible, a correlation between mass and the time elapsed from the start of the scan may be used as the basis of a table containing an entry for each time in the magnet scan at which the electrostatic sector voltage should be altered by one digital step [57]. Alternatively, the table can be based on a correlation between mass and the magnet reference voltage [63]. The accelerating voltage and the electrostatic sector voltage which are normally maintained in a fixed ratio in a double-focusing instrument are decoupled during linked scan operation.

An appreciation of the resolution available in the various forms of linked scan may be obtained from Fig. 7.5. In this figure the peak representing the dissociation $226^+ \rightarrow 184^+ + 42$ is broadened due to internal energy release since the resultant range of ion energies requires a range of B and E values for transmission to the detector. Change in momentum is proportional to the change in velocity so that the magnetic field must change from B to $B(1 + \Delta v/v)$. Energy is proportional to velocity squared so that the electrostatic field must change from E to $E(1 + \Delta v/v)^2$.

These precise changes are provided by a linked scan at constant B^2/E so that peak broadening is reproduced without discrimination along the B^2/E curve connecting m/z 184 to the origin. Thus, the displayed (precursor ion) resolution in a B^2/E scan is low like that in a MIKE scan, and energy release data may be determined from a B^2/E scan just as from accelerating voltage and MIKE scans. Figure 7.5 also shows that adjacent B^2/E scans barely intersect the peak corresponding to the transition $226^+ \rightarrow 184^+$. Thus, the resolution of selected fragment ions is relatively high but ion intensity can also be markedly attenuated or even lost if the values of B and E are not accurately correlated during B^2/E scanning. An alternative precursor ion scan, which displays improved resolution but no energy release information, is the linked scan at constant B^2E [64]. This scan is, however, only applicable to fragmentations that take place in the second field free region of a reversed geometry instrument.

Inspection of Fig. 7.5 also demonstrates that linked B/E and constant neutral loss scans discriminate against those ions that would contribute to the sides of a B^2/E peak so that the displayed resolution in these scans is considerably higher (e.g. Fig. 7.6). Porter *et al.* [65] have shown that the resolution displayed in B/E scans depends on the magnitude of the kinetic energy release accompanying fragmentation and

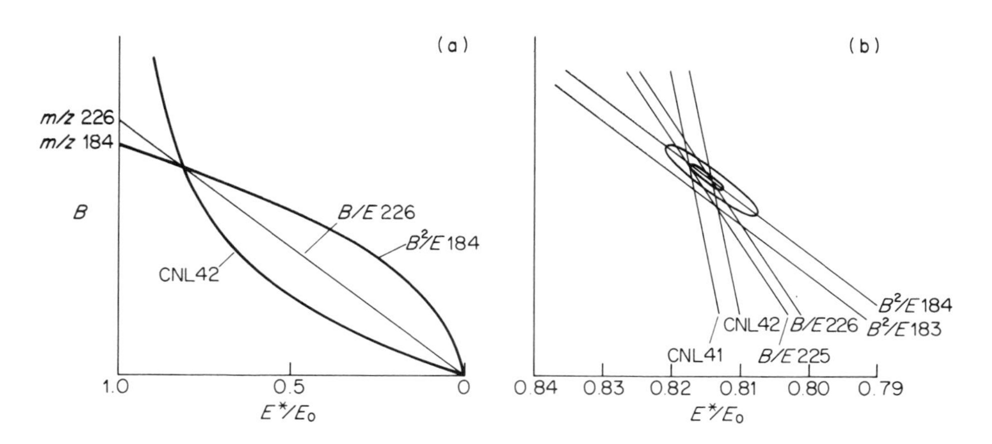

Fig. 7.5 (a) Scan lines for constant neutral loss, B/E, and B^2/E scans, and (b) intersection of these and other scan lines with the transition $226^+ \rightarrow 184^+ + 42$. Ion intensity is represented by contour mapping

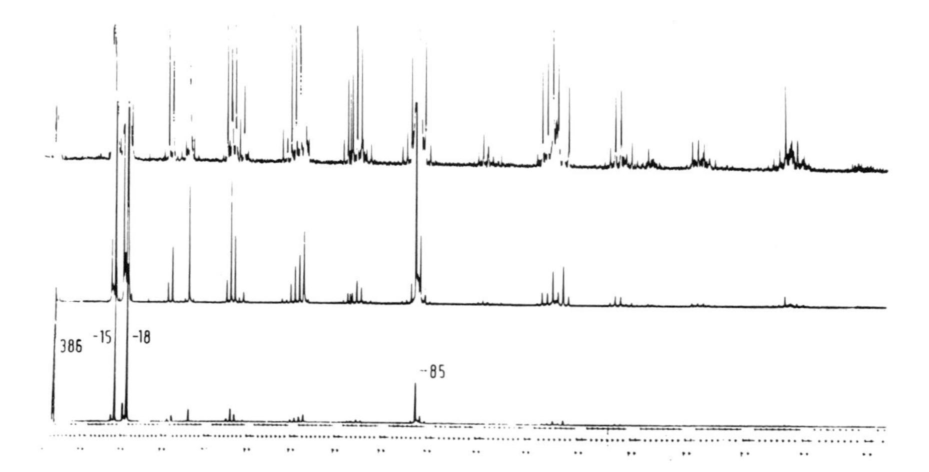

Fig. 7.6 B/E scan fom m/z 386 in cholesterol (EI spectrum). (Courtesy Kratos Analytical Instruments)

have itemized instrumental conditions for maximizing this resolution. Figure 7.5 also shows that adjacent B/E and constant neutral loss scans intersect the transition $226^+ \rightarrow 184^+ \rightarrow 42^+$ so that the resolution of selected precursor ions or neutral losses is much lower and may lead to ambiguities in interpretation (section 7.3.1.3).

7.3.1.3 Ambiguities and errors in MIKE and linked scans

Ambiguities in the interpretation of linked scan data from double-focusing sector instruments of either geometry can arise because of the limited precursor ion resolution in B/E and constant neutral loss scans. Thus, a B/E scan from a precursor ion at m/z 500 may also record transitions from ions at adjacent masses (m/z 501 and 499). Scans where the displayed resolution is low, e.g. B^2/E and MIKE, can present problems in the assignment of mass values, although this is less of a problem when the spectrum is merely used as a characteristic fingerprint for compound identification.

A more serious source of error in mass assignment in MIKE scans is the energy shift observed for fragment ions from high mass precursors (section 7.2.1). The size of this shift depends upon the experimental method employed [66] as well as parameters such as the mass of the fragment ion [67] and the choice of collision gas [31]. The same energy loss shift can also result in a sensitivity loss in linked scans [68].

Ambiguities may also arise during the study of fragmentations occurring in a particular field free region since fragmentations elsewhere in the instrument may provide ions with the correct momentum and energy to be transmitted to the detector. For example, several scanning modes for instruments of reversed geometry can transmit fragment ions formed in both the first field free region and the second field free region [69,70]. Thus, while fragmentations within the second field free region of an instrument of conventional geometry do not interfere with the analysis of fragmentations within the first field free region, this is not the case for an instrument of reversed geometry [71].

Similarly, a reversed geometry instrument tuned to study the fragmentations of ion m_1^+ (equation 7.2) in the second field free region by MIKE scanning may also record fragmentations in the first field free region corresponding to the transition

$$m_4^+ \longrightarrow m_5^+ + m_6^0 \tag{7.22}$$

if $m_5^2/m_4 = m_1$ (for transmission by the magnet) and $m_5/m_4 = m_2/m_1$ (for transmission by the electrostatic sector) [72]. Fragmentation within the fields along the flight path can cause additional interference with instruments of either geometry when low intensity ions are being studied [69,73]. The most satisfactory method of eliminating interferences is to use one of the specially designed instrument geometries described in the following sections.

7.3.2 Triple-quadrupole instruments

A number of instruments specifically designed for the study of ion fragmentations are now available. The basic design of these instruments consists of two independent mass analysers arranged in tandem and separated by a collision cell

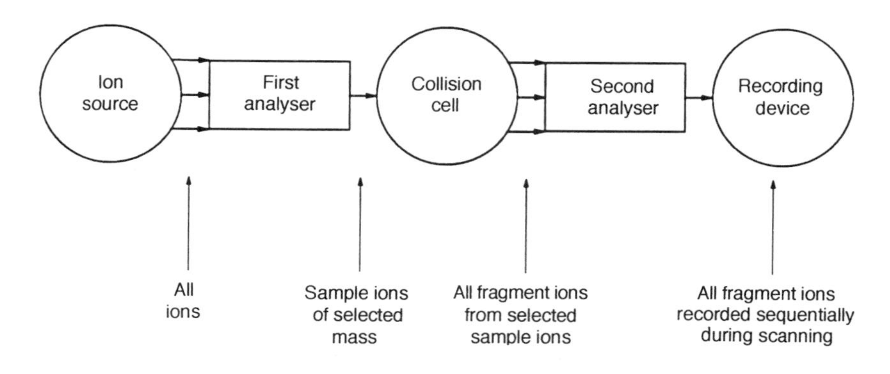

Fig. 7.7 A basic tandem mass spectrometry (MS–MS) experiment

(Fig. 7.7). These instruments are known as tandem mass spectrometers or MS–MS instruments. In general, tandem mass spectrometers can offer advantages in resolution and sensitivity for the study of ion fragmentations as well as freedom from artifact peaks compared with the use of double focusing magnetic sector instruments.

The triple quadrupole (TQ) [6] is a particularly effective instrument of this type. The instrument (Fig. 7.8) comprises a standard quadrupole mass filter (Q_1), an r.f.-only quadrupole collision cell (Q_2), and a second quadrupole mass filter (Q_3). As a result, independent mass analysers are available that can select precursor and fragment ions for a fragmentation taking place in the collision cell.

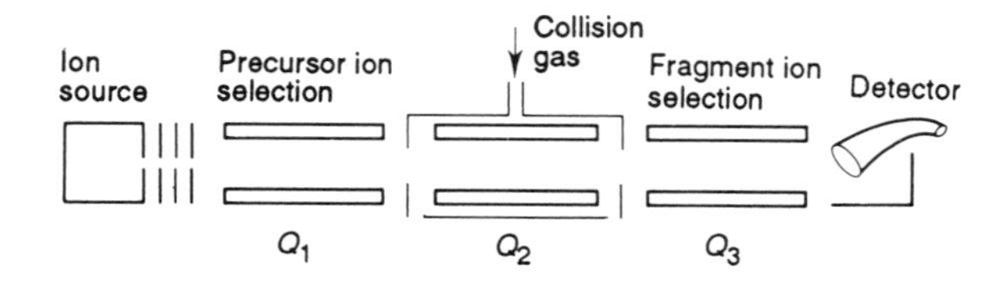

Fig. 7.8 Schematic diagram of a triple quadrupole analyser system

The mass filtering property of a quadrupole analyser is independent of ion velocity so that the peak broadening on fragmentation seen with sector instruments is absent when a quadrupole mass filter is used. Thus, a triple quadrupole is able to provide unit mass resolution for both precursor and fragment ions and the recorded spectra are free from artefacts.

An important advantage of the triple quadrupole is its ease of operation with the use of simple scanning procedures for each type of MS–MS experiment [74,75]. In addition, the whole instrument is very amenable to data system control. The simplicity of the scan laws arises mainly from the fact that a quadrupole mass filter recognizes only the mass-to-charge ratio of an ion. Thus, unlike a magnetic sector, a quadrupole analyses precursor and fragment ions, with the same mass-to-charge ratio, under identical conditions.

The operating modes available for triple-quadrupole instruments are equivalent to those that have been described for double-focusing sector instruments (section 7.3.1):

(a) A scan of the last quadrupole while the first quadrupole is set to transmit a selected mass produces a spectrum of all fragment ions from the selected precursor ion (e.g. Fig. 7.21).
(b) The spectrum of all precursor ions that fragment to produce a given fragment ion is recorded by scanning the first quadrupole with the last quadrupole set to transmit the selected fragment ion (e.g. Fig. 7.9).

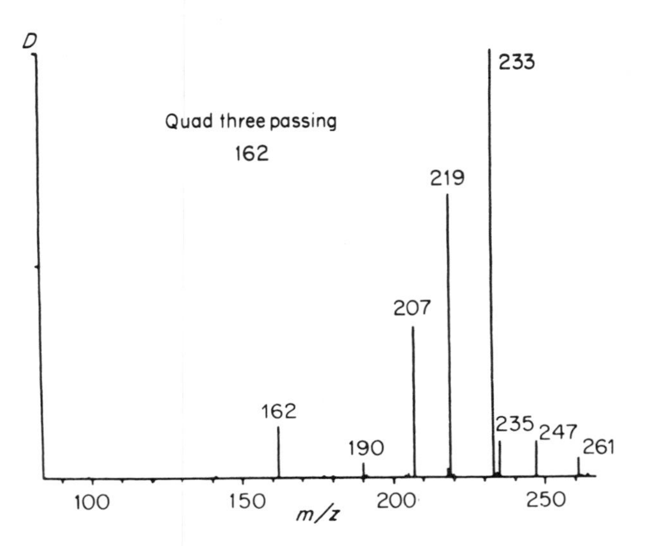

Fig. 7.9 Precursor ion spectrum of the abundant *m/z* 162 fragment ion from primidone (MW 218) and its metabolites, phenylethylmalonamide (MW 206) and phenobarbital (MW 232) present in plasma. Methane chemical ionization conditions. (Reprinted with permission from R.J. Perchalski *et al.*, *Anal. Chem.*, **54**, 1466[79]. Copyright (1982) American Chemical Society)

(c) The recording of a constant neutral loss spectrum is simply achieved by scanning both mass filters with a fixed difference in mass (e.g. Fig. 7.10).
(d) Selected fragmentations may be monitored by setting both the first and last quadrupoles to the appropriate mass values.

In low collision energy experiments, an r.f.-only quadrupole is able to refocus most of the ions that have been scattered by the collision process. This feature

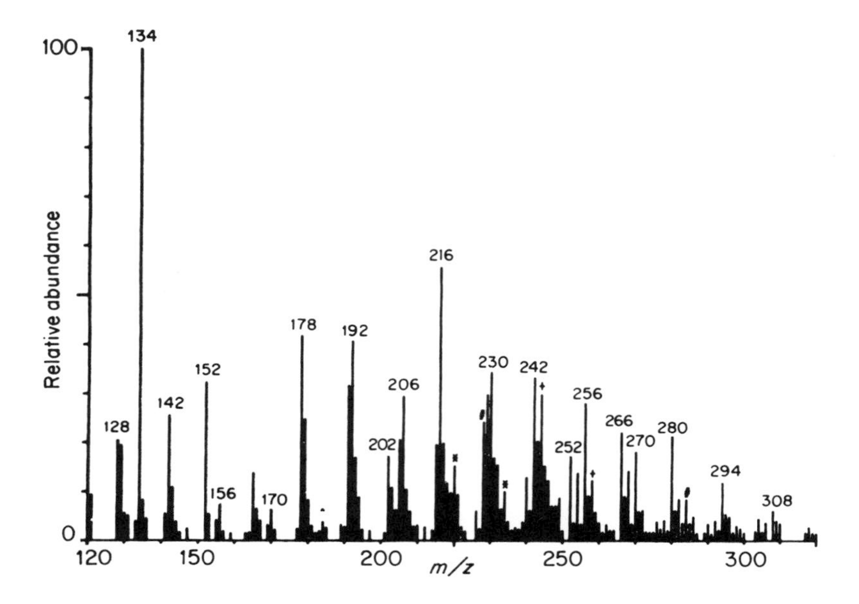

Fig. 7.10 Constant neutral loss spectrum produced by loss of SH (33 mass units) from $(M+H)^+$ ions of organosulphur components of South Swan Hills crude oil. (Reprinted with permission from D.F. Hunt and J. Shabanowitz, *Anal. Chem.*, **54**, 574[198]. Copyright (1982) American Chemical Society)

permits the use of more massive gases, which increase the centre-of-mass collision energy (equation 7.10), at pressures suitable for multiple collisions, conditions that might otherwise result in substantial scattering losses. In this r.f.-only mode, the collision quadrupole transmits all ions with a mass greater than a low mass cut-off determined by the amplitude of the r.f. voltage. In practice, the r.f.-only quadrupole is not a perfect collision cell. Ion transmission depends on a number of parameters [76] and it is difficult to optimize the transmission of precursor and fragment ions simultaneously [77].

Sensitivity in MS–MS experiments, at least for lower mass (<600 Da) ions, is high in TQ instruments. This is a result of the efficient collection of fragment ions generated in the r.f.-only collision cell and the relatively long residence time, of the order of some tens of microseconds, in the collision cell for low energy experiments [78] (section 7.2). This sensitivity and the ease of control using a data system make the triple quadrupole well suited to fast scanning experiments and to more complex analyses in which parameters may be changed during the course of the experiment [79,80].

These advantages have to be set against a more limited performance at higher mass (but see section 6.3.2.2) and less flexibility in the type of data that can be obtained, e.g. high energy CID experiments are not practical. The triple-quadrupole system also cannot perform MS–MS–MS experiments in which, for example, a fragment ion from one collision-induced dissociation is dissociated a second time to provide second-generation fragment ions.

A general problem encountered with low collision energy MS–MS is a lack of spectrum reproducibility [81]. The relative abundance of fragment ions in a low collision energy MS–MS spectrum depends on the internal energy transferred to the precursor ion so that parameters such as collision energy and target gas pressure and composition must be carefully controlled [82]. A further contribution to non-reproducibility is the fact that ion transmission through instruments that incorporate an r.f.-only collision quadrupole depends in a complex way on a number of different parameters [77,83].

7.3.3 Three- and four-sector instruments

A natural development of magnetic sector MS–MS instrumentation was the addition of a further sector to an existing double-focusing instrument. The first of these instruments was of *EBE* geometry (Fig. 7.11), obtained by the addition of an electrostatic sector to an instrument of conventional geometry [3,84]. With this configuration, a precursor ion may be selected at high resolution using the first two sectors and fragmentation products formed in reaction region 3 (Fig. 7.11), recorded by scanning the second electrostatic analyser as in a MIKE scan. Linked scans of an appropriate pair of sectors may be used to record precursor ion or neutral loss information for a fragmentation which occurs in any one of the available reaction regions. An instrument of *BEE* geometry, which has many similarities to the *EBE* geometry, has also been described [85].

In an instrument of *BEB* geometry [86,87] (Fig. 7.11), the first two sectors may be used to select a precursor ion, at high resolution if necessary, just as with an instrument of *EBE* or *BEE* geometry. In this case, however, scanning the second magnet provides a fragment ion spectrum of improved resolution. A further advantage of the *BEB* geometry lies in its ability to provide good resolution of fragment ions using a *B*/*E* linked scan with the second and third sectors following precursor ion selection with the first magnet.

The provision of reaction regions between successive analysers in multisector instruments allows the observation of sequential fragmentation reactions. Thus, a given fragment ion from a collision-induced process in the first collision region of an *EBE* instrument (Fig. 7.11) may be selected using the first electrostatic analyser and magnet and then further dissociated in the third region. The products of this second dissociation are analysed using the second electrostatic analyser [88].

The next stage in the evolution of tandem sector instruments was the construction of four-sector instruments of *EBEB* (Fig. 7.12) and *BEEB* geometry [89,98]. In a four-sector instrument, the first two sectors are typically operated as a conventional double-focusing mass spectrometer to select a suitable precursor ion for collision induced dissociation. The fragment ions from this dissociation are then analysed

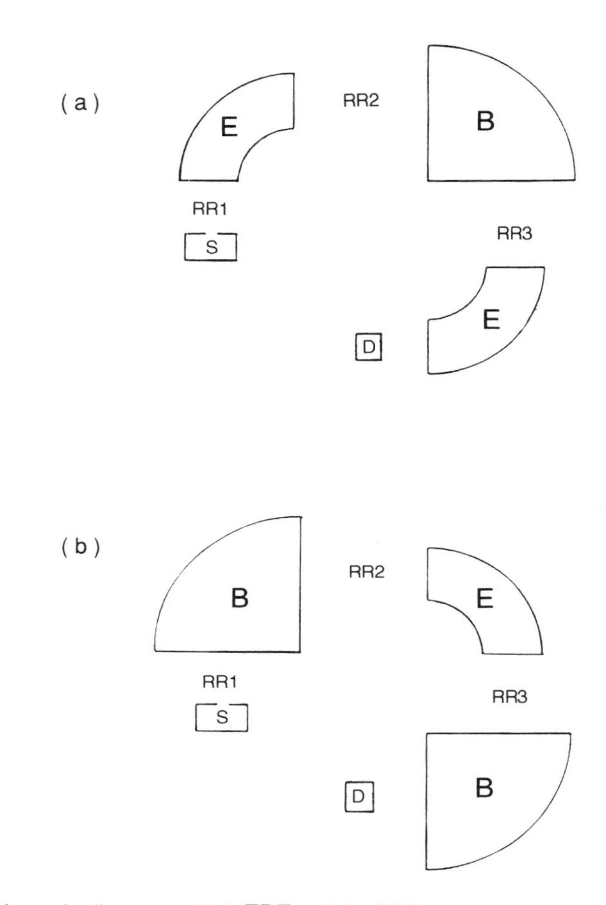

Fig. 7.11 Schematic diagrams of (a) *EBE* and (b) *BEB* geometry instruments showing ion source (S), detector (D), and reaction regions (RR). (Reproduced with permission from ref. 139)

using a *B*/*E* linked scan of the magnet and electrostatic sector of the second double-focusing mass spectrometer.

In contrast to experiments with a triple-quadrupole instrument, MS–MS experiments with a three- and four-sector instruments generally involve kiloelectronvolt energy precursor ions that experience one or a very few collisions. Only fragmentations which can take place in microseconds or less will be recorded. The collision cell which separates the two double-focusing mass spectrometers in a four-sector instrument can be held at any voltage between earth and ion source potential. This ability to float the collision cell has a number of advantages on sector instruments in general, when they are used for MS–MS experiments [68]. For

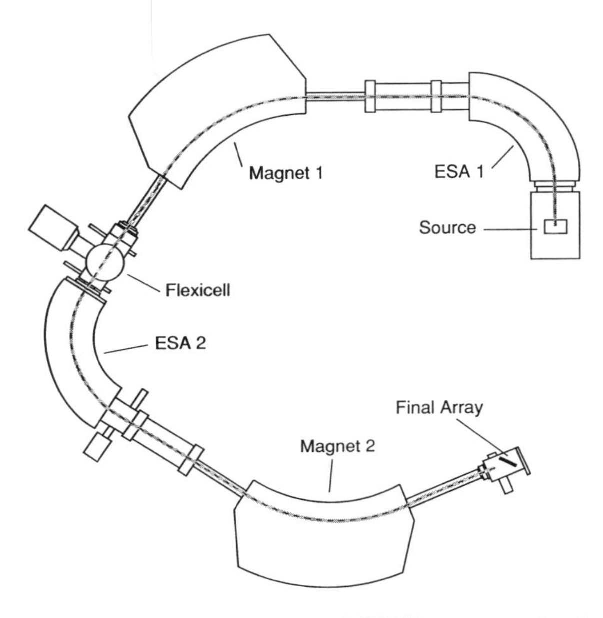

Fig. 7.12 Schematic diagram of an instrument of *EBEB* geometry showing intermediate collision region ('Flexicell') and final array detector. (Courtesy Kratos Analytical Instruments)

example, fragment ions formed by collision-induced dissociation can be separated from those formed by unimolecular dissociation, since the latter are produced in the surrounding field free region at a different potential from ions produced within the collision cell [90,91].

A floated collision cell also improves fragment ion resolution since peak broadening due to energy release (section 7.3.1.1) on fragmentation is reduced as the kinetic energy of ions entering the collision cell is reduced [87]. Similarly, the mass shift due to energy loss on fragmentation ('Derrick shift'—section 7.2.1) is also reduced when a floated collision cell is used [68]. A further beneficial effect of floating the collision cell away from ground potential is to improve the detection of low mass fragment ions [35,68]. Naturally enough, as the collision cell is floated to a higher potential, the collision energy decreases, with attendant qualitative changes in the fragmentation spectrum.

Four-sector instruments provide the highest performance in tandem mass spectrometry, including a capability for very high resolution precursor ion selection and high resolution recording of fragment ion spectra [89]. Again, MS–MS–MS experiments are readily performed with an instrument of four-sector geometry [92]. The same instrument operated as a single four-sector mass spectrometer can also offer a useful increase in resolution without a significant decrease in sensitivity [93]. Four-sector tandem MS instruments have, so far, provided the most practical

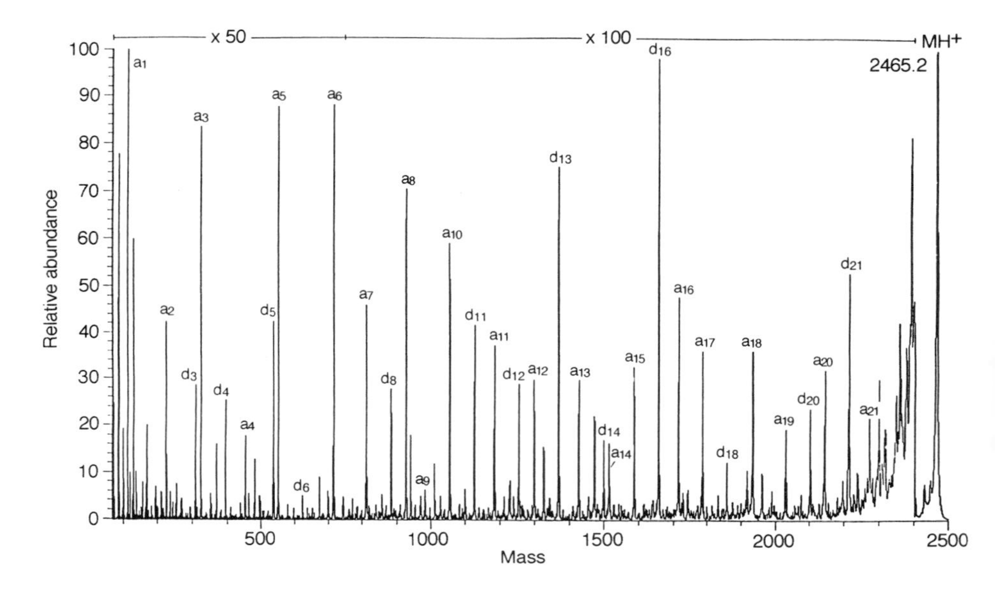

Fig. 7.13 Fragment ion spectrum from the peptide ACTH (amino acids 18–39). Series of prominent main chain (a_n) and side chain (d_n) fragment ions are seen. Fragment ion nomenclature is that of ref. 196. (Courtesy Kratos Analytical Instruments)

approach to the study of the collision-induced fragmentation of higher molecular weight peptides, i.e. much above MW 1000 [94]. The high energy CID spectra recorded with four-sector instruments have been shown to offer good reproducibility from day to day and between instruments [95]. In this respect, performance is superior to that offered by a triple-quadrupole instrument [35].

Further improvements in four-sector performance can be effected by the use of an array detector (section 1.4.4) located on a focal plane at the end of the instrument (Fig. 7.12). Although such detectors simultaneously record only a limited mass range, a complete tandem mass spectrum is available by taking a number of overlapping exposures with the array as B_2 and E_2 are stepped to new values. The low level of chemical noise in MS–MS spectra means that the benefits of array detectors in such experiments have so far been realized as a corresponding increase in useful signal intensity (Fig. 7.13), although such spectra are not completely free from interferences [96]. Instrumental developments that further increase the mass range of array detectors have been reported more recently [97,98].

7.3.4 Hybrid instruments

MS–MS instruments of hybrid geometry are based on the addition of an r.f.-quadrupole collision cell and a quadrupole mass filter to a double-focusing instrument of conventional or reversed geometry. Instruments of hybrid geometry offer high precursor ion resolution by virtue of the double focusing sectors together with unit mass resolution of fragment ions in the analytical quadrupole. The use of two independent mass analysers which both have good resolution also reduces the problem of artefact peaks. Hybrid instruments are, in addition, less complex and less expensive than four-sector instruments.

A particular feature of hybrid instruments is the fact that ions may be caused to undergo collisions at low or high (kiloelectronvolt) energies. Instruments of hybrid geometry may also be used to record ions formed by sequential fragmentations. Details of both of these facilities are provided later in this section.

The two major hybrid geometries, *EBqQ* [99] and *BEqQ* [100], are shown schematically in Fig. 7.14 with the reaction regions that can be used for MS–MS experiments. A hybrid instrument in which the quadrupole analyser precedes the double-focusing sectors has also been described [101]. Taking the *EBqQ* instrument as an example, the ion beam, which has an initial energy up to 8 keV, is decelerated before it enters the collision quadrupole (q). This quadrupole is maintained at a d.c. voltage which is variable from 8 kV down to 7.5 kV for 8 keV operation to give a collision energy within the range 0–500 eV. The final analytical quadrupole (Q) is also floated at a similar d.c. level.

A precursor ion is conventionally selected with the sector portion of the instrument and the mass selected ion beam decelerated into the collision cell. A scan of the analytical quadrupole can then be used to separate the fragment ions formed. Many such experiments with hybrid instruments have used collision energies that are more similar to those used in triple-quadrupole instruments, i.e. some tens of electronvolts of collision energy [102]. For precursor ion scans, the magnetic sector is scanned while the analytical quadrupole is set to pass the desired fragment ion with deceleration and CID as before. A linked scan of the magnet and analytical quadrupole such that B^2/Q = constant, where Q represents the r.f. voltage applied to the analytical quadrupole, can be used to record neutral loss spectra.

With an *EBqQ* instrument, high energy CID takes place in the first reaction region (Fig. 7.14) and the spectrum is recorded at a separate detector, positioned after the magnet, by linked scanning of the sectors. A *BEqQ* instrument can be operated in an equivalent mode, but in addition offers the possibility of high energy CID between the magnetic and electric sectors. This is, in principle, a MIKE scan experiment, but by also scanning the analytical quadrupole together with the electrostatic sector so as to maintain constant E/Q, unit mass resolution of the fragment ions is achieved [103] (Fig. 7.15). The yield of fragment ions in E/Q scans is, however, extremely low [103,104]. Sequential fragmentations may readily be studied on a hybrid instrument by following the high energy fragmentation, carried out in either the first field free region [36] or between the magnetic and electric sectors in a *BEqQ* instrument [105], with a low energy fragmentation in the collision quadrupole.

An alternative approach to higher energy collisions with a hybrid instrument is to make use of an increased collision energy in the r.f.-only quadrupole, using either

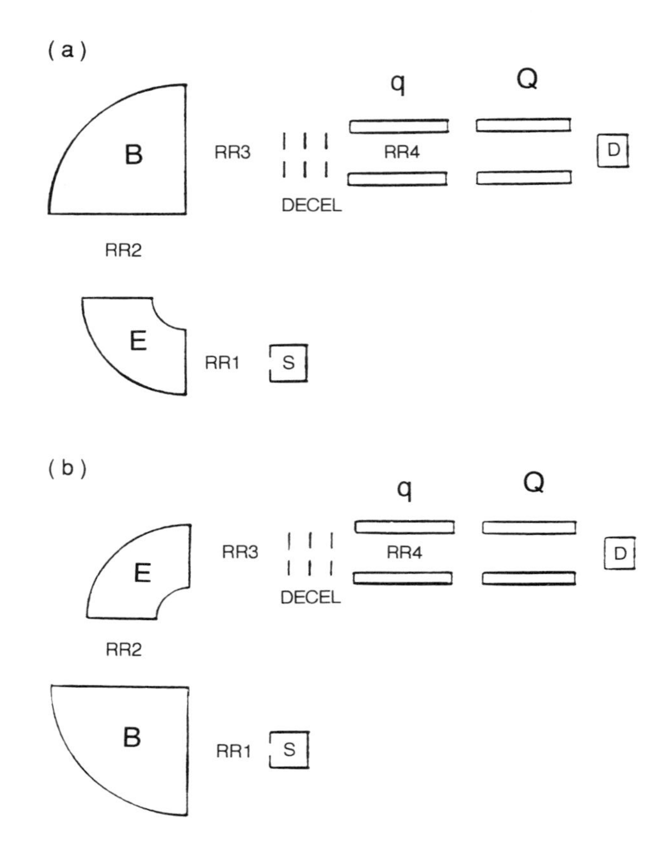

Fig. 7.14 Schematic diagrams of (a) *EBqQ* and (b) *BEqQ* geometry instruments showing ion source (S), detector (D), collision quadrupole (q), analytical quadrupole (Q), and reaction regions (RR). (Reproduced with permission from ref. 139)

EBqQ or *BEqQ* geometries. There are a number of application areas where this may be desirable. For example, while the overall efficiency of the CID process with precursor ions of relatively low mass, e.g. less than 500–600 Da, can be extremely high with a hybrid instrument, this level of performance is not seen with higher mass precursor ions [26,102,106].

Early attempts to overcome this problem were simply based on the use of a higher value of $E_{LAB.}$ For example, an E_{CM} value of 10 eV, used in a four-sector instrument (calculated from E_{LAB}= 5 kV, m = 2000, and m_g = 4(helium) in equation 7.10) would require a hybrid instrument to use an $E_{LAB.}$ value of 500 eV (calculated from E_{CM} = 10 eV, m = 2000, and m_g = 40(argon) in equation 7.10) to achieve the same collision energy. These first efforts were, however, largely unsuccessful [107]. Davis and Wright [77] have since shown that energetic (>100 eV) high mass

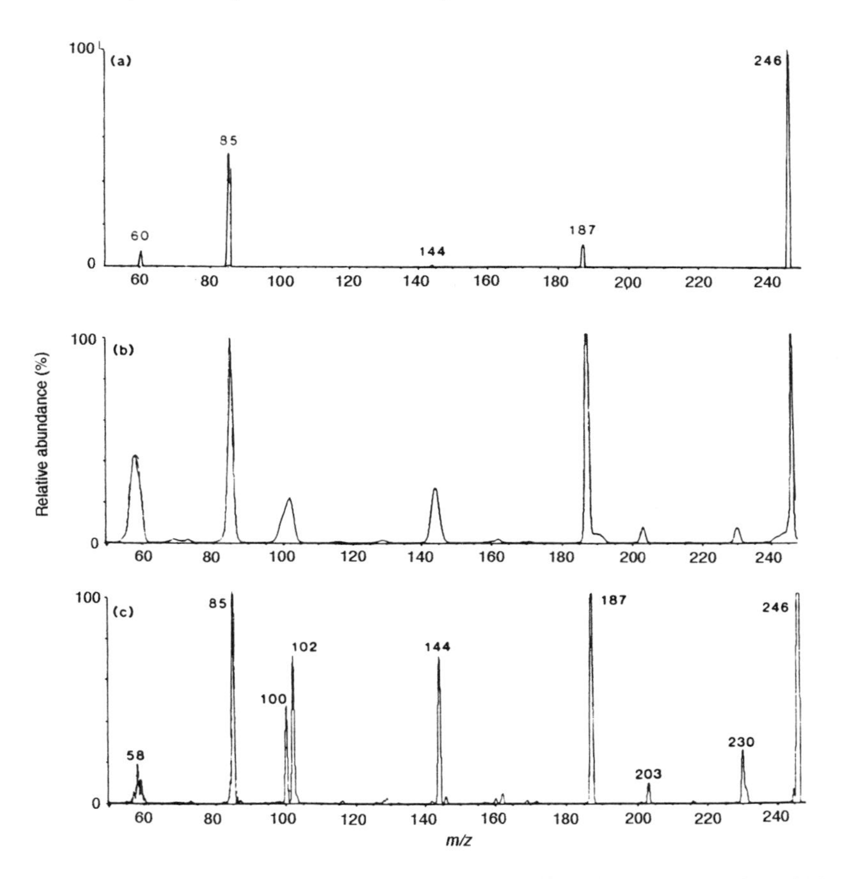

Fig. 7.15 Fragment ion spectra obtained on a *BEqQ* instrument by collision-induced dissociation of isovaleryl carnitine $(M+H)^+$ ions : (a) low energy (20 eV) activation in *q* : spectrum obtained by scanning *Q*; (b) high energy (8 keV) activation in the region between *B* and *E* ; spectrum obtained by scanning *E*; (c) activation as in (b), spectrum obtained by linked scanning of *E* and *Q*. (Reproduced with permission from ref. 104)

precursor ions need higher amplitude r.f. voltages to achieve focusing in an r.f. collision cell. Improved utility for higher molecular weight (>1000 Da) precursor ions has been demonstrated by this means [102,108].

7.3.5 Time-of-flight instruments

In a linear time-of-flight instrument, metastable ions are not observed separately. This is because an ion that results from a fragmentation in the flight tube following acceleration continues with approximately the same velocity as the precursor ion (equation 7.5) and therefore arrives at the detector at approximately the same time as an unfragmented precursor ion.

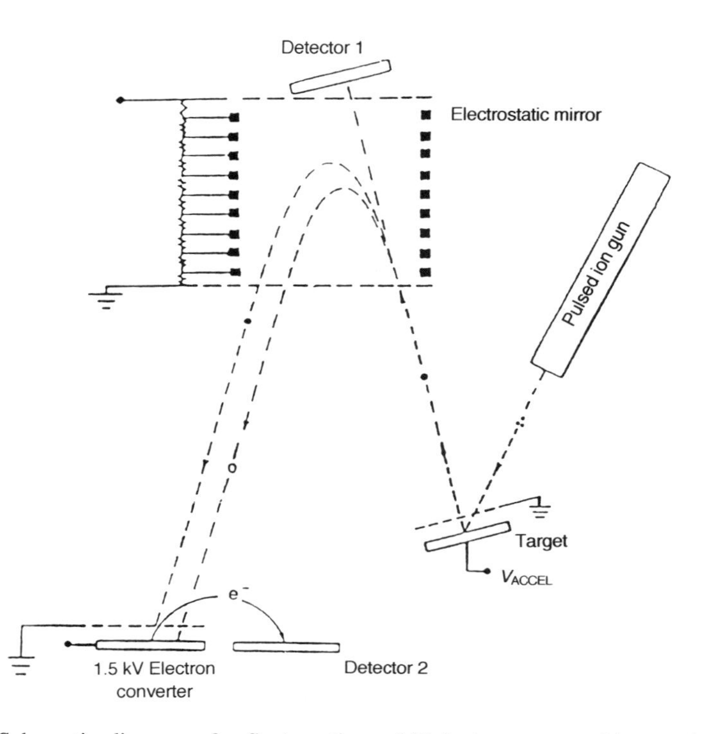

Fig. 7.16 Schematic diagram of reflectron time-of-flight instrument with two detectors. (Adapted from ref. 194)

In a reflectron instrument, if an ion fragments after acceleration and before the entrance to the electrostatic mirror, the neutral product is not reflected, but continues with approximately the same velocity as the precursor. The time of arrival of the neutral fragment at a detector behind the mirror therefore identifies the precursor ion [109,110], while the charged fragment from the decomposition is reflected by the mirror into a second detector (Fig. 7.16). Measurement of the arrival times of fragment ions correlated with a particular precursor ion then defines the routes for unimolecular fragmentation of this ion [111]. A drawback of MS–MS experiments with this type of instrument is that a very low ion flux must be used to make unambiguous coincidence measurements. The use of a reflectron instrument has also been proposed for laser photodissociation experiments with high mass ions [112].

A time-of-flight analyser can be advantageous as the second stage of an MS–MS instrument since its integrating capability provides a useful increase of sensitivity for these experiments. MS–MS instruments in which a time-of-flight analyser follows a quadrupole [113,114] or a double-focusing magnetic sector [47] instrument have been described. Tandem TOF–TOF instruments have been used for the study of surface-induced [115] and laser-induced [116,117] dissociations.

7.3.6 Ion traps

MS–MS experiments with an ion trap (ITMS) or with a Fourier transform (FTMS) instrument (section 7.3.7) involve the manipulation of an ion population that is trapped in a single physical location. In these instruments, all the steps of an MS–MS experiment are separated in time but not in space. This is in contrast to the instruments discussed so far in this chapter, where the various steps are separated in space but occur on a time scale that is determined by instrumental characteristics.

The sequence of events for an MS–MS experiment which records fragment ions from a selected precursor ion, using an ITMS, is as follows. Ions are formed within the trap or injected from an external source during period A (Fig. 7.17). Subsequently, ions other than the required precursor ion are ejected from the trap during period B and a supplementary alternating voltage of appropriate frequency is applied to the end cap electrodes of the trap to excite the precursor ion during period C. This supplementary voltage, referred to as a 'tickle' voltage [118], serves to increase the motion of the precursor ions so that they undergo energetic collisions with neutral gas molecules, although recent observations have thrown some doubt on the detailed explanation of the mechanism [119]. A final scan of the r.f. voltage during period D then causes the fragment ions formed to leave the trap, mass by mass, whereupon they are recorded at the detector and a fragment ion spectrum is the result [120].

Isolation of a selected precursor ion has been accomplished by means of a d.c. voltage applied to the ring electrode in conjunction with a ramp of the r.f. voltage

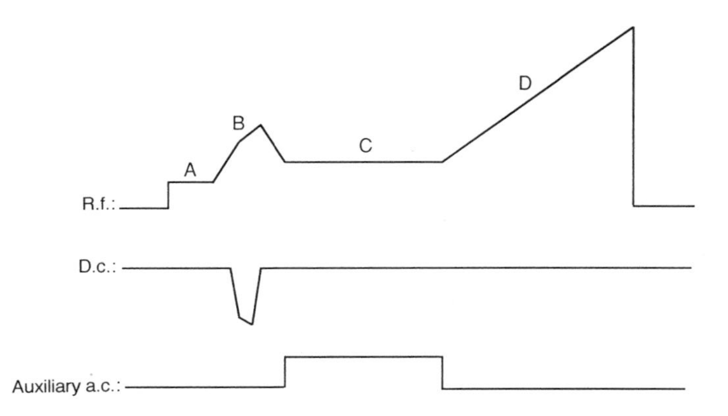

Fig. 7.17 Fragment ion scan using an ion trap. The ion formation period (A) is followed by precursor ion selection (B), collision-induced dissociation (C), and a final mass scan (D). (Adapted from ref. 195)

[121]. By this means it is possible to define a_z and q_z values (section 1.3.5) so that only a very few m/z values, or even a single m/z value, are stable within the trap at the time. More recently, precursor ion isolation, which offers higher resolution [122], has been accomplished by a combination of forward and reverse r.f. voltage scans while applying an additional alternating voltage to the end-cap electrodes for resonant ejection of ions at the frequency of the alternating voltage [122,123].

An increase in the amplitude of the 'tickle' voltage used in the ion trap leads to more energetic collisions with the buffer gas [36,124]. The reaction time between ion excitation and ejection of the fragment ions is also under experimental control and may be extended for periods as long as several hundred milliseconds [123]. An increase in the reaction time also leads to an increased energy deposition because of the increased number of energetic collisions [124]. Nevertheless, under normal operating conditions, the maximum energy deposited by this method is somewhat less than that available in a TQ mass spectrometer [119,124].

The use of an ion trap for MS–MS experiments can provide very high overall CID efficiencies [120]. This high efficiency leads to one of the most important advantages of MS–MS with an ion trap, i.e. the ability to perform sequential CID, by the use of successive ejection and excitation stages, on the ions generated from a single ionization event [121]. This technique is referred to as $(MS)^n$ (Fig. 7.18). Sequential reactions with as many as eleven steps $(MS)^{12}$ have been accomplished [121]. The long residence times also make the ion trap very suitable for photodissociation experiments [46].

One drawback of the ion trap in MS–MS experiments is that it has so far only been used to record fragment ion data since there is presently no simple method of recording precursor ion or neutral loss data [125]. Another potential difficulty is the possibility of unwanted ion–molecule reactions, e.g. involving neutral analyte molecules within the trap.

7.3.7 Fourier transform instruments

Just as with the ion trap, the trapping facilities offered by Fourier transform instruments make them well suited to the manipulations of ion populations required for MS–MS experiments. Thus, Fourier transform instruments also offer reaction periods that are separated in time but not in space and that may last for several seconds or even minutes. All the steps of an MS–MS experiment with a Fourier transform instrument involve excitation of the same ion cyclotron frequency; the difference is in the amount of energy absorbed.

In practice, all ions except those of a selected m/z value are ejected from the cell by means of a frequency sweep that is of higher amplitude and/or longer duration than that used for ion detection and that also lacks a narrow band of frequencies. The application of an r.f. potential at the cyclotron frequency, but with a slightly lower amplitude, then energizes the selected ion prior to dissociation. Dissociation itself is induced by collision with a target gas which may be introduced

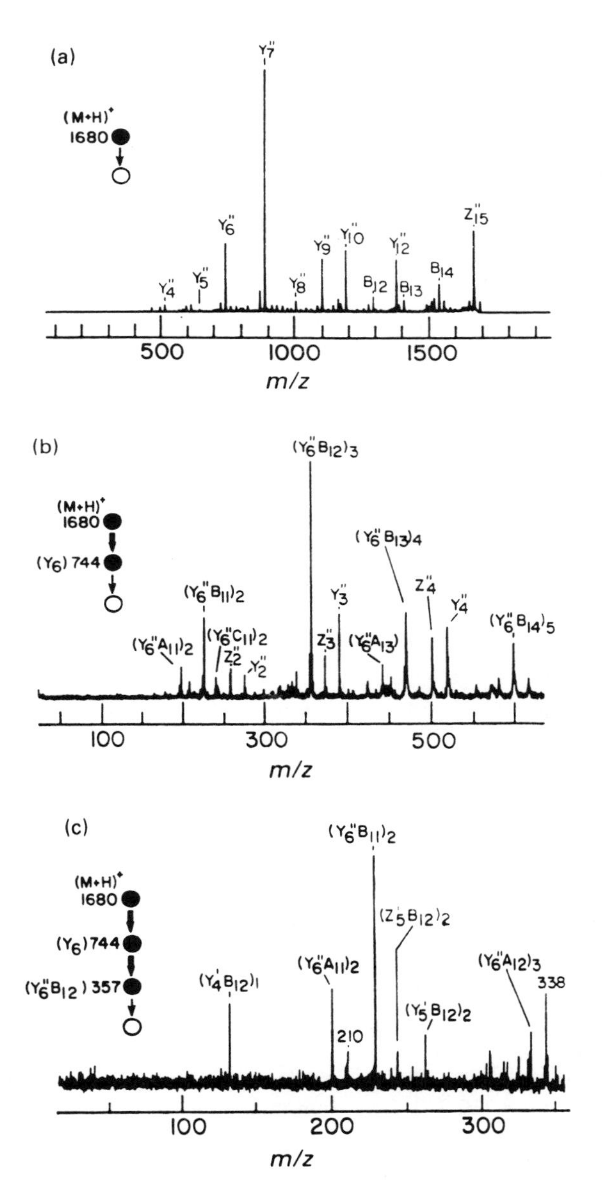

Fig. 7.18 (a) MS–MS fragment ion spectrum from a peptide of molecular weight 1680. (b) MS^3 spectrum of *m/z* 744 (Y_6) isolated from spectrum (a). The sequence ion nomenclature, based on that of ref. 196, expresses their relation- ship to the molecular ion (*m/z* 1680). (c) MS^4 spectrum of *m/z* 357 (Y''_6B_{12}) isolated from spectrum (b). (Reproduced with permission from ref. 123)

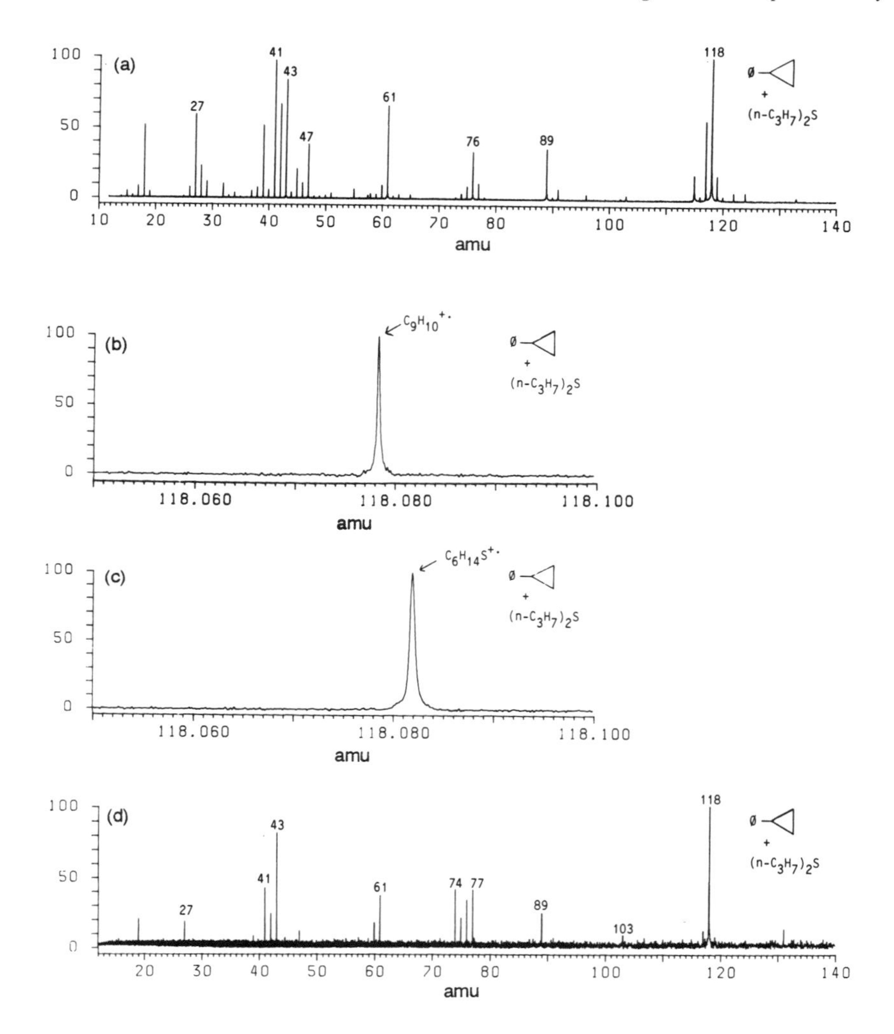

Fig. 7.19 (a) Electron impact FTMS spectrum of a 1:1 mixture of cyclopropylbenzene (MW 118) and di-n-propylsulphide (MW 118). Selective isolation of the molecular ions of (b) cyclopropylbenzene and (c) di-n-propylsulphide from the mixture. (d) Collision induced dissociation spectrum of *m/z* 118 from from di-n-propylsulphide after selective isolation. (Reproduced with permission from ref. 132)

as a pulse into the FTMS cell so as not to compromise vacuum requirements. An appropriate waveform for selective activation or ejection may be prepared by the stored waveform inverse Fourier transform (SWIFT) method [126]. The fragment ions resulting from CID are then detected in the usual manner to give the fragment ion MS–MS spectrum.

The kinetic energy acquired by an ion on activation depends on the amplitude and length of the r.f. pulse used. Naturally, there is a limit to this kinetic energy since an excessive amount will simply remove the ion from the cell. The maximum activation energy increases with cell size and magnetic field strength but decreases with mass [127]. MS–MS experiments with a Fourier transform instrument are usually in the low energy range.

Just as with the ITMS, ion trapping in a Fourier transform instrument lends itself very readily to the implementation of sequential steps in $(MS)^n$ experiments [128,129]. More recently, experiments have been reported in which activation is used to promote ion-molecule reactions as well as subsequent dissociation steps [129–131]. Another feature of FTMS in MS–MS experiments is its high resolution capabilities. Thus, highly selective ion excitation may be achieved by the use of tailored waveforms [126,129]. Again, Nibbering's group have demonstrated high precursor ion resolution (Fig. 7.19) and high fragment ion resolution with accurate mass measurement in CID experiments [132].

The disadvantages of using Fourier transform instruments for MS–MS experiments parallel those encountered with ITMS instruments. For example, precursor ion and neutral loss scans cannot be carried out in a simple manner with trapping instruments. Another difficulty is again the possibility of unwanted ion–molecule reactions which involve neutral analyte or background molecules, particularly when an external ion source is not available.

Fourier transform instruments have been extensively used with alternative methods of ion activation, especially those in which the operating pressure is not raised. Photodissociation has been widely used [45,133–137] since trapping allows an extended ion–photon interaction time. Electron impact activation has also been demonstrated with a dual-cell instrument [51].

7.4 Applications of Tandem Mass Spectrometry

The use of tandem mass spectrometry techniques provides information regarding the origin and fate of ions observed in a mass spectrometer. Such information should be related to the structure of these ions and therefore to the structure of the compounds from which they originate. Since, however, the recording of a routine EI spectrum is so straightforward and since comprehensive databases of EI spectra are available for comparison [138], the normal spectrum is the preferred basis for structural investigations in cases where electron impact ionization is applicable.

Thus, in general, tandem mass spectrometry has been used for the systematic search for specific compounds or compound classes, particularly in unseparated mixtures (section 7.4.1), rather than as a stand-alone technique for structural analysis. An important extension of this approach is the use of selected reaction monitoring (section 8.3.2) as a highly selective technique for quantitative analysis.

This is not to say that tandem mass spectrometry cannot be used as a complementary technique in structural elucidation. Thus, it can provide a more sensitive probe of chemical structure, e.g. for the differentiation of isomers. Again, tandem mass spectrometry used in conjunction with collision-induced dissociation can provide fragment ion information when soft ionization techniques, which otherwise only provide molecular ions, are used. Applications to structural analysis are described in more detail in sections 7.4.2 and 7.4.3, and particularly in two more specialized books [2,139].

7.4.1 Targeted analysis

Mixtures may be analysed for the presence of specific compound types by the selection of specific fragment ions or of specific neutral losses. For example, dansyl derivatives of amines show an intense ion at m/z 171 in their EI spectra, so that a scan of the precursors of this ion records the molecular ions of the dansyl derivatives present [140]. The direct analysis of terpanes and steranes in an unseparated shale oil using characteristic fragment ions at m/z 191 and m/z 217 respectively is another excellent application of this technique (Table 7.3).

Table 7.3 Terpanes and steranes in the saturate portion of Green River shale. Comparison of GCMS and accelerating voltage scan (AVS) methods. (Reprinted with permission from E.J. Gallegos *Anal. Chem.*, **48**, 1348[197]. Copyright (1976) American Chemical Society)

Terpanes	C_{20}	C_{21}	C_{29}	C_{30}	C_{31}	Total
GC–MS	1.4	0.4	2.3	13.6	1.0	18.7
AVS method	1.7	0.4	2.2	13.2	1.1	18.6
Steranes	C_{22}	C_{27}	C_{28}	C_{29}		Total
GC–MS	—	1.9	7.6	10.3		19.8
AVS method	0.6	1.8	6.3	10.1		18.8

A practical example of constant neutral loss scanning is the screening of crude mixtures for chlorinated biphenyls by monitoring the loss of one or two chlorine atoms [141]. Again, the formation of $(M-H)^-$ ions from carboxylic acids under CI conditions allows these compounds to be detected in mixtures by monitoring the subsequent loss of 44 mass units (CO_2) [142]. Constant neutral loss scans following FAB ionization can be used to detect glucuronide (176 mass unit loss) and sulphate (80 mass unit loss) conjugates present as drug metabolites [143]. Precursor ion scans of fragment ions characteristic of the original drug structure were also used to locate metabolites in this investigation. Precursor ion and neutral loss scans have also been employed in the direct analysis of complex fuel constituents [144,145].

When the presence of distinct compounds in a mixture is presumed then the selection of the corresponding molecular ions for tandem mass spectrometric analysis is an appropriate technique. For example, it has been demonstrated that steroid mixtures can be analysed using MIKE scans of molecular ions produced in the EI mode [146] (Fig. 7.4). Collisional activation was not used in this case. A change to chemical ionization can, however, offer a potentially more discriminatory and sensitive technique since the number of competing fragment ions is greatly reduced. Chemical ionization followed by a MIKE scan analysis of the quasi-molecular ions formed has been promoted as a method of mixture analysis by Cooks and co-workers, who have recorded a number of interesting applications, especially in the alkaloid field [147,148]. The method has subsequently been extended to other ionization techniques, e.g. negative ion CI applied to organic acid detection [149], desorption EI (section 5.2.1) applied to the detection of anabolics in tissue [150], and especially FAB applied to antibiotics [151], acylcarnitines [152] and ionic surfactants [153], amongst others.

An excellent example of the comprehensive use of tandem mass spectrometric techniques for the analysis of targeted compounds in complex mixtures is given by Startin and co-workers [154]. Using an instrument of hybrid geometry with chemical ionization, these workers were able to screen for sulphonamide residues by searching for the precursor ions of m/z 156 and then confirm the identity of the sulphonamides found by recording fragment ion scans from the precursor ions. These fragment ion scans could be compared with those recorded for standard compounds. As a further extension of this work, selected reaction monitoring (section 8.3.2) was used where detection of sulphonamides with the maximum sensitivity was required.

The analysis of selected precursor ions has also been implemented using triple quadrupole analysers. For example, Hunt [155] has been able to demonstrate scanning detection limits of 10 ppb for the direct analysis of *p*-nitrophenol and 2,4-dinitrophenol in an industrial sludge using OH^- reagent ions to form quasi-molecular ions from both compounds. Triple quadrupole analysers are well suited to more complex analyses which can be carried out completely under data system control. Thus, fragment ions shown to be characteristic of a pure drug, and therefore indicative of chemical substructures that might be found in its metabolites, can be selected and scans showing the associated precursor ions recorded (Fig. 7.9). The structure of each metabolite located in this way can then be established from a complete fragment ion spectrum either by interpretation or by comparison with standard spectra [79].

Another advantage of a triple quadrupole instrument is the absence of artefact peaks in the spectra. For this reason, the triple quadrupole is generally more suitable than conventional sector instruments for direct analysis at trace levels in unseparated mixtures, e.g. by selected reaction monitoring (section 8.3.2). A similar advantage is offered by instruments of hybrid (section 7.3.4) or extended (section 7.3.3) geometry, which also provide high precursor ion resolution permitting the separation and subsequent analysis of isobaric ions (Fig. 7.20).

High precursor ion resolution is of no avail for the analysis of isomeric ions. Some isomeric mixtures may, however, be resolved by the observation of

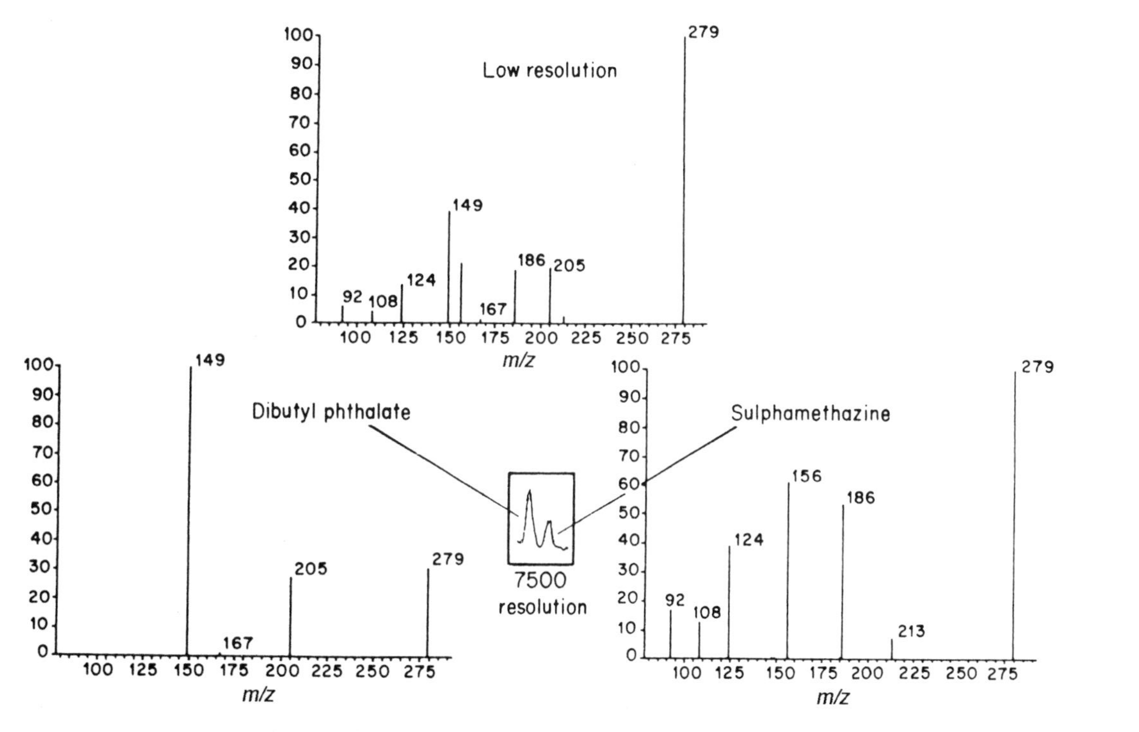

Fig. 7.20 Separation and MS–MS analysis of isobaric ions at *m/z* 279 from sulphamethazine (279.0916) and dibutylphthalate (279.1596) using a hybrid geometry (BEqQ) instrument with a primary beam resolution of 7500. (Courtesy Kratos Analytical Instruments)

consecutive reactions in instruments of extended or hybrid geometry with collision cells in appropriate regions. Thus, benzyl acetate, which shows a major fragment ion at m/z 108 ($C_6H_5CH_2OH^{+\cdot}$) may be identified in the presence of a large excess of *o*-cresol (molecular ion $CH_3C_6H_4OH^{+\cdot}$ also at m/z 108) if an instrument is set to record only fragment ions formed by the reactions $150^+ \rightarrow 108^+ \rightarrow (\text{fragments})^+$ in separate collision regions [88].

Direct analysis techniques, such as those described in this section, have often been advanced as an alternative to more conventional methods in which a chromatographic separation is coupled to the mass spectrometer, or as a substitute for extensive sample work-up procedures [147,156]. The principal advantages of direct methods are their speed and their obvious suitability for compounds that are not amenable to chromatography. Clearly, however, such methods may not have the same discriminating power as chromatography, particularly when artefact peaks can be present, and their use may benefit from a combination with some simple form of chromatographic purification or separation. Again, while direct methods can often provide very suitable screening procedures for selected compounds, quantitative data from unpurified samples may be unreliable because of matrix effects (section 8.1.1).

7.4.2 Structural analysis

The structures of relatively low mass ions in a spectrum may be determined directly by matching collision-induced spectra with the spectra of ions of known structure [10,157]. More complex ions may be associated with these lower mass ions by the observation of fragmentations and so provide a further stage in the analysis of the spectra of unknowns [158,159]. With some exceptions, however, this logical approach has not been particularly rewarding as a method of structural analysis and may be subject to misinterpretation if high resolution mass measurement data are not also available [160].

The most successful application of tandem mass spectrometry to structural analysis has been the sequencing of natural oligomers such as peptides where both the modes of fragmentation and the composition of possible fragment ions are well documented. Thus, collision-induced dissociation of peptide molecular ions generated by FAB/LSIMS has become a well-recognized method of establishing a complete amino acid sequence for these materials [94,161]. The high energy fragment ion spectra from peptides (e.g. Fig. 7.13) exhibit fairly complete series of peaks that correspond to an initial cleavage at successive amide bonds in the peptide backbone, whereas low energy spectra are more complex and less well understood [162]. In addition, the use of high energy CID, available with four-sector instruments, also produces abundant, structurally informative fragment ions that result from the cleavage of amino acid side chains and that are absent from low energy spectra. A particularly valuable feature of tandem mass spectra is the lack of chemical background and matrix contributions that make any

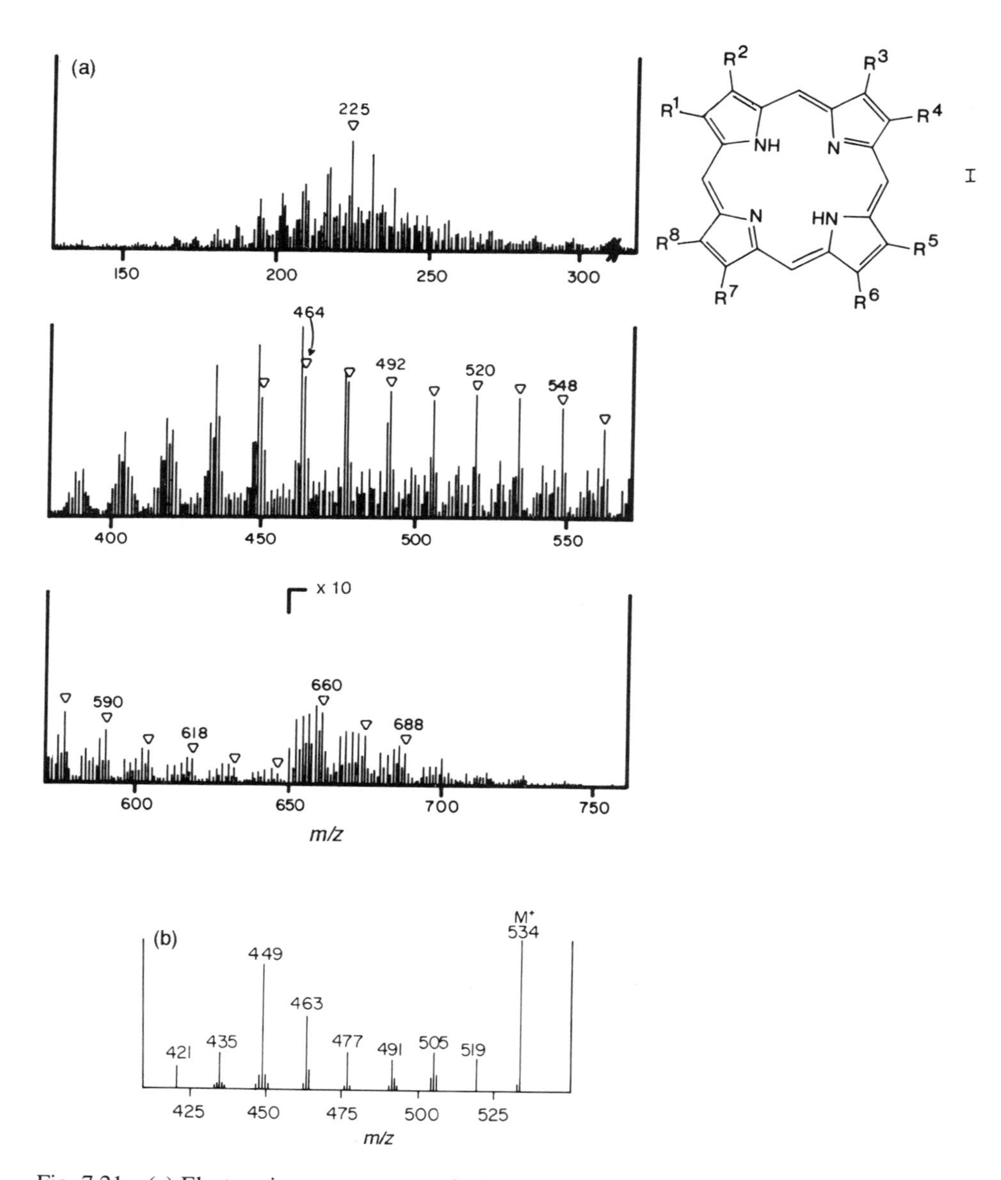

Fig. 7.21 (a) Electron impact spectrum of etioporphyrin (structure I) fraction from Boscan oil. (b) Collision induced fragment ion spectrum from *m/z* 534 in the same fraction, recorded using a triple quadrupole instrument. (Reprinted with permission from J.V. Johnson *et al.*, *Anal. Chem.*, **58**, 1325[168]. Copyright (1986) American Chemical Society)

fragmentation in conventional FAB/LSIMS spectra difficult to interpret.

A limited number of studies have used fragment ion scans to elicit structural information from other compound types. For example, CID of a prominent fragment ion in the FAB spectra of some polyether antibiotics gives information on the disposition and type of substituent groups present [163]. Again, dissociation of the m/z 174 fragment ion found in the EI spectra of β-lactam antibiotics produces spectra characteristic of β-lactam ring size and of substituent type and position [164]. Information about the carbohydrate units in cyclic glycopeptides is also available from fragment ion scans [165].

The same methods can be used to obtain structural information from unpurified samples. For example, the fatty acid composition of mixtures of phospholipids [166] or higher molecular weight esters [167] are available following FAB and FD ionization respectively. Other studies have applied these methods to determine alkyl substitution patterns in porphyrin mixtures [168] (Fig. 7.21) and general substitution patterns in fatty acid mixtures [169].

Spectra generated by other ionization techniques such as chemical ionization, electrospray ionization, or thermospray ionization may often contain insufficient fragment ion data to properly characterize the compounds being analysed, and under these conditions a collision-induced spectrum can provide significantly more structural information [170,171] (e.g. Figs. 2.18 and 6.20). The use of tandem mass spectrometry techniques also allows the direct observation of characteristic ions in the low mass region which may be obscured in the normal spectrum, e.g. in LC–MS operation.

Metastable techniques may be used to record molecular ions that fragment before they reach the detector. For example, the B^2/E scan in Fig. 7.22 quite clearly indicates the formation and subsequent decomposition of protonated dimer ions not seen in the conventional chemical ionization spectrum [172]. An analogous method allows the identification of metastable molecular ions formed from phospholipids under EI conditions [173].

7.4.3 The analysis of isomeric compounds

Most peaks in the mass spectra of isomeric compounds which have only minor differences in structure are common to all the spectra and any more specific differences are often obscured. Tandem mass spectrometry provides a method by which these specific ions, which might serve to distinguish isomeric compounds, may be studied in isolation. Thus, unimolecular fragmentations have been used to distinguish isomeric cyclic diols [174,175] and stereoisomeric permethylated disaccharides [176] amongst many other compound types. Puzo *et al.* [177] have demonstrated that competition between the unimolecular reactions

$$\text{(hexose-cation-matrix)}^+ \rightarrow \text{(hexose-cation)}^+ + \text{matrix} \quad (7.23)$$

$$\text{(hexose-cation-matrix)}^+ \rightarrow \text{(matrix-cation)}^+ + \text{hexose} \quad (7.24)$$

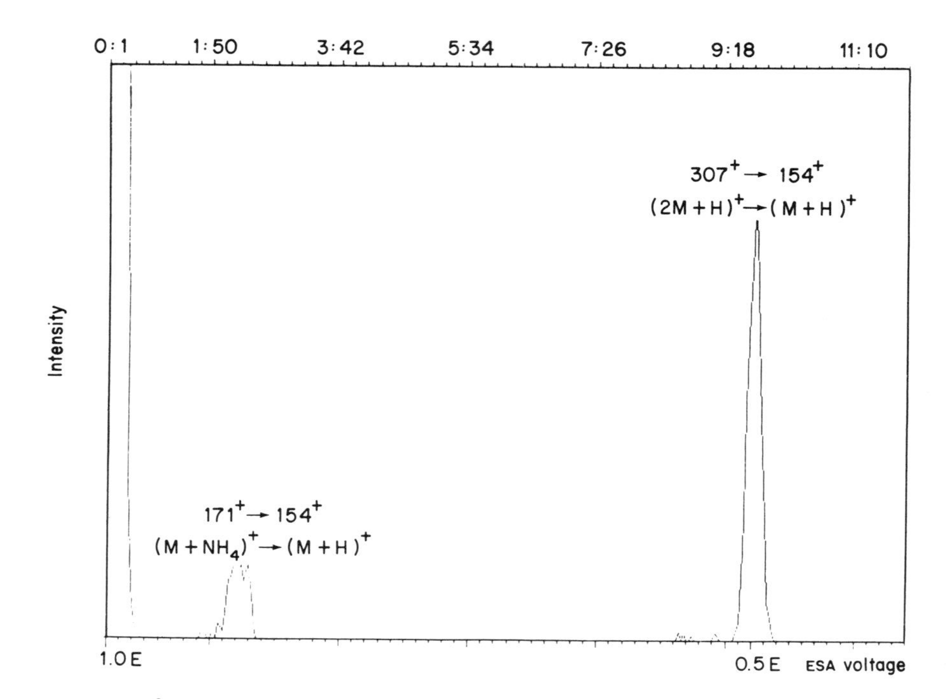

Fig. 7.22 B^2/E scan from dopamine. The origin of the ion at *m/z* 154 from both the $(M+NH_4)^+$ and $(2M+H)^+$ ions is illustrated. (Courtesy Kratos Analytical Instruments)

is very dependent on hexose geometry, so that isomeric hexoses may be distinguished using MIKE scans of the [hexose-cation-matrix]$^+$ ion formed by fast ion bombardment. A similar method is applicable to hexosamine isomers [178].

The combination of FAB/LSIMS and CID has been used in a large number of published examples of isomer differentiation. For example, fragment ion spectra from such experiments have been used to determine the location of double bonds in fatty acids [179] and to distinguish branched and linear chain oligosaccharides [180,181], isomeric cyclic nucleotides [182,183], isomeric acyl carnitines [184], steroid epimers [182,185], and β-lactam epimers [185] (Fig. 7.23). Similar experiments with steroid epimers have used electron impact ionization [186]. A study using isomeric steroid glucuronides has sought to determine the effect of a range of experimental conditions on peak intensity in fragment ion spectra and therefore the practical utility of this technique [91].

The energy released in a metastable decomposition can also be a sensitive indicator of stereochemical differences. Thus, the widths of metastable peaks corresponding to the unimolecular loss of bromine from the molecular ions of

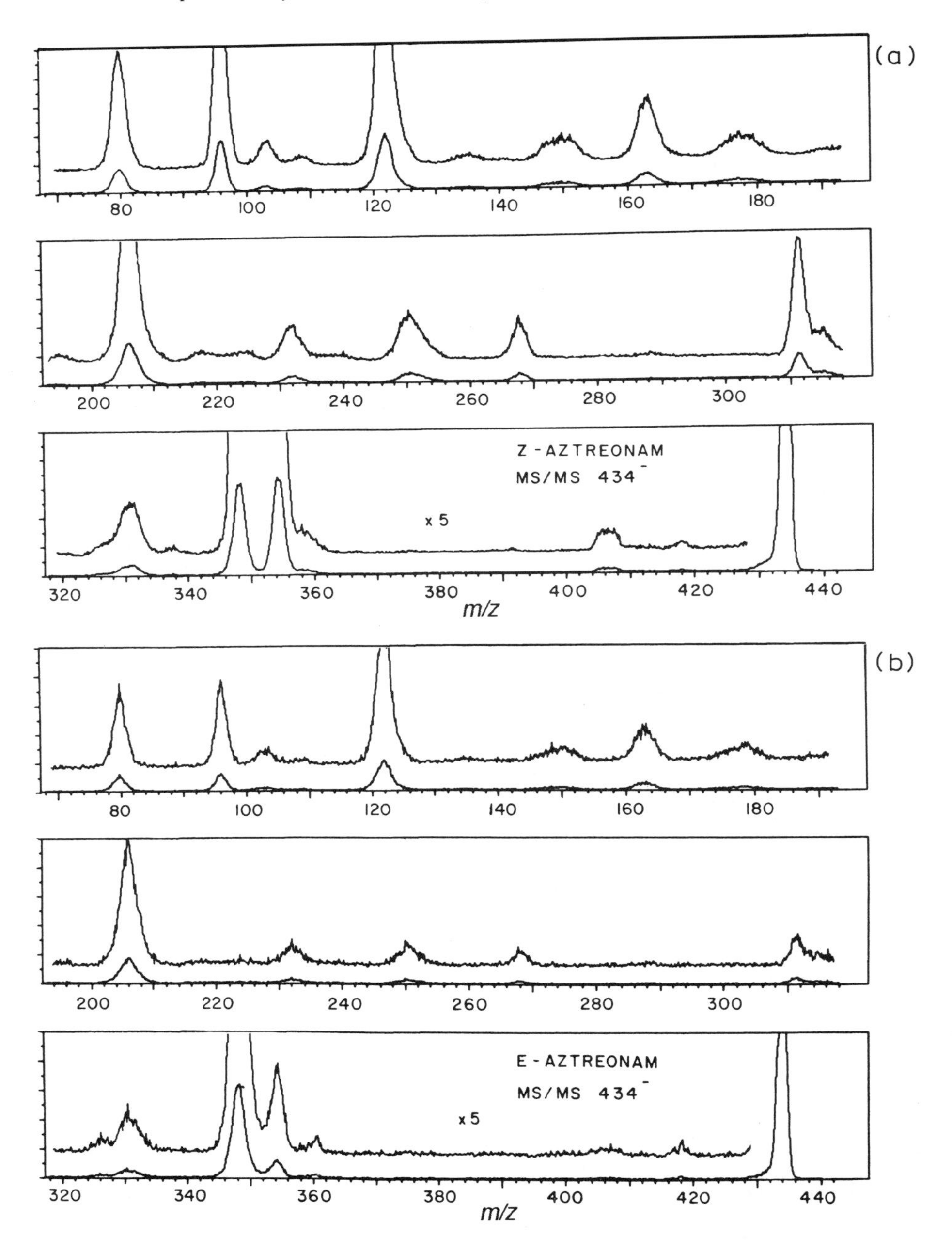

Fig. 7.23 MIKE scans of the collision-induced fragment ions from the deprotonated molecular ions produced by fast atom bombardment of (a) the Z- and (b) the E-isomer of aztreonam (structure I, overleaf). (Reproduced with permission from ref. 185)

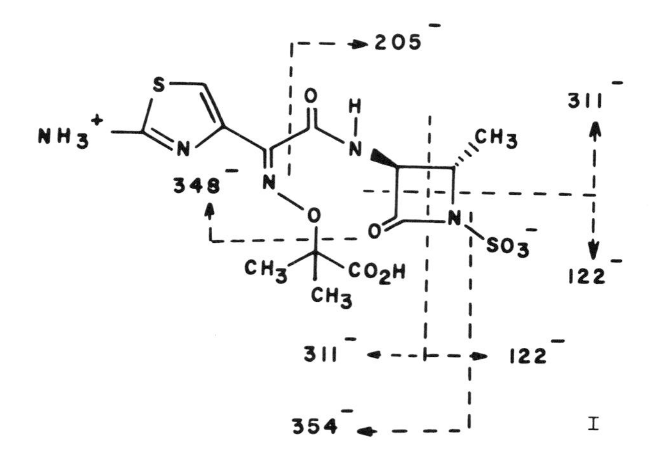

the three isomeric bromotoluenes and benzyl bromide show significant and reproducible differences [187] (Table 7.4). Differences in both translational energy releases and metastable ion intensity ratios have been reported in a study of steroid epimers [188].

Table 7.4 Kinetic energy release for the loss of Br from the molecular ion of four isomeric aromatic bromo-compounds. (Reprinted with permission from J.R. Hass, Y. Tondeur, and R.D. Voyksner, *Anal. Chem.*, **55**, 295[187]. Copyright (1983) American Chemical Society)

Compound	Kinetic energy release (meV)	Standard deviation
Benzyl bromide	1.7	0.3
o-Bromotoluene	73.2	3.7
m-Bromotoluene	271	22
p-Bromotoluene	162	14

References

1. Howe, I., Williams, D.H., and Bowen, R.D. (1981), *Mass Spectrometry—Principles and Applications* (2nd Edition), McGraw-Hill.
2. McLafferty, F.W. (Ed.) (1983), *Tandem Mass Spectrometry*, John Wiley, New York.
3. Gross, M.L., Chess, E.K., Lyon, P.A., Crow, F.W., Evans, S., and Tudge, H. (1982), *Int. J. Mass Spectrom. Ion Processes*, **42**, 243.
4. Beynon, J.H., Cooks, R.G., and Keough, T. (1974), *Int. J. Mass Spectrom. Ion Processes*, **14**, 437.
5. Glish, G.L., and Todd, P.J. (1982), *Anal. Chem.*, **54**, 842.
6. Yost, R.A., and Enke, C.G. (1970), *Anal. Chem.*, **51**, 1251A.

7. Boitnott, C.A., Steiner, U., Story, M.S., and Smith, R.D. (1981), *Dynamic Mass Spectrometry*, **6**, 71.
8. Siegel, M.E. (1980), *Anal. Chem.*, **52**, 1790.
9. Yost, R.A., Enke, C.G., McGilvery, D.C., Smith, D., and Morrison, J.D. (1979), *Int. J. Mass Spectrom. Ion Processes*, **30**, 127.
10. McLafferty, F.W., Kornfeld, R., Haddon, W.F, Levsen, K., Sakai, I., Bente, P.F.,III, Tsai, S.-C., and Schuddemage, H.D.R. (1973), *J. Am. Chem. Soc.*, **95**, 3886.
11. Alexander, A.J., Thibault, P., Boyd, R.K., Curtis, J.M., and Rinehart, K.L. (1990), *Int. J. Mass Spectrom. Ion Processes*, **98**, 107.
12. Massey, H.S.W. (1949), *Rep. Progr. Phys.*, **12**, 248.
13. Mahan, B.H. (1970), *J. Chem. Phys.*, **52**, 5221.
14. Jennings, K.R. (1968), *Int. J. Mass Spectrom. Ion Processes*, **1**, 227.
15. Horning, S.R., Vincenti, M., and Cooks, R.G. (1990), *J. Am. Chem. Soc.*, **112**, 119.
16. Douglas D.J., (1982), *J. Phys. Chem.*, **86**, 185.
17. Busch, K.L., Glish, G.L., and McLuckey, S.A. (1988), *Mass Spectrometry/Mass Spectrometry: Techniques and Applications of Tandem Mass Spectrometry*, p. 81, VCH, New York.
18. Smith R.D., and Barinaga, C.J. (1990), *Rapid Commun. Mass Spectrom.*, **4**, 54.
19. Busch, K.L., Glish, G.L., and McLuckey, S.A. (1988), *Mass Spectrometry/Mass Spectrometry: Techniques and Applications of Tandem Mass Spectrometry*, pp.78–80, VCH, New York.
20. Wysocki, V.H., Kenttamaa, H.I., and Cooks, R.G. (1987), *Int. J. Mass Spectrom. Ion Processes*, **75**, 181.
21. Reiner, E.J., Harrison, A.G., and Bowen, R.D. (1989), *Can.J. Chem.*, **67**, 2081.
22. Adams, J. (1990), *Mass Spectrom. Rev.*, **9**, 141.
23. Zakett, D., Ciupek, J.D., and Cooks, R.G. (1981), *Anal. Chem.*, **53**, 723.
24. Ciupek, J.D., Zakett, D., Cooks, R.G., and Wood, K.V. (1982), *Anal. Chem.*, **54**, 2215.
25. Bursey, M.M., Nystrom, J.A., and Hass, J.R. (1984), *Anal. Chim. Acta*, **159**, 275.
26. Alexander, A.J., and Boyd, R.K. (1989), *Int. J. Mass Spectrom. Ion Processes*, **90**, 211.
27. Harvan, D.J., Nystrom, J.A., Grady, W.L., Voyksner, R.D., Cerny, R.L., Bursey, M.M., Siegel, M.W., Yinon, J., and Hass, J.R. (1981), *Anal. Lett.*, **14**, 985.
28. Schey, K.L., Kenttamaa, H.I., Wysocki, V.H., and Cooks, R.G. (1989), *Int. J. Mass Spectrom. Ion Processes*, **90**, 71.
29. Ouwerkerk, C.D., McLuckey, S.A., Kistemaker, P.G., and Boerboom, A.J. (1984), *Int. J. Mass Spectrom. Ion Processes*, **56**, 11.
30. Laramee, J.A., Cameron, D., and Cooks, R.G. (1981), *J. Am. Chem. Soc.*, **103**, 12.
31. Alexander, A.J., Thibault, P., and Boyd, R.K. (1990), *J. Am. Chem. Soc.*, **112**, 2484.
32. Uggerud, E., and Derrick, P.J., (1989), *Z. Naturforsch.*, **44A**, 245.
33. Biemann, K. (1987), *Anal. Chem.*, **59**, 125R.
34. Bordas-Nagy, J., Despeyroux, D., and Jennings, K.R. (1992), *J. Am. Soc. Mass Spectrom.*, **3**, 502.
35. Bordas-Nagy, J., Despeyroux, D., Jennings, K.R., and Gaskell, S.J. (1992), *Org. Mass Spectrom.*, **27**, 406.
36. Brodbelt-Lustig, J.S., and Cooks, R.G. (1988), *Int. J. Mass Spectrom. Ion Processes*, **86**, 253.
37. Mabud, M.A., Dekrey, M.A., and Cooks, R.G. (1985), *Int. J. Mass Spectrom. Ion Processes*, **67**, 285.
38. Bier, M.E., Amy, J.W., Cooks, R.G., Syka, J.E.P., Ceja, P., and Stafford, G. (1987), *Int. J. Mass Spectrom. Ion Processes*, **77**, 31.
39. Schey, K., Cooks, R.G., Grix, R., and Wollnik, H. (1987), *Int. J. Mass Spectrom. Ion Processes*, **77**, 49.

40. DeKrey, M.J., Kenttämaa, H.I., Wysocki, V.H., and Cooks, R.G. (1986), *Org. Mass Spectrom.*, **21**, 193
41. Horning, S.R., Bier, M.E., Cooks, R.G., Brusini, G., Traldi, P., Guiotto, A., and Rodighiero, P. (1989), *Biomed. Environ. Mass Spectrom.*, **18**, 927.
42. Schey, K.L., Cooks, R.G., Kraft, A., Grix, R., and Wollnik, H. (1989), *Int. J. Mass Spectrom. Ion Processes*, **94**, 1.
43. Martens, J., Ens, W., Standing, K.G., and Verentchikov, A. (1992), *Rapid Commun. Mass Spectrom.*, **6**, 147.
44. Wright, A.D., Despeyroux, D., Jennings, K.R., Evans, S., and Riddoch, A. (1992), *Org. Mass Spectrom.*, **27**, 525.
45. Williams, E.R., Furlong, J.J.P., and McLafferty, F.W. (1990), *J. Am. Soc. Mass Spectrom.*, **1**, 288.
46. Louris, J.N., Brodbelt, J.S., and Cooks, R.G. (1987), *Int. J. Mass Spectrom. Ion Processes*, **75**, 345.
47. Tecklenburg, R.E., Jr, and Russell, D.H. (1990), *Mass Spectrom. Rev.*, **9**, 405.
48. Tecklenburg, R.E., Jr, Miller, M.N., and Russell, D.H. (1989), *J. Am. Chem. Soc.*, **111**, 1161.
49. Fukuda, E.K., and Campana, J.E. (1985), *Int. J. Mass Spectrom. Ion Processes*, **65**, 321.
50. Watson, C.H., Baykut, G., and Eyler, J.R. (1987), *Anal. Chem.*, **59**, 1133.
51. Cody, R.B., and Freiser, B.S. (1987), *Anal. Chem.*, **59**, 1054.
52. Aberth, W., and Burlingame, A.L. (1990), in *Biological Mass Spectrometry* (Eds. A.L.Burlingame and J.A.McCloskey), Ch.13, Elsevier, Amsterdam.
53. Sen-Sharma, D.K., Jennings, K.R., and Beynon, J.H. (1976), *Org. Mass Spectrom.*, **11**, 319.
54. Cooks, R.G., Beynon, J.H., Caprioli, R.M., and Lester, G.R. (1973), *Metastable Ions*, Elsevier, Amsterdam.
55. Beynon, J.H., Cooks, R.G., Amy, J.W., Baitinger, W.E., and Ridley, T.Y. (1973), *Anal. Chem.*, **45**, 1023A.
56. Stahl, D., and Tabet, J.C. (1979), *Chimia*, **33**, 287.
57. Haddon, W.F. (1979), *Anal. Chem.*, **51**, 983.
58. Shushan, B., Bunce, N.J., Boyd, R.K., and Corke, C.T. (1981), *Biomed. Mass Spectrom.*, **8**, 225.
59. Longstaff, C., and Rose, M.E. (1982), *Org. Mass Spectrom.*, **17**, 508.
60. Davies, N.W., and Bignall, J.C. (1982), *Org. Mass Spectrom.*, **17**, 451.
61. Trott, G.W., Porter, C.J., Beynon, J.H., Brenton, A.G., and Morgan, R.P. (1979), *J. Phys. E: Sci. Instrum.*, **12**, 979.
62. Friedli, F., and Beck, E. (1982), *Org. Mass Spectrom.*, **17**, 646.
63. Kassel, D.B., Musselman, B.D., and Smith, J.A. (1991), *Anal. Chem.*, **63**, 1091.
64. Boyd, R.K., Porter, C.J., and Beynon, J.H. (1982), *Org. Mass Spectrom.*, **16**, 490.
65. Porter, C.J., Brenton, A.G., and Beynon, J.H. (1980), *Int. J. Mass Spectrom. Ion Processes*, **36**, 69.
66. Guevremont, R., and Boyd, R.K. (1988), *Rapid Commun. Mass Spectrom.*, **2**, 1.
67. Sheil, M.M., and Derrick, P.J. (1988), *Org. Mass Spectrom.*, **23**, 429.
68. Boyd, R.K., (1987), *Int. J. Mass Spectrom. Ion Processes*, **75**, 243.
69. Lacey, M., and Macdonald, C. (1980), *Org. Mass Spectrom.*, **15**, 484.
70. Morgan, R.P., Porter, C.J., and Beynon, J.H. (1977), *Org. Mass Spectrom.*, **12**, 735.
71. Lacey, M.J., Macdonald, C.G., Donichi, K.F., and Derrick, P.J. (1981), *Org. Mass Spectrom.*, **16**, 351
72. Ast, T., Bozorgzadeh, M.H., Wiebers, J.L., Beynon, J.H., and Brenton, A.G. (1979), *Org. Mass Spectrom.*, **14**, 313.

73. Lacey, M.J., and Macdonald, C.G. (1978), *Aust. J. Chem.*, **31**, 2161.
74. Dawson, P.H., French, J.B., Buckley, J.A., Douglas, D.J., and Simmons, D. (1982), *Org. Mass Spectrom.*, **17**, 205.
75. Dawson, P.H., French, J.B., Buckley, J.A., Douglas, D.J., and Simmons, D. (1982), *Org. Mass Spectrom.*, **17**, 212.
76. Miller, P.E., and Bonner Denton, M. (1986), *Int.J. Mass Spectrom. Ion Processes*, **72**, 223.
77. Davis, S.C., and Wright, B. (1990), *Rapid Commun. Mass Spectrom.*, **4**, 186.
78. Boyd, R.K., Bott, B.A., Beynon, J.H., Harvan, D.J., and Hass, J.R. (1985), *Int. J. Mass Spectrom. Ion Processes*, **66**, 253.
79. Perchalski, R.J., Yost, R.A., and Wilder, B.J. (1982), *Anal. Chem.*, **54**, 1466.
80. Brotherton, H.O., and Yost, R.A. (1983), *Anal. Chem.*, **55**, 549.
81. Dawson, P.H., and Sun, W.-F. (1983/1984), *Int. J. Mass Spectrom. Ion Processes*, **55**, 155.
82. Hayes, R.N., and Gross, M.L. (1990), in *Methods in Enzymology*, Vol. 193: *Mass Spectrometry* (Ed. J.A. McCloskey), Academic Press, San Diego, pp. 255–6.
83. Yost, R.A., and Boyd, R.K. (1990), in *Methods in Enzymology*, Vol. 193: *Mass Spectrometry* (Ed. J.A. McCloskey), Ch.7, Academic Press, San Diego.
84. Russell, D.H., Smith, D.H., Warmack, R.J., and Betram, L.K. (1980), *Int. J. Mass Spectrom. Ion Processes*, **35**, 381.
85. Rabrenovic, M., Brenton, A.G., and Beynon, J.H. (1983), *Int. J. Mass Spectrom. Ion Processes*, **52**, 175.
86. Gilliam, J.M., and Occolowitz, J.L., (1983), in *Proceedings of the 31st ASMS Conference on Mass Spectrometry and Allied Topics*, Boston, Massachusetts, p. 193.
87. Beynon, J.H., Harris, F.M., Green, B.N., and Bateman, R.H. (1982), *Org. Mass Spectrom.*, **17**, 55.
88. Burinsky, D.J., Cooks, R.G., Chess, E.K., and Gross, M.L. (1982), *Anal. Chem.*, **54**, 295.
89. Hass, J.R., Green, B.N., Bott, P.A., and Bateman, R.H. (1984), in *Proceedings of the 32nd ASMS Conference on Mass Spectrometry and Allied Topics*, San Antonio, Texas, p. 380.
90. Morgan, R.P., Brenton, A.G., and Beynon, J.H. (1979), *Int. J. Mass Spectrom. Ion Processes*, **29**, 195.
91. Cole, R.B., Guenat, C.R., and Gaskell, S.J. (1987), *Anal. Chem.*, **59**, 1139.
92. Unger, S.E., McCormick, T.J., Treher, E.N., and Nunn, A.D. (1987), *Anal. Chem.*, **59**, 1145.
93. Matsuo, T., Ishihara, M., Martin, S.A., and Biemann, K. (1983), *Int. J. Mass Spectrom. Ion Processes*, **86**, 83.
94. Ashcroft, A.E., and Derrick, P.J. (1990), in *Mass Spectrometry of Peptides* (Ed. D.M. Desiderio), CRC Press, Boca Raton, Fla.
95. Hayes, R.N., and Gross, M.L. (1990), in *Methods in Enzymology*, Vol. 193: *Mass Spectrometry* (Ed. J.A. McCloskey), Academic Press, San Diego, pp. 247–8.
96. Falick, A.M., Medzihradszky, K.F., and Walls, F.C. (1990), *Rapid Commun. Mass Spectrom.*, **4**, 318.
97. Hill, J.A., Biller, J.E., and Biemann, K. (1991), *Int. J. Mass Spectrom. Ion Processes*, **111**, 1.
98. Gross, M.L. (1990), in *Methods in Enzymology*, Vol. 193: *Mass Spectrometry* (Ed. J.A. McCloskey), Ch.6, Academic Press, San Diego.
99. Bateman, R.H., Green, B.N., and Smith, D.C., (1982), *Proceedings of the 30th ASMS Conference on Mass Spectrometry and Allied Topics*, Honolulu, p. 516.

100. Schoen, A.E., Amy, J.W., Ciupek, J.D., Cooks, R.G., Doberstein, P., and Jung, G., (1985), *Int. J. Mass Spectrom. Ion Processes*, **65**, 125.
101. Glish, G.L., McLuckey, S.A., McBay, E.H., and Bertram, L.K., (1986), *Int. J. Mass Spectrom. Ion Processes*, **70**, 321.
102. Alexander, A.J., Dyer, E.W., and Boyd, R.K. (1989), *Rapid Commun. Mass Spectrom.*, **3**, 364.
103. Boyd, R.K., Dyer, E.W., and Guevremont, R., (1989), *Int. J. Mass Spectrom. Ion Processes*, **88**, 147.
104. Gaskell, S.J., and Reilly, M.H. (1988), *Rapid Commun. Mass Spectrom.*, **2**, 139.
105. Thorne, G.C., and Gaskell, S.J. (1989), *Rapid Commun. Mass Spectrom.*, **3**, 217.
106. Poulter, L., and Taylor, L.C.E. (1989), *Int. J. Mass Spectrom. Ion Processes*, **91**, 183.
107. Alexander, A.J., Thibault, P., and Boyd, R.K. (1989), *Rapid Commun. Mass Spectrom.*, **3**, 30.
108. Bradley, C.D., Curtis, J.M., Derrick, P.J., and Wright, B. (1992), *Anal. Chem.*, **64**, 2628.
109. Tang, X., Beavis, R., Ens, W., Lafortune, F., Schueler, B., and Standing, K.G. (1988), *Int. J. Mass Spectrom. Ion Processes*, **85**, 43.
110. Della-Negra, A., and LeBeyec, Y. (1985), *Anal. Chem.*, **57**, 2035.
111. Tang, X., Ens, W., Standing, K.G., and Westmore, J.B. (1988), *Anal. Chem.*, **60**, 1791.
112. LaiHing, K., Cheng, P.Y., Taylor, T.G., Willey, K.F., Peschke, M., and Duncan, M.A. (1989), *Anal. Chem.*, **61**, 1460.
113. Glish, G.L., and Goeringe, D.E. (1984), *Anal. Chem.*, **56**, 2291.
114. Glish, G.L., McLuckey, S.A., and McKown, H.S. (1987), *Anal. Instrum.*, **16**, 191.
115. Schey, K., Cooks, R.G., Grix, R., and Wollnik, H. (1987), *Int. J. Mass Spectrom. Ion Processes*, **77**, 49.
116. Geusic, M.E., Jarrold, M.F., McIlrath, T.J., Freeman, R.R., and Brown, W.L. (1987), *J. Chem. Phys.*, **86**, 3862.
117. Liu, Y., Zhang, Q.L, Tittel, F.K., Curl, R.F., and Smalley, R.E.J. (1986), *J. Chem. Phys.*, **85**, 7434.
118. Todd, J.F.J. (1991), *Mass Spectrom. Rev.*, **10**, 3.
119. Curcuruto, O., Fontana, S., and Traldi, P. (1992), *Rapid Commun. Mass Spectrom.*, **6**, 322.
120. Louris, J.N., Cooks, R.G., Syka, J.E.P., Kelley, P.E., Stafford, G.C., Jr, and Todd, J.F.J. (1987), *Anal. Chem.*, **59**, 1677.
121. Louris, J.N., Brodbelt-Lustig, J.S., Cooks, R.G., Glish, G.L., Van Berkel, G.J., and McLuckey, S.A. (1990), *Int. J. Mass Spectrom. Ion Processes*, **96**, 117.
122. Schwartz, J.C., and Jardine, I. (1992), *Rapid Commun. Mass Spectrom.*, **6**, 313.
123. Kaiser, R.E., Jr, Cooks, R.G., Syka, J.E.P., and Stafford, J.C., Jr (1990), *Rapid Commun. Mass Spectrom.*, **4**, 30.
124. Brodbelt, J.S., Kenttämaa H.I., and Cooks, R.G. (1988), *Org. Mass Spectrom.*, **23**, 6.
125. Todd, J.F.J., Penman, A.D., Thorner, D.A., and Smith, R.D. (1990), *Rapid Commun. Mass Spectrom.*, **4**, 108.
126. Marshall, A.J., Wang, T. -C., and Ricca, T.L. (1985), *J. Am. Chem. Soc.*, **107**, 7893.
127. Freiser, B.S. (1988), in *Techniques for the Study of Ion Molecule Reactions*, Vol. 20 (Eds. J.M. Farrar and W.H. Saunders Jr), pp. 61–118, John Wiley, New York.
128. Carlin, T.J., Sailans, L., Cassady, C.J., Jacobson, D.B., and Freiser, B.S. (1983), *J. Am. Chem. Soc.*, **105**, 6320.
129. Forbes, R.A., Laukien, F.H., and Wronka, J. (1988), *Int. J. Mass Spectrom. Ion Processes*, **83**, 23.
130. Bensimon, M., and Houriet, R. (1986), *Int. J. Mass Spectrom. Ion Processes*, **72**, 93.
131. Forbes, R.A., Lech, L.M., and Freiser, B.S. (1987), *Int. J. Mass Spectrom. Ion Processes*, **77**, 107.

132. De Koning, L.J., Fokkens, R.H., Pinkse, F.A., and Nibbering, N.M.M. (1987), *Int. J. Mass Spectrom. Ion Processes*, **77**, 95.
133. Cassady, C.J., and Freiser, B.S. (1984), *J. Am. Chem. Soc.*, **106**, 6176.
134. Bowers, W.D., Delbert, S., and McIver, R.T., Jr (1986), *Anal. Chem.*, **58**, 969.
135. Bensimon, M., Rapin, J., and Gäumann, T. (1986), *Int. J. Mass Spectrom. Ion Processes*, **72**, 125.
136. Nuwaysir, L.M., and Wilkins, C.L. (1989), *Anal. Chem.*, **61**, 689.
137. Hunt, D.F., Shabanowitz, J., and Yates, J.R., III (1987), *J. Chem. Soc. Chem. Commun.*, **1987**, 548.
138. See, for example, McLafferty, F.W., and Stauffer, D.B. (1991), *The Important Index of the Registry of Mass Spectral Data*, John Wiley, New York.
139. Busch, K.L., Glish, G.L., and McLuckey, S.A. (1988), *Mass Spectrometry/Mass Spectrometry: Techniques and Applications of Tandem Mass Spectrometry*, VCH, New York.
140. Addeo, F., Malorni, A., and Marino, G. (1975), *Anal.Biochem.*, **64**, 98.
141. Shushan, B., and Boyd, R.K. (1981), *Anal. Chem.*, **53**, 421.
142. Zakett, D., Schoen, A.E., Kondrat, R.W., and Cooks, R.G. (1979), *J. Am. Chem. Soc.*, **101**, 6781.
143. Rudewicz, P., and Straub, K.M. (1986), *Anal. Chem.*, **58**, 2928.
144. Wood, K.V., Schmidt, C.E., Cooks, R.G., and Batts, B.D. (1984), *Anal. Chem.*, **56**, 1335.
145. Wood, K.V., Cooks, R.G., Laugal, J.A., and Benkesser, R.A. (1985), *Anal. Chem.*, **57**, 692.
146. Maquestiau, A., van Haverbeke, Y., Flammang, R., Mispreuve, H., Kaisin, M., Braekman, J.C., Daloze, D., and Tursch, B. (1978), *Steroids*, **31**, 31.
147. Kondrat, R.W., and Cooks, R.G. (1978), *Anal. Chem.*, **50**, 81A.
148. Roush, R.H., and Cooks, R.G. (1984), *J. Natl. Prod.*, **47**, 197.
149. Rinaldo, P., Miolo, G., Chiandetti, L., Zacchello, F., Daolio, S., and Traldi, P. (1985), *Biomed. Mass Spectrom.*, **12**, 570.
150. Facino, R.M., Carini, M., Da Forno, A., Traldi, P., and Pompa G. (1986), *Biomed. Environ. Mass Spectrom.*, **13**, 121.
151. Tondeur, Y., Shorter, M., Gustafson, M.E., and Pandey, R.C. (1984), *Biomed. Mass Spectrom.*, **11**, 622.
152. Millington, D.S., Roe, C.R., and Maltby, D.A. (1984), *Biomed. Mass Spectrom.*, **11**, 236.
153. Lyon, P.A., Crow, F.W., Tomer, K.B., and Gross, M.L. (1984), *Anal. Chem.*, **56**, 2278.
154. Finlay, E.M.H., Games, D.E., Startin, J.R., and Gilbert, J. (1986), *Biomed. Environ. Mass Spectrom.*, **13**, 633.
155. Hunt, D.F., Shabanowitz, J., and Giordani, A.B. (1980), *Anal. Chem.*, **52**, 386.
156. Cooks, R.G., and Glish, G.L. (1981), *Chem. Engng. News*, **59**, No.48, 40.
157. Levsen, K., and Schwarz, K. (1976), *Angew. Chem. Int. Ed. Engl.*, **15**, 509.
158. Cheng, M.T., Kruppa, G.H., McLafferty, F.W., and Cooper, D.A. (1982), *Anal. Chem.*, **54**, 2204.
159. Cheng, M.T., Barbalas, M.P., Pegues, R.F., and McLafferty, F.W. (1983), *J. Am. Chem. Soc.*, **105**, 1510.
160. Rose, M.E. (1981), *Org. Mass Spectrom.*, **16**, 323.
161. Johnson, R.S., Martin, S.A., and Biemann, K. (1988), *Int. J. Mass Spectrom. Ion Processes*, **86**, 137.
162. Burlet, O., Orkiszewski, R.S., Ballard, K.D., and Gaskell, S.J. (1992), *Rapid Commun. Mass Spectrom.*, **6**, 658.

163. Siegel, M.M., McGahren, W.J., Tomer, K.B., and Chang, T.T. (1987), *Biomed. Environ. Mass Spectrom.*, **14**, 29.
164. Barbalas, M.P., McLafferty, F.W., and Occolowitz, J.L. (1983), *Biomed. Mass Spectrom.*, **10**, 258.
165. Roberts, G.D., Carr, S.A., and Christensen, S.B., (1987), in *Proceedings of the 35th ASMS Conference on Mass Spectrometry and Allied Topics*, Denver, Colorado, p. 933.
166. Sherman, W.R., Ackermann, K.E., Bateman, R.H., Green, B.N., and Lewis, I. (1985), *Biomed.Mass Spectrom.*, **12**, 409.
167. Niwa, Y., and Ishikawa, K., (1987), in *Proceedings of the 35th ASMS Conference on Mass Spectrometry and Allied Topics*, Denver, Colorado, p. 975.
168. Johnson, J.V., Britton.E.V., Yost, R.A., Quirke, J.M.E., and Cuesta, L.L. (1986), *Anal. Chem.*, **58**, 1325.
169. Tomer, K.B., Jensen, N.J., and Gross, M.L. (1986), *Anal. Chem.*, **58**, 2429.
170. Henion, J.D., Thomson, B.A., and Dawson, P.H. (1982), *Anal. Chem.*, **54**, 451.
171. McFadden, W.H., and Lammert, S.A. (1987), *J. Chromatogr.*, **385**, 201.
172. Cosgrove, M., Hazelby, D., Warburton, G.A., Stradling, R.S., and Chapman, C.J. (1983), *Int. J. Mass Spectrom. Ion Processes*, **46**, 89.
173. Klein, R.A. (1971), *J. Lipid Res.*, **12**, 628.
174. Winkler, F.J., and Stahl, D. (1978), *J. Am. Chem. Soc.*, **100**, 6779.
175. Houriet, R., Stahl, D., and Winkler, F.J. (1980), *EHP, Environ. Health Perspect.*, **36**, 63.
176. DeJong, E.G., Heerma, W., and Dijkstra, G. (1980), *Biomed. Mass Spectrom.*, **7**, 127.
177. Puzo, G., Fournie, J.-J., and Prome, J.-C. (1985), *Anal. Chem.*, **57**, 892.
178. Fournie, J.-J., and Puzo, G. (1985), *Anal. Chem.*, **57**, 2287.
179. Jensen, N.J., Tomer, K.B., and Gross, M.L. (1985), *Anal. Chem.*, **57**, 2018.
180. Carr, S.A., Reinhold, V.N., Green, B.N., and Hass, J.R. (1985), *Biomed. Mass Spectrom.*, **12**, 288.
181. Chen.Y., Chen, N., Li, H., Zhao, F., and Chen, N. (1987), *Biomed. Environ. Mass Spectrom.*, **14**, 9.
182. Kingston, E.E., Beynon, J.H., and Newton, R.P. (1984), *Biomed. Mass Spectrom.*, **11**, 367.
183. Kingston, E.E., Beynon, J.H., Newton, R.P., and Liehr, J.G. (1985), *Biomed. Mass Spectrom.*, **12**, 525.
184. Gaskell, S.J., Guenat, C., Millington, D.S., Maltby, D.A., and Roe, C.R. (1986), *Anal. Chem.*, **58**, 2801.
185. Unger, S.E. (1985), *Int. J. Mass Spectrom. Ion Processes*, **66**, 195.
186. Pelli, B., Traldi, P., Gargano, N., Ravasio, N., and Rossi, M. (1987), *Org. Mass Spectrom.*, **22**, 183.
187. Hass, J.R., Tondeur, Y., and Voyksner, R.D. (1983), *Anal. Chem.*, **55**, 295.
188. Zaretskii, Z.V.I., Dan, P., Kustanovich, Z., Larka, E.A., Herbert, C.G., Beynon, J.H., and Djerassi, C. (1984), *Org. Mass Spectrom.*, **19**, 321.
189. Boyd, R.K., and Beynon, J.H. (1977), *Org. Mass Spectrom.*, **12**, 163.
190. Barber, M., and Elliott, R.M. (1964), in *Proceedings of the 12th Annual Conference on Mass Spectrometry and Allied Topics*, Montreal.
191. Bruins, A.P., Jennings, K.R., and Evans, S. (1978), *Int. J. Mass Spectrom. Ion Phys.*, **26**, 395.
192. Haddon, W.F. (1980), *Org. Mass Spectrom.*, **15**, 539.
193. Lacey, M.J., and Macdonald, C.G. (1979), *Anal. Chem.*, **51**, 691.
194. Standing, K.G., Ens, W., Mayer, F., Tang, X., and Westmore, J.B. (1989) in *Ion Formation from Organic Solids*, IFOS V (Eds. A. Hedin, B.U.R. Sundquist and A. Benninghoven), John Wiley, Chichester.

195. Nourse, B.D., and Cooks, R.G. (1990), *Anal. Chim. Acta*, **228**, 1.
196. Roepstorff, P., and Fohlman, J. (1984), *Biomed. Mass Spectrom.*, **11**, 601.
197. Gallegos, E.F. (1976), *Anal. Chem.*, **48**, 1348.
198. Hunt, D.F., and Shabanowitz, J. (1982), *Anal. Chem.*, **54**, 574.

Chapter 8

Quantitative Analysis

8.1 Introduction

Although mass spectrometry has been recognized as a technique for quantitative analysis since its inception, the greatest impetus to its use in the field of quantitative organic analysis has come from the coupling of mass spectrometry with gas chromatography as a separation technique. Following a chromatographic separation, specific compounds can be determined in complex mixtures with minimal interference by monitoring only selected m/z values that are characteristic of the compounds under study. An important early report of selected ion monitoring (SIM) using a combined GC–MS instrument was presented by Sweeley *et al.* [1] using a so-called accelerating voltage alternator [1]. This device repeatedly switched the accelerating voltage of a magnetic sector instrument between the values needed to focus two ions, and was used to effect the separate monitoring of d_0-trimethylsilyl

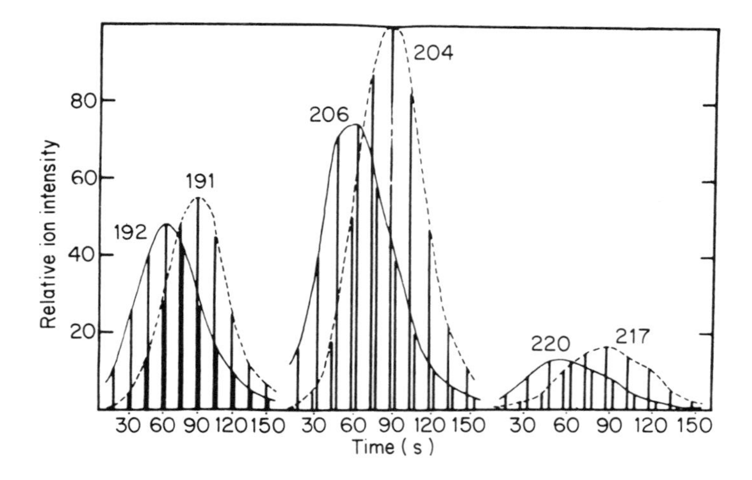

Fig. 8.1 Changes in fragment ion intensities during the elution of d_0- and d_7-trimethylsilyl glucose from a packed column. (Reprinted with permission from C.C. Sweeley *et al.*, *Anal. Chem.*, **38**, 1549[1]. Copyright (1966) American Chemical Society)

glucose and d_7-trimethylsilyl glucose which eluted from the gas chromatograph with closely similar retention times (Fig. 8.1).

Since Sweeley's original publication, technology and methodology have progressed and the number of applications of selected ion monitoring to quantitative analysis has increased very rapidly. Present day technology has made available laminated electromagnets which can be switched rapidly and precisely under data system control. Previously, magnetic sector instruments were less well adapted to selected ion monitoring than were quadrupole instruments since magnet switching, which offers constant sensitivity, was not practical and accelerating voltage switching is limited to a mass range of perhaps 30% if dramatic changes in sensitivity are not to ensue. Quadrupole analysers, on the other hand, have enjoyed a continuous reputation for ease of data system control with switching possible at high speeds over wide mass ranges without loss of sensitivity. Most organic quantitative mass spectrometry experiments are still based on selected ion monitoring following a gas chromatographic separation into an EI/CI source, although other inlet systems such as an LC or a direct insertion probe and other ionization techniques (section 8.3.3) are also used. Sections 8.1.1 to 8.2.4 in this chapter offer a more detailed consideration of various aspects of selected ion monitoring.

Quantitative data can also be obtained by repetitive scanning, e.g. during elution of a sample from the GC, although detection limits are generally much poorer than with selected ion monitoring since the instrument spends very little time at each m/z value during scanning (section 8.1.2). Such scans have either been scans of the entire mass range or scans of a limited mass range for increased sensitivity. This technique has been used, for example, in the analysis of carbohydrate and acidic metabolites in urine, quantitative data being calculated from areas of GC peaks displayed as selected mass chromatograms [2].

Repetitive scanning is also applied to the measurement of a large number of target compounds in complex environmental samples. One such report [3] describes a scanning method for the determination of polychlorinated biphenyls (PCBs), chlorinated dibenzo-*p*-dioxins (CDDs), and chlorinated dibenzofurans (CDFs). Automated data processing for PCBs, which incorporates a library search of the recorded spectra, has been found to be as reliable and accurate and considerably more cost-effective than operator-mediated methods [3,4]. A more recent report [5] describes the determination of a range of involatile organic compounds in aqueous environmental samples using scanning thermospray LC–MS. The authors of these reports suggest [6] that although improved sensitivity and precision should be available with selected ion monitoring, the loss of qualitative information is significant and usually unacceptable.

Scanning as a quantitative technique is also used in the analysis of unfractionated mixtures, e.g. in oil analysis, where the sample is admitted as a steady flow from a reservoir inlet (section 2.3.1). In order to carry out the quantitative analysis of an unfractionated mixture, it is necessary to record spectra of the pure components to determine sensitivity coefficients. A sensitivity coefficient is the response at an

m/z value of interest from a pure component at unit pressure. The analysis then requires the solution of a series of simultaneous equations which involve sensitivity coefficients for each component [7].

8.1.1 Specificity

The specificity of the mass spectrometer as an analytical instrument has correctly been stressed by many authors. Comparison of mass spectrometry with a number of other analytical techniques has shown that mass spectrometry is routinely able to provide more reliable data because of high specificity combined with instrumental sensitivity. The major disadvantages of mass spectrometry are the high cost and the relatively low sample throughput. Thus, the use of mass spectrometry in quantitative analysis is indicated where sensitivity and specificity rather than throughput are of major importance.

Despite the foregoing comments, mass spectrometry does not offer absolute specificity and the degree of specificity can depend on the manner in which the instrument is used. Generally, a recording of signals at two or three selected masses that shows the correct relative intensities is considered sufficient to diagnose the presence of one particular compound rather than one of several thousand others. A molecular ion, provided it is sufficiently intense, is usually the most suitable ion. Fragment ions of high m/z ratio, again provided they are sufficiently intense, are preferred to those of low m/z ratio for diagnostic purposes. Selected ion monitoring is particularly specific when the sample is introduced chromatographically and the signal at each mass then maximizes at the correct retention time for the particular compound.

Figure 8.2 illustrates a number of possible results from the search for a particular compound by selected ion monitoring. The examples given in Fig.8.2(a) to (c) seem straightforward, and even the example in Fig. 8.2(d) where two of the three ions

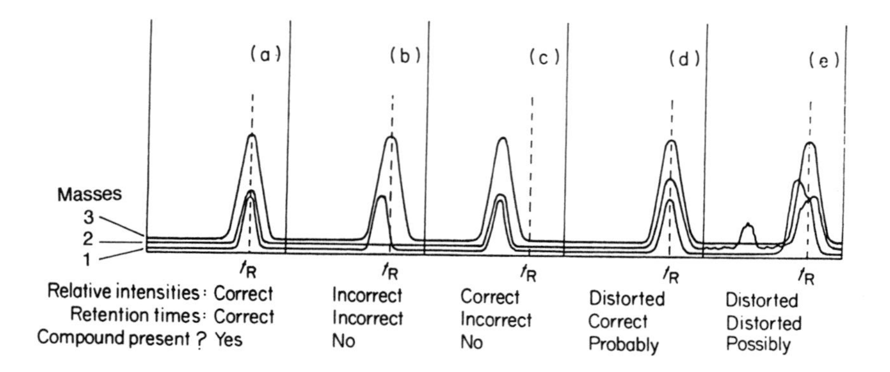

Fig. 8.2 Selected ion monitoring with GC separation

have the correct relative intensity may constitute an acceptable identification. Once we begin to look for compounds at lower concentrations, however, much higher levels of interference may be expected and the situation can become much more complex. For example, in Fig. 8.2(e) the response at mass 1 is only an inflection on a poorly resolved peak. The response at mass 2 sits on top of a continuous response (e.g. from column bleed) whose inherent variation makes the trace noisy and perhaps more difficult to interpret. The whole chromatogram is much 'busier', and while it appears that the expected compound may be present, our confidence in this assessment is somewhat lower than in the previous examples.

There are a number of non-mass-spectrometric means by which this situation may be improved. Investment in simple preliminary purification or fractionation procedures, e.g. liquid chromatographic separations using small scale columns [8,9], extraction into less polar solvents, and acid–base separations, should be considered as a matter of course for any routine analysis. The use of fused silica capillary columns should be considered in all cases where the sample is amenable to gas chromatography. The sample is introduced over a very short time interval from these columns so that the sample flow measured in grams per second is high, the background from the columns is exceedingly low, and many impurities in the sample will hopefully be resolved out. The column length and operating conditions may be chosen to give the shortest possible analysis time consistent with useful resolution. Fused silica columns also minimize the losses of polar compounds by adsorption, particularly when the columns are introduced directly into the ion source [10].

Table 8.1 Qualitative criteria used to establish specificity for the determination of 2,3,7,8-TCDD by high resolution selected ion monitoring. (Reproduced with permission from ref. 11)

(a)	Retention time for unlabelled 2,3,7,8-TCDD must be within -1 to +3 s of that of the isotopically labelled internal standard. Retention times of other TCDDs must fall within the retention time window established by the GC column performance mixture.
(b)	Ion current responses for *m/z* 258.930, 319.897, and 321.894 from the unlabelled TCDDs must maximize simultaneously (± 1 scan) and all ion current intensities must be at least 2.5 times the noise level.
(c)	The response ratios for *m/z* 319.897/321.894, and for 331.937/333.934 must be between 0.67 and 0.90 (0.77 theoretical).
(d)	Ion current responses for *m/z* 331.937 and 333.934 from ^{13}C-labelled TCDDs must maximize simultaneously (± 1 scan).

The alternative approach is to make the mass spectrometer more selective. A number of methods of increasing instrumental specificity are discussed in detail in later sections, viz. high resolution selected ion monitoring (section 8.3.1), selected reaction monitoring (section 8.3.2), and the use of alternative ionization techniques (section 8.3.3). Another possibility is to monitor a much higher m/z value where the background is lower, perhaps following derivative formation to increase the molecular weight of the sample (section 8.3.3). All of these techniques may be used in conjunction with chromatographic separation methods. The data in Table 8.1 give an example of the qualitative criteria used to establish specificity for 2,3,7,8-tetrachlorodibenzo-*p*-dioxin (2,3,7,8-TCDD) when this is determined by high resolution selected ion monitoring.

Increased mass spectrometric specificity has also been promoted as an alternative to extensive sample work-up procedures, particularly as a basis for screening methods [12], and is usually mandatory if the sample cannot be separated chromatographically. The direct quantitative analysis of unpurified samples can, however, be subject to matrix effects. Considerable variations in the response to analytes can be ascribed to the presence of larger quantities of matrix material which have a significant effect on the ion source sensitivity [13,14].

8.1.2 Sensitivity and limits of detection

During scanning, the mass spectrometer spends very little time at each mass so that the number of ions in a mass spectral peak can be very small. Statistically, a single or two ion peak at a given mass might be considered to be an unreliable event since there is a 37% chance that a single ion peak would be a zero ion peak, i.e. not recorded. Similarly, there is a 14% chance that a two ion peak would not be recorded in a subsequent scan. Therefore, we take a four ion peak, which has only a 2% chance of not being re-recorded, as a reasonable detection limit.

A suitable dynamic range of intensity for an identifiable spectrum may be taken as 20 to 1, i.e. base peak down to 5% peak, as this range covers most of the entries in the 'Eight Peak Index of Mass Spectrometry' [15], itself a very suitable basis for compound identification. Therefore, a useful spectrum should contain a minimum of 80 ions in the base peak.

For an exponentially scanning magnetic sector instrument with a scan speed of 1 s per decade in mass and a resolution of 1000, equation (1.6) shows that any peak in the spectrum will have a width, $t_p = 435\,\mu s$. The minimum of 80 ions contained in a triangular peak of this width implies an ion arrival rate of $n = (80 \times 2)/(435 \times 10^{-6}) = 3.7 \times 10^5$ ions s^{-1}. The sample flow rate (F $\mu g\,s^{-1}$) which corresponds to this ion arrival rate is given by

$$F = (3.7 \times 10^5)(1.6 \times 10^{-19})/S \tag{8.1}$$

where S is the sensitivity in coulombs per microgram. If, for example, we were to use an EI source with a sensitivity of $10^{-7}\,C\,\mu g^{-1}$ at the base peak, then the appropriate value of F is $6 \times 10^{-7}\,\mu g\,s^{-1}$. As an example, an identifiable spectrum could be achieved by scanning at the maximum of a GC peak, which has a base width of 12 s, corresponding to approximately 4 pg (4×10^{-12}g) entering the ion source. Better detection limits might be obtained by reducing the scan speed although a slower scan speed would not be compatible with GC operation.

If the analysis is now designed specifically for one compound so that only a limited number of masses are monitored, the potential sensitivity is greatly increased. A simple comparison suggests that, whereas during the scanning experiments described above less than one-thousandth of the total time is spent at any one mass, half the time is available when only two selected ions are monitored and the sensitivity increases accordingly. A calculation of the sample level needed to give a well-defined peak in selected ion monitoring suitable for quantitative measurements may be made as follows. Assume that the chromatographic peak requires twelve separate data points to establish its profile and that the data point located at the outset of the peak corresponds to our minimum detection limit of four ions. Then the total peak contains approximately 170 ions so that the detection limit is 170 ions or $170n$ ions when n separate masses are being monitored. With a sensitivity of 10^{-7} C μg^{-1}, a 170 ion peak corresponds to approximately 0.3 fg (3×10^{-16}g) entering the ion source.

Instrumental detection limits for GC–MS calculated in this way are idealized, and generally do not reflect the detection limits that can be achieved with real samples. This is especially true of the very low limit calculated for selected ion monitoring which, in practice, is usually determined by factors other than analyte ion statistics, although techniques that efficiently discriminate against background (e.g. high resolution selected ion monitoring, section 8.3.1, or negative ion chemical ionization, section 8.3.3) can approach these theoretical limits. A limit of detection (LOD) that is a generally more appropriate indication of analytical capability also takes into account factors such as instrumental background, background from the sample matrix, recovery from the matrix and other analytical variables (cf. section 2.2.5.3).

A practical estimate [16] of the limit of detection for a selected ion monitoring GC–MS method is based on the peak-to-peak noise measured on the baseline close to the analyte peak. The limit of detection is then defined using the amount by which the analyte peak height measured above the base line ($X_D - X_B$) exceeds the baseline variability (σ_B):

$$X_D - X_B = K_D \sigma_B \tag{8.2}$$

A minimum value of $K_D = 3$ is recommended for detection [16]. It will be obvious, for example, that, if the analyte is being measured at an m/z value where there is already a high background ion current, fluctuations in this ion current will provide a substantial contribution to the baseline noise level.

The limit of detection expressed in concentration units (C_D) is derived from the calibration (section 8.1.4) which relates peak height to analyte concentration. Thus, if concentration C is a simple linear function of peak height X, i.e. $X - X_B = mC$, then the limit of detection, in concentration units, is given by

$$C_D = K_D \frac{\sigma_B}{m} \tag{8.3}$$

The LOD should not be extrapolated from the analysis of more concentrated samples that give a very much greater than background response. A satisfactory determination of LOD would be based on the analysis of dilute samples where the S/N of GC–MS peaks is no greater than approximately 10:1 [17].

The limit of quantitation (LOQ) is defined in a similar manner to the LOD:

$$X_Q - X_B = K_Q \sigma_B \tag{8.4}$$

$$C_Q = K_Q \frac{\sigma_B}{m} \tag{8.5}$$

A minimum value of $K_Q = 10$ is recommended for quantitation [16]. The various regions of analyte measurement are summarized in Table 8.2. More sophisticated statistical analyses of detection limits are also available [18–20].

Table 8.2 Regions of analyte measurement. (Reprinted with permission from ACS Committee on Environmental Improvement, *Anal. Chem.*, **52**, 2242[16]. Copyright (1980) American Chemical Society)

Analyte signal	Recommended inference
$< 3\sigma$	Analyte not detected
3σ to 10σ	Region of detection
$> 10\sigma$	Region of quantitation

8.1.3 Internal standards

Quantitative measurements in mass spectrometry are usually referred to an internal standard. The internal standard provides its own characteristic ion or ions which are monitored together with those characteristic of the sample by rapidly switching between the appropriate mass values. Typically the internal standard is an isotopically labelled analogue of the sample, although isomers, homologues, or structurally analogous compounds are alternative possibilities [21]. Ideally, sources of variation should affect both sample and standard in a similar manner and for this reason the stable isotope labelled analogue, which comes closest to meeting this requirement, is generally the most suitable form of internal standard. Stable isotope labelled internal standards may also be used in multicomponent GC–MS analyses, but economic considerations often dictate that only a limited number of internal standards, chosen to elute at intervals throughout the analysis, or even a single internal standard, will be used. Thus, $^{2}H_{10}$-anthracene has been used as a single quantitation standard for the determination of a range of polycyclic aromatic hydrocarbons (PAHs) in waste water [22,23].

Deuterium-labelled standards are the most readily prepared with good isotopic purity [24,25] and are therefore the most commonly used. A possible disadvantage of deuterium labelling, particularly where it involves a larger number of atoms, is that the labelled analogue is at least partially separated from the unlabelled analyte when a capillary GC column is used and therefore may no longer act as an ideal internal standard. ^{13}C-labelled analogues are much better in this respect [26] but are less readily available and generally of lower isotopic purity. ^{18}O analogues have also been used in a limited number of cases [27–29] and similarly do not show the chromatographic separation experienced with deuterated analogues [30].

An inability to prepare a labelled analogue in high purity is a disadvantage if the labelled material is to be used as an internal standard. The addition of the internal standard will then create an interference at the m/z value used to monitor the analyte. It is also possible for the natural isotope signals from the unlabelled analyte to interfere at the m/z value used to monitor the internal standard, although this effect can be minimized by ensuring that the internal standard molecule contains at least three heavy atoms [32]. The consequences of these interferences are considered in the following section on calibration (section 8.1.4).

A large excess of a co-eluting, stable isotope labelled analogue has traditionally been used to reduce analyte losses due to adsorption via the so-called 'carrier effect' [31]. The mechanism of this carrier effect has been questioned [25,31,32] and in some cases, e.g. the analysis of octopamine with a deuterated internal standard [33], the effect has been shown not to exist. Other workers have, however, reported a carrier effect even when a homologue, rather than a labelled analogue, is used as an internal standard [34].

When the use of a carrier effect is not anticipated, the amount of labelled material added to each unknown sample should be only a few times greater than the lowest expected level of unlabelled material. This will lead to improved limits of detection, since the amount of unlabelled material which originates from the internal standard will be almost negligible.

8.1.4 Calibration [35,36]

Calibration is a prerequisite for quantitation. With the use of an internal standard, calibration relates the ratio of the response to the analyte (X) and the response to the internal standard (X_S) to the quantities of analyte (C) and internal standard (C_S) present.

$$\frac{X}{X_S} = f(C, C_S)$$

The graph of this function, determined from a number of known mixtures of the analyte and internal standard, forms the basis of the calibration curve (e.g. Fig. 8.3). The ion abundance ratios for unknown samples spiked with known amounts of internal standard are subsequently measured and compared with this calibration curve to determine the quantity of analyte present in the unknowns.

The form of the calibration curve when an isotopically labelled standard is used can be expressed as follows. Let the mass at which the analyte is measured be M_1 and that at which the standard is measured be M_2. Let the abundances of the contributions at M_1 and M_2 from the pure analyte be a_1 and a_2 respectively, and those from an equal amount of the standard be b_1 and b_2, respectively. If, in any

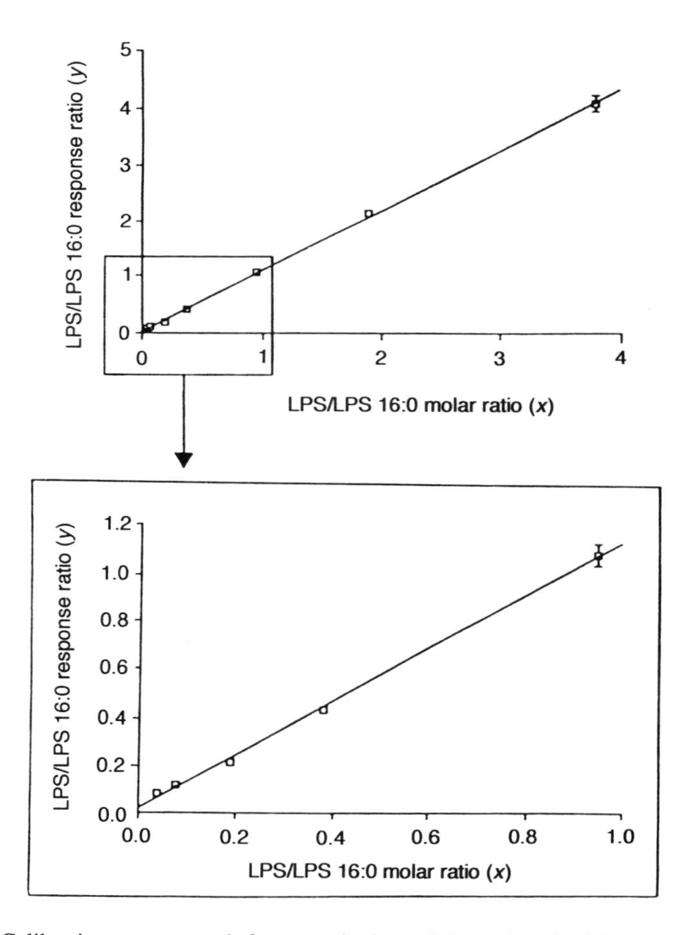

Fig. 8.3 Calibration curve used for quantitation of lyso-phosphotidylserine (LPS) by fast atom bombardment using 1-hexadecanoyl-*sn*-glycero-3-phospho-L-serine (LPS 16:0) as internal standard. (Reproduced with permission from ref. 120)

analysis, the respective molar amounts of analyte and standard are C and C_S, the measured ion abundance ratio, X/X_S, is then given by

$$\frac{X}{X_S} = \frac{a_1 C + b_1 C_S}{a_2 C + b_2 C_S} \tag{8.6}$$

and dividing through by C_S gives

$$\frac{X}{X_S} = \frac{a_1 (C/C_S) + b_1}{a_2 (C/C_S) + b_2} \tag{8.7}$$

There are four general cases of equation (8.7). The first is the ideal case where the analyte does not interfere at mass M_2 and the standard does not interfere at mass M_1. In this case, $a_2 = b_1 = 0$ and the calibration is a straight line. The slope will approach 1, but deviation will occur if, for example, there is an isotope effect that influences the fragmentation of the standard [41]. The second case occurs when the standard interferes at mass M_1 but the analyte does not interfere at mass M_2. In this case, $a_2 = 0$ and the calibration is again linear but the intercept is not zero. If the analyte interferes at mass M_2 or, as is most often the case, the analyte interferes at mass M_2 and the standard interferes at mass M_1, then the calibration is no longer a straight line but is hyperbolic, as may be seen by rearranging (8.7) to the form:

$$\frac{X}{X_S} = P + Q\left(\frac{C}{C_S}\right) + R\left(\frac{C}{C_S}\right)\left(\frac{X}{X_S}\right) \tag{8.8}$$

where P, Q, and R are constants.

For linear data, the calibration is constructed by the use of a straightforward least squares method [41]. The resulting regression line is that which minimizes the sum of the squares of the distances between the line and the observed data. As we have seen previously, the use of a labelled internal standard can introduce cross-contributions at each mass monitored and consequently lead to a non-linear calibration. Non-linearity necessitates either discarding data in the non-linear portion of the calibration, linearization of the data through some transformation [37], or fitting the data to a non-linear equation. Use of the latter method has become more popular with the availability of non-linear statistical programs. Most investigators have chosen to fit the data to a hyperbolic form [38,39] while others have used a polynomial fitting routine [40].

At least four different concentrations of the analyte together with the internal standard should be prepared for calibration. These solutions should include both extremes of the concentration range expected in unknown samples. In order to construct a calibration curve, a regression of ion abundance ratios on concentration ratios is generally carried out (e.g. the sum of the squares of the deviations along the y axis is minimized in Fig. 8.3). In this case, repeat determinations of the ion abundance ratio are carried out on each of the solutions in order to improve the precision of the calibration. The use of this regression line assumes that while the concentration ratios are not known with perfect accuracy, they exhibit only a small variance compared with the variance in the mass spectrometric determinations. In practice, this is an acceptable assumption that has been discussed in more detail by Falkner and other authors [36,38].

When a calibration covers a wide range of concentrations, the corresponding ion abundance ratios will show a wide range of absolute standard deviations since the standard deviation increases as the ion abundance ratio increases (section 8.2.3). Thus, standard regression techniques, which assume a constant standard deviation, are no longer applicable. A solution to this problem is the use of a weighted regression that is less sensitive to the less precise points at one extreme of the calibration [41,42].

When there is little interference at each mass monitored ($a_2 = b_1 = 0$ in equation (8.6)), the calibration may be abbreviated to the use of a single response factor. An example is in dioxin analysis where a typical internal standard is a $^{13}C_{12}$ analogue which is monitored at an m/z value 12 mass units above that used for the analyte itself [11,43]. In this case, if the relative response for the analyte and its labelled analogue has a coefficient of variation which is less than 20% over the calibration range, an average relative response may be used to calculate analyte concentrations. Otherwise, the complete calibration curve is used [44]. The relative response (RR) takes the form given by

$$RR = \frac{XC_S}{X_S C} \tag{8.9}$$

where the other variables are defined as before.

A drawback of the calibration curve method is that measured ion abundance ratios can change with changes in instrumental operating conditions. Despite this, an existing calibration is often used for a considerable period of time before a new calibration is prepared. The validity of an existing calibration may be checked by the analysis of a sample of known mole ratio with a statistical assessment of the difference between the measured and expected ratios [42]. Schoeller [45] has also discussed the effect of data reduction methods in minimizing such errors.

A more laborious alternative, which avoids the difficulties associated with a previously established calibration curve, is bracketing. Thus, each sample measurement is bracketed both in time and abundance ratio by the measurement of calibration standards. The standards selected for bracketing are a pair whose ion abundance ratios closely surround the ion abundance ratio of the sample. Interpolation between the two standards is then used to find the molar ratio of the sample. Bracketing is only used when the highest accuracy and precision are required [26].

8.2 Sources of Error

8.2.1 Sample handling errors

A typical sample handling procedure is illustrated in Fig. 8.4. Errors may be introduced at any stage of this handling procedure. The principal causes of these errors will be errors in the measurement of weights and volumes and analyte losses, e.g. due to evaporation of volatile materials or incomplete extraction or derivatization. The use of a suitable internal standard (surrogate) which is added at the beginning of the work-up procedure as shown, and which is therefore subject to all the same processes as the analyte, can obviate most of these sources of error and bring precision and accuracy to a very acceptable level [17].

Obviously, care must still be exercised in the work-up procedure, especially if an isotopically labelled standard is not used, since some processes, such as extraction, will vary in effectiveness from compound to compound. Overall, depending on the

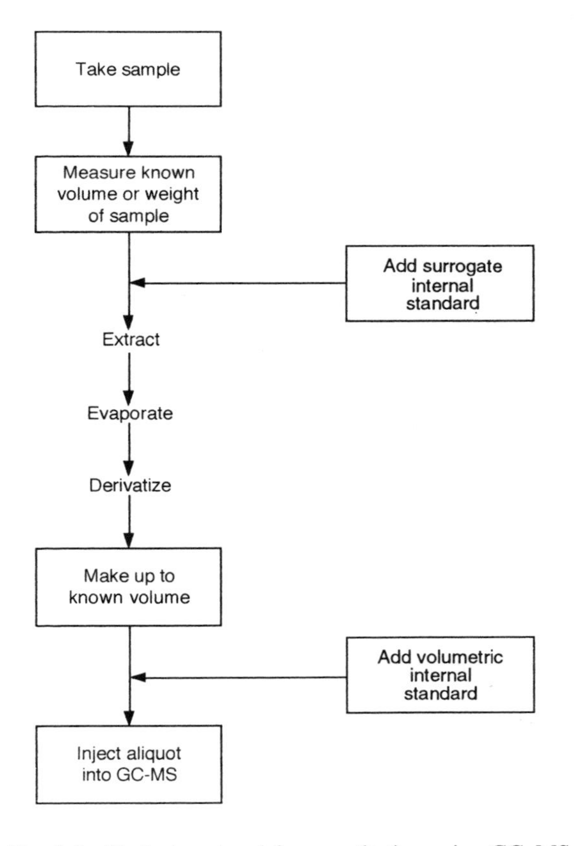

Fig. 8.4 Typical protocol for quantitation using GC–MS

sample type, it is very often a factor such as sample heterogeneity, for which the use of an internal standard does not compensate, that is the major contributor to error. For example, Eicemann *et al.* [46] have demonstrated that in the analysis of di-(2-ethylhexyl)phthalate in sludge-aided soils, more than 90% of the imprecision originates from sample heterogeneity and less than 10% from GC–MS or sample work-up.

Errors in gravimetric and volumetric procedures which follow the addition of an internal standard (IS) in Fig. 8.4 will be compensated for by the use of the standard. Gravimetric and volumetric procedures involved in making up solutions of analyte and IS for calibration, however, include no such compensation. Therefore these procedures must be carried out as accurately as possible.

8.2.2 Instrumental errors

It is important that an instrument is set to monitor the peak top for each mass selected. The ideal instrumental peak shape is flat topped so that any error in the mass setting has the minimum effect on the recorded signal intensity. Instrumental stability is also important so that any mass setting once effected will be maintained throughout the analysis. There is, as Schoeller [45] has pointed out, no escaping the need for short-term instrumental stability [47] if within-run accuracy and precision are to be maintained.

Ideally, instrumental sensitivity is optimised at the beginning of an analysis and is then maintained throughout the analysis. Source parameters such as the electron energy and the ion repeller voltage are generally set on broad maxima (cf. Fig. 1.2) where small changes in parameter value have little effect on ion intensity. Sometimes, however, other values are preferred. For example, the electron energy for dioxin GC–MS analysis is optimally set at *ca.* 30 eV when sensitivity is only marginally reduced by a high helium pressure in the ion source [48].

Sensitivity is affected by the build-up of contamination within the ion source during use. Again, the competitive ionization of materials that co-elute from a GC or LC can suppress sensitivity towards the analyte. A co-eluting internal standard is able to compensate effectively for such changes in instrumental sensitivity and provide improved precision and accuracy in quantitative analyses [49].

8.2.3 Measurement procedure errors

Ion statistics provides a fundamental limit to the precision of quantitative measurements. The standard deviation for the measurement of peak area is given by

$$\sigma = N^{1/2} \tag{8.10}$$

where N is the number of ions contained in the peak. Alternatively, this measure may be expressed as a percentage coefficient of variation:

$$\nu = 100\left(\frac{1}{N}\right)^{1/2} \tag{8.11}$$

Thus, a peak that contains 400 ions can, at best, be measured with a coefficient of variation equal to 5%.

When, as is usual in quantitative mass spectrometry, a response ratio is measured, the standard deviation for a ratio $R(= N_1/N_2)$ is given by

$$\sigma = R\left(\frac{1}{N_1} + \frac{1}{N_2}\right)^{1/2} \tag{8.12}$$

where N_1 and N_2 are the numbers of ions in the peaks ratioed [42]. This measure can also be expressed as a percentage coefficient of variation:

$$\nu = 100\left(\frac{1}{N_1} + \frac{1}{N_2}\right)^{1/2} \tag{8.13}$$

These statistics may be improved, for example, with a weaker ion current, by increasing the relative dwell time at the appropriate m/z value. Naturally, this improvement is realized at the expense of the other measurements if the whole experiment has to be conducted in a limited time, as, for example, when the sample is introduced chromatographically.

In practice, the accuracy and precision of peak height measurements can exceed those of peak area measurements due primarily to interfering compounds eluting close to the analyte and problems in assigning the beginning and end of a peak. The use of peak areas usually requires baseline separation, whereas poorer separations can be tolerated with peak height measurements. Peak height measurements may be applied after smoothing the data or after fitting the data to a suitable peak profile model. For example, Fukuda and co-workers [50] have demonstrated that the use of peak heights calculated using a modified Gaussian peak shape model offers better precision than a direct manual measurement of peak heights.

An effective means of reducing imprecision in ion intensity ratio determinations, especially from the weaker signals recorded in high resolution selected ion monitoring (section 8.3.1), is the use of a peak profile method [39,51]. Figure 8.5 shows a plot of the response at m/z 416.257 versus the response at m/z 419.276, recorded during a high resolution experiment. The elliptical appearance of the plot is due to the earlier gas chromatographic elution of the internal standard. Application of an empirical shift correction reduces the plot to a linear form. Treatment of the corrected data by linear regression then gives a slope equal to the

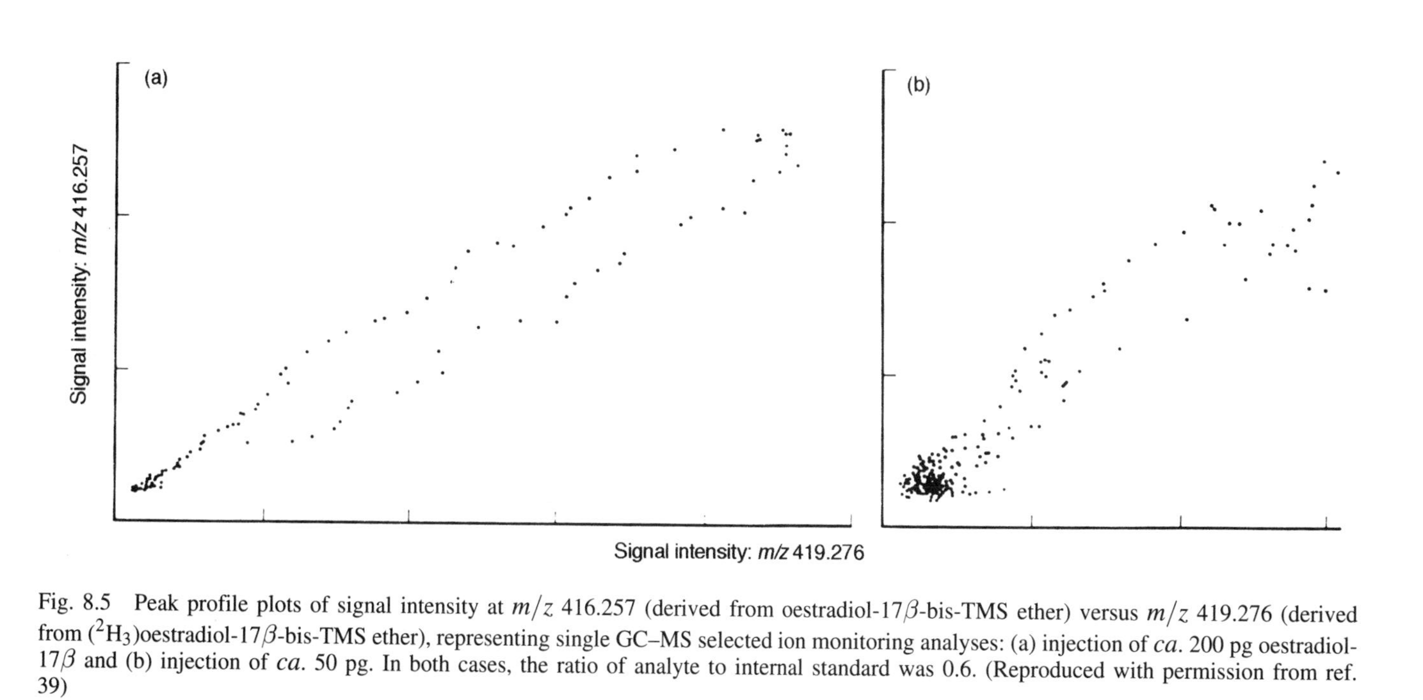

Fig. 8.5 Peak profile plots of signal intensity at m/z 416.257 (derived from oestradiol-17β-bis-TMS ether) versus m/z 419.276 (derived from ($^{2}H_{3}$)oestradiol-17β-bis-TMS ether), representing single GC–MS selected ion monitoring analyses: (a) injection of *ca.* 200 pg oestradiol-17β and (b) injection of *ca.* 50 pg. In both cases, the ratio of analyte to internal standard was 0.6. (Reproduced with permission from ref. 39)

ion intensity ratio m/z 416.257/419.276. Gaskell and co-authors [39] attribute the improvement in the precision of ratio determination using this procedure to two features: (a) regression analysis is equivalent to smoothing of the intensity data; (b) the baseline and peak-top values at each m/z value need not be determined.

8.2.4 Measurement quality

There are two basic indicators of measurement quality: precision and accuracy. Systematic errors, such as loss of volatile samples during concentration of a solution, incomplete extraction of a sample from the matrix, or the presence of a co-eluting material which also provides a signal at the m/z value being monitored, affect the accuracy by introducing a net bias. These systematic errors will, however, also contain a random contribution that affects precision and may even be larger than the purely random error. In general, all practicable steps will have been taken to eliminate or compensate for systematic errors before the adoption of an analytical method.

Purely random errors originate from sources such as measuring uncertainties, sample heterogeneity, and ion statistics, many of which have been discussed in previous sections. In this context, it should be noted that imprecision in the mass spectrometric stage of the determination of analyte concentration will arise from uncertainty in the calibration (section 8.1.4) as well as in the mass spectrometric measurements made on the sample itself. Thus, replicate analyses of the unknown, which only include the latter source of error, will underestimate the total uncertainty [42].

With a conventional GC–MS selected ion monitoring procedure and a suitable internal standard, a short-term (within-run) coefficient of variation of the order of 1–2% should be attainable at intermediate concentrations (ng–mg cm^{-3}) [21,52]. However, precision worsens rapidly to 10–20% or more for the analysis of lower analyte levels when adsorption and other non-reproducible losses occur and when the specificity may be much poorer because of background interference. As an example of the improvement that can be effected at higher sample levels by careful attention to analysis conditions, a definitive method for serum cholesterol which uses bracketing (section 8.1.4) for quantitation achieved a coefficient of variation for single measurements of 0.22% [26]. The use of a capillary GC–MS instrument specifically modified to produce high precision results and a rigid measurement protocol that uses the instrument under optimal conditions are cited as important criteria in achieving this level of precision.

The most effective procedure for the elimination of bias, and therefore the provision of accurate results, is the use of some form of internal standard which is carried through the whole procedure from initial extraction to final mass spectrometric determination. The use of an isotopically labelled analogue as the internal standard ensures that, as far as possible, the standard and analyte are affected to the same extent by each step of the analytical procedure. Naturally, such precautions will not compensate for the effects of co-eluting materials which also provide a signal at the m/z value being monitored. As far as possible, a knowledge of likely interfering materials should be established beforehand to facilitate assessment and control of these problems during method development [16,17,26,53].

The use of surrogate standards in conjunction with a conventional (volumetric) internal standard, which is added just before the GC–MS analysis (Fig. 8.4), allows a separate recovery figure to be determined for the sample handling process. For example, in the determination of 2,3,7,8-TCDD [11], $^{13}C_{12}$-2,3,7,8-TCDD is added, as a surrogate, to the original matrix before extraction, while $^{13}C_{12}$-1,2,3,4-TCDD is added, as an internal standard, to the cleaned-up extract just before GC–MS analysis. All three compounds are monitored and, using a prior calibration, the recovery of the $^{13}C_{12}$-2,3,7,8-TCDD may be determined (Eqn. 8.14) by reference to the level of $^{13}C_{12}$-1,2,3,4-TCDD:

$$\text{Recovery}(\%) = 100\left(\frac{\text{amount of surrogate measured}}{\text{amount of surrogate added}}\right) \tag{8.14}$$

The quantity of native 2,3,7,8-TCDD in the extract is also measured, as usual, by reference to $^{13}C_{12}$-1,2,3,4-TCDD.

Alternatively, a prior estimate of the recovery of the sample work-up procedure for each analyte can be made by spiking suitable sample matrices with different, known levels of each analyte before analysis. The increase in GC–MS response due to the spike, which is measured relative to an internal standard added just before the GC–MS stage, is then used to calculate a recovery figure [16,22].

Data that are associated with recoveries that fall within an acceptable range may be corrected using the appropriate recovery figure in order to report accurate results [16,17,22], although it should be recognized that analyte spiked into a blank sample may behave differently (typically showing higher recovery) from native analyte [16,17]. Data for which recoveries are outside the acceptable range will generally not be corrected since, for example, a low recovery implies a greater possibility of variability [16,17]. Alternatively, low recovery data may be discarded. As an example, before quantifying a response in 2,3,7,8-TCDD analysis, the recovery of the surrogate standard ($^{13}C_{12}$-2,3,7,8-TCDD) must be between 40 and 120% [11].

Once an analytical method has been adopted for the measurement of unknowns, it should be periodically retested by analyses of blanks, standards, and spiked samples. Whenever practicable, analytical accuracy should be assessed by the analysis of homogeneous, certified reference materials. The provision of such standard reference materials is an important aspect of the development of trace analytical methods [16,26,54].

8.3 Alternative Methods

8.3.1 High resolution selected ion monitoring

If the resolution of the mass spectrometer is increased, the cross-contribution between one mass and another is reduced (Fig. 8.6) so that the interference from background or other sample components which are close to the mass being monitored is reduced accordingly. Operation at increased resolution forms the basis of a commonly accepted method of increasing performance in selected ion monitoring. A resolution of 5000–10 000 is typically employed so that the method is restricted to magnetic sector instruments which can provide a higher resolution with an acceptable sensitivity.

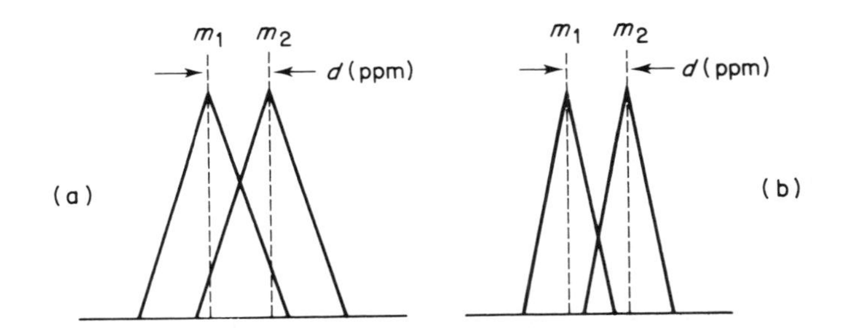

Fig. 8.6 Cross-contribution at different resolutions. (a) Low resolution; significant cross-contribution from m_1 to m_2 and vice versa. (b) Higher resolution and higher multiplier gain. No cross-contribution from m_1 to m_2 or vice versa

To set against the improved discrimination available at higher resolution, the absolute sensitivity of the instrument will be lower. This does not necessarily result in detection limits that are poorer than those obtained at low resolution. For example, if a resolution of 10 000, at which the sensitivity is approximately 10 times less than at a standard resolution of 1000, is sufficient to reduce the cross-contribution to a negligible level in a particular case then the detection limit of a 170*n* ion peak (section 8.1.2) is now provided by 3*n* fg ($3n \times 10^{-15}$g) in the ion source. In practice, it has been possible to achieve a close approximation to these theoretical detection limits (Fig. 8.7) because the background level is so much lower under high resolution conditions whereas theoretical detection limits calculated at low resolving power are rarely attained in practice.

In addition to the improved detection limits which may be realized, the use of higher resolution also makes any assay more specific. Thus, the observation of a response at the mass being monitored is less likely to be a contribution from an adjacent mass. For example, Table 8.3 [53] lists common chemical interferences in the GC–MS determination of 2,3,7,8-TCDD, a number of which can be separated at a resolution of 10 000. Unfortunately, it has been customary in high resolution

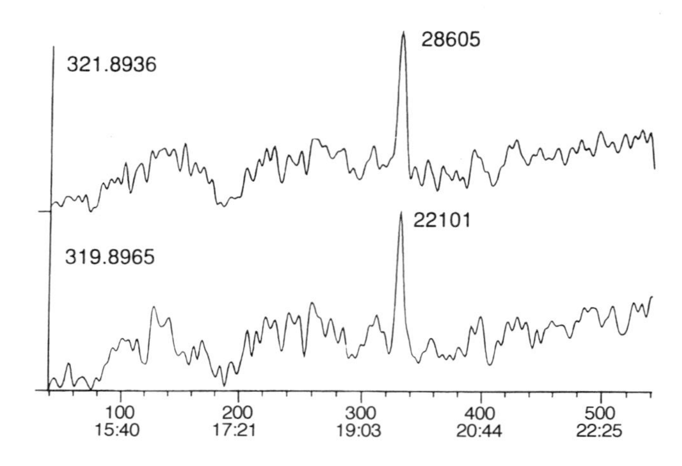

Fig. 8.7 Detection of 10 fg TCDD at 10 000 resolving power monitoring ten ions plus lock mass. (Courtesy Kratos Analytical Instruments)

selected ion monitoring to state the exact mass at which the instrument has been set up, e.g. 321.8936, perhaps giving the erroneous impression that the only response recorded would be from ions having this exact mass. In fact, even at high resolution, the instrument can still register the signal from a closely adjacent mass by reading the side of this peak rather than the top of the expected peak. In addition, interferences that are present at much higher concentrations than that of the analyte are likely to contribute at the recorded mass even when using a mass resolution that is apparently sufficient.

Table 8.3 Common chemical interferences in the GC–MS determination of 2,3,7,8-TCDD. (Reproduced with permission from ref. 53)

Compound or compound type	Molecular ion	Interfering ion	Exact mass	Resolution needed for separation
Dioxin quantification ion				
2,3,7,8-TCDD	$C_{12}H_4{}^{35}Cl_3{}^{37}ClO_2$	—	321.8936	—
Biphenyl	$C_{12}H^{35}Cl_8{}^{37}Cl$	$M^{+\cdot} - 4^{35}Cl$	321.8491	7223
Biphenyl	$C_{12}H_3{}^{35}Cl_7$	$M^{+\cdot} - 2^{35}Cl$	321.8678	12476
Xanthene	$C_{13}H_6O^{35}Cl_2{}^{37}Cl_2$	$M^{+\cdot}$	321.9114	18084
Benzylphenylether	$C_{13}H_7O^{35}Cl_3{}^{37}Cl_2$	$M^{+\cdot} - H^{35}Cl$	321.9114	18084
DDT	$C_{14}H_9{}^{35}Cl_2{}^{37}Cl_3$	$M^{+\cdot} - H^{35}Cl$	321.9292	9042
DDE	$C_{14}H_8{}^{35}Cl^{37}Cl_3$	$M^{+\cdot}$	321.9292	9042
Benzylphenylether	$C_{13}H_8{}^{35}Cl_3{}^{37}Cl$	$M^{+\cdot}$	321.9299	8868
Methoxybiphenyl	$C_{13}H_8{}^{35}Cl_3{}^{37}Cl$	$M^{+\cdot}$	321.9299	8868

Good instrumental stability is a particular prerequisite for high resolution selected ion monitoring since any change in the mass focused at the detector will result in a greater proportional change in the intensity of the recorded signal at high resolution than at low resolution. If the instrumental stability is inadequate, some short-term instability may be compensated for by integration of the output from a small mass sweep superimposed on the monitoring process, although this method leads to some loss of sensitivity and selectivity [55].

Long-term drift may be compensated for by monitoring an additional signal of known mass (lock mass) from a material that is introduced into the ion source at a constant rate. In the high resolution analysis of polychlorinated dibenzodioxins and dibenzofurans, a perfluorokerosene (PFK) lock mass chosen to be in the same mass range as the analyte ions being monitored is recommended [43,44]. The level of the reference compound introduced during the analysis should be adjusted to give a lock mass signal which is not subject to large statistical variations, while ensuring that there is no interference from PFK at the analyte masses. The lock mass signal can also be used to monitor changes in instrumental sensitivity during the analysis [44,56,57]. It is also necessary to use a mass sweep to locate the lock mass so that its position on the mass scale can be measured. Any change in the position of the lock mass will be ascribed to drift, which is assumed to have affected all the other masses being monitored so that an appropriate correction is then applied to these mass values.

In an alternative method designed to further increase the selectivity of high resolution selected ion monitoring, a wider, calibrated mass sweep is synchronized with the start of data acquisition at each mass being monitored and the data stored as peak profiles [58–61]. If these data are then displayed by integrating the results from each mass sweep, a conventional selected ion chromatogram is obtained. If, instead, all of the sweep data are summed over the period of sample elution but not integrated over the mass range, a mass profile of improved signal-to-noise ratio is obtained (Fig. 8.8). Since the sweep is calibrated, mass measurement may be carried out using the peak profile data. The use of a wider sweep does mean that sensitivity is sacrificed, but this is compensated for by the very much higher selectivity available when mass measurement is employed. Thus, the identity of any GC peak may be confirmed by mass measurement from the profile data. In addition, the profile data will display any interferences that are close to the mass being monitored (Fig. 8.8). Good instrumental stability is, of course, essential if this process of data summation and mass measurement is to be a viable one.

High resolution selected ion monitoring has most notably been applied to the quantitative determination by GC–MS of trace levels of polychlorinated dibenzodioxins and dibenzofurans in complex environmental matrices [11,17,43,54]. A comprehensive protocol for this method has been published by the US Environmental Protection Agency [44]. High resolution GC–MS operation has also proved valuable for the determination of hormonal steroids [39,62–64], phenylacetic acids [65], drugs and their metabolites in body fluids [66–69], and other xenobiotics such

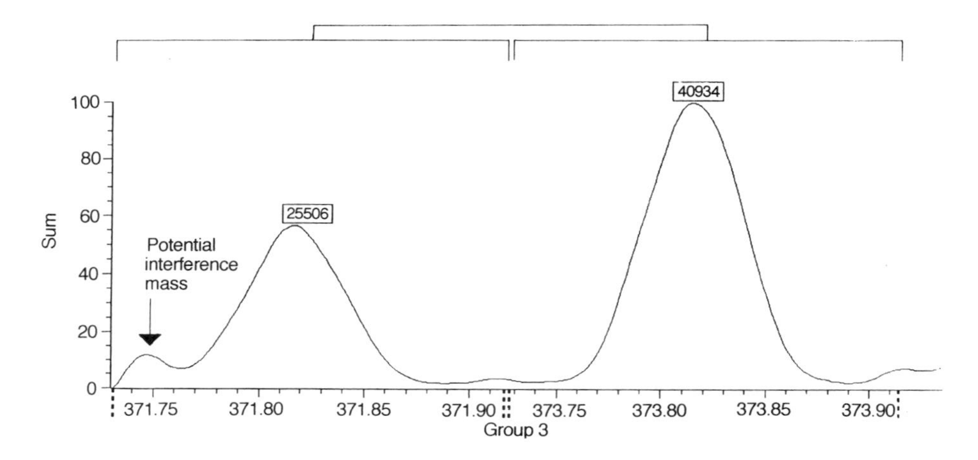

Fig. 8.8 Mass profiles for *m/z* 372 and 374, summed over the elution of hexachlorodibenzofuran. (Courtesy Kratos Analytical Instruments)

as *N*-nitrosodimethylamine [70] in complex matrices. Quantitation at high resolution, in conjunction with FAB ionization, although using narrow range scanning rather than selected ion monitoring, has been demonstrated as an effective method for cationic surfactants in environmental matrices [71]. Monitoring the intensity of the CH_2N^+ fragment ion at a resolution of 3000 has been proposed as an alternative specific detection method for nitrogen-containing compounds [72].

8.3.2 Selected reaction monitoring

An alternative method of making ion monitoring more selective is to monitor ions of a given mass which have been formed by the unimolecular or collision-induced decomposition of a particular precursor ion. This method is often referred to as selected reaction monitoring. Selected reaction monitoring increases selectivity by accessing finer resolution elements just as high resolution selected ion monitoring does. For example, at mass 300, fractional masses range from approximately 299.8 to 300.35 amongst organic compounds. When high resolution selected ion monitoring is carried out at a resolving power of 10 000, this fractional mass range is effectively divided into 100 ppm wide elements, of which there are approximately eighteen. With selected reaction monitoring, assuming, for example, a precursor ion resolution of 100 on the same sector instrument, rather more than eighteen resolution elements would be achieved even if the highest mass precursor ion for m/z 300 was only m/z 400. Thus selected reaction monitoring can be highly selective, although this selectivity is dependent on the instrumentation and the manner in which it is used [73].

A good example of selected reaction monitoring is provided by steroid *t*-butyldimethylsilyl ethers which show particularly simple fragmentation patterns and intense metastable ions associated with the formation or further reaction of $(M\text{-}C_4H_9)^+$ ions under electron impact ionization conditions:

$$ROH \rightarrow ROSi(CH_3)_2Bu^t \qquad (8.15)$$

The use of these derivatives was pioneered by Gaskell and co-workers [62,64,74] who demonstrated that, using a double-focusing magnetic sector instrument, metastable ion monitoring can provide sensitivity and selectivity for the assay of a range of steroid hormones in crude extracts at least equal to that provided by high resolution selected ion monitoring (Table 8.4).

Table 8.4 Precision of determination of testosterone in low and high concentration plasma samples by GC–MS. (Reproduced with permission from ref. 62)

		Testosterone (nmol/litre^{-1})	
		Low	High[a]
GCMS/high resolution selected ion monitoring			
Intra-assay	Mean	10.47	26.49
	SD	0.54	0.74
	n	6	6
	CV (%)	5.2	2.8
Inter-assay	Mean	10.40	26.49
	SD	0.72	0.82
	n	5	5
	CV (%)	6.9	3.1
GCMS/metastable ion monitoring			
Intra-assay	Mean	9.02	25.69
	SD	0.24	0.72
	n	6	6
	CV (%)	2.7	2.8
Inter-assay	Mean	9.99	25.24
	SD	0.87	1.44
	n	6	5
	CV (%)	8.7	5.7

[a] Prepared by supplementing the low concentration pool with testosterone 15.60 nmol/litre^{-1}.

Another example of metastable ion monitoring is a direct insertion probe method for the analysis of the miticide hexythiazox in crude fruit extracts [75]. This comparative study demonstrated the greater selectivity of selected reaction monitoring and achieved detection limits of 0.1, 0.1, and 0.02 mg kg^{-1} for low resolution selected ion monitoring, high resolution selected ion monitoring, and selected reaction monitoring respectively. Similar conclusions regarding the sensitivity and selectivity of selected reaction monitoring have been reached by other workers [76,77].

In practice, an ion of mass m_2 formed from a precursor of mass m_1 in the first field free region of a magnetic sector instrument (effective mass m_2^2/m_1 (Equation 7.8) and energy m_2/m_1) is monitored by first setting the magnetic field to focus m_1 at the detector and then reducing both the electrostatic analyser voltage and the magnetic field strength in the same ratio (m_2/m_1). A simpler but less general method that is useful if the magnet current cannot be switched rapidly allows a number of fragmentations with the same product ion m_2^+ to be monitored. In this case the electrostatic sector and accelerating voltages are again decoupled, the magnet tuned to mass m_2, and the accelerating voltage raised by a factor (m_1/m_2) to focus the fragmentation product (Fig. 8.9) [78].

In favourable cases a metastable ion may have an intensity of as much as 10% relative to the corresponding fragment ion, but in related compounds the same metastable ion may be orders of magnitude weaker. Metastable ion intensity is determined by energetic considerations both for the reaction being monitored and for any competing reactions (section 7.1) and may be affected by relatively minor changes in chemical structure. Such effects may then constitute a further source of selectivity. For example, while lanosterol esters may be measured quantitatively using an intense metastable ion which corresponds to the loss of a fatty acid moiety, the metastable ion for the corresponding reaction in cholesterol esters is two orders of magnitude less intense [79]:

$$(M - CH_3)^+ \rightarrow (M - CH_3 - RCO_2H)^+ + RCO_2H \qquad (8.16)$$

In an instrument of reversed geometry, the fragmentation in the second field free region of an ion selected by the magnetic sector can be monitored by setting an appropriate value for the electrostatic sector voltage. The use of a reversed geometry was elaborated as a method for the quantitative analysis of low level components of unpurified mixtures by Cooks and co-workers [12,80–82]. Their method was based on the use of chemical ionization as a means of simplifying the original mixture spectrum followed by the monitoring of a suitable collision-induced fragmentation reaction for the appropriate $(M + H)^+$ ion. Subsequently, other ionization techniques, such as FAB for the assay of nucleotides [83], have been used with this method.

Selected reaction monitoring using a triple quadrupole instrument is a straightforward process in which the first quadrupole is set to pass the precursor ion and the third quadrupole is set to pass the appropriate product ion. The second quadrupole, operated in the r.f.-only mode, acts as a collision cell. With this instrumentation, ions created by a variety of ionization techniques, e.g. EI [28,84–86] and positive

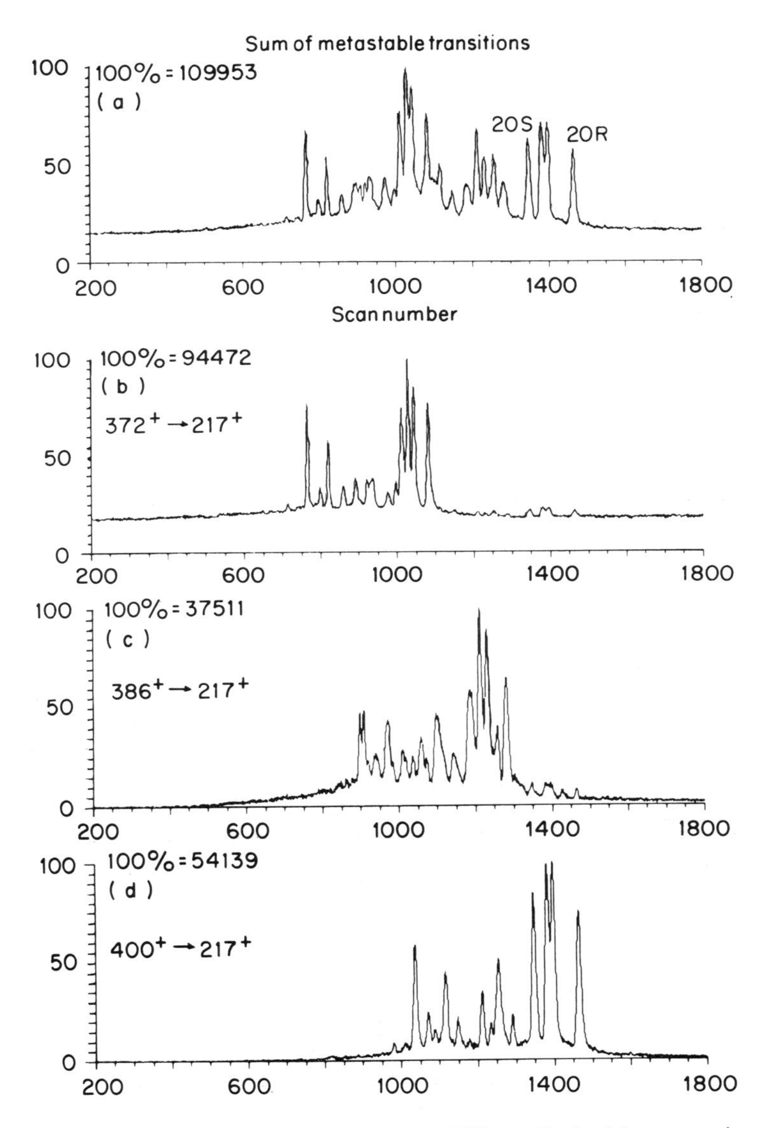

Fig. 8.9 Sterane distributions in a crude oil from Williston Basin, Montana, using selected metastable ion monitoring. Individual traces represent (b) C_{27}, (c) C_{28}, and (d) C_{29} steranes. (Reprinted with permission from G.A. Warburton and J.E. Zumberge, *Anal. Chem.*, **55**, 123[78]. Copyright (1983) American Chemical Society)

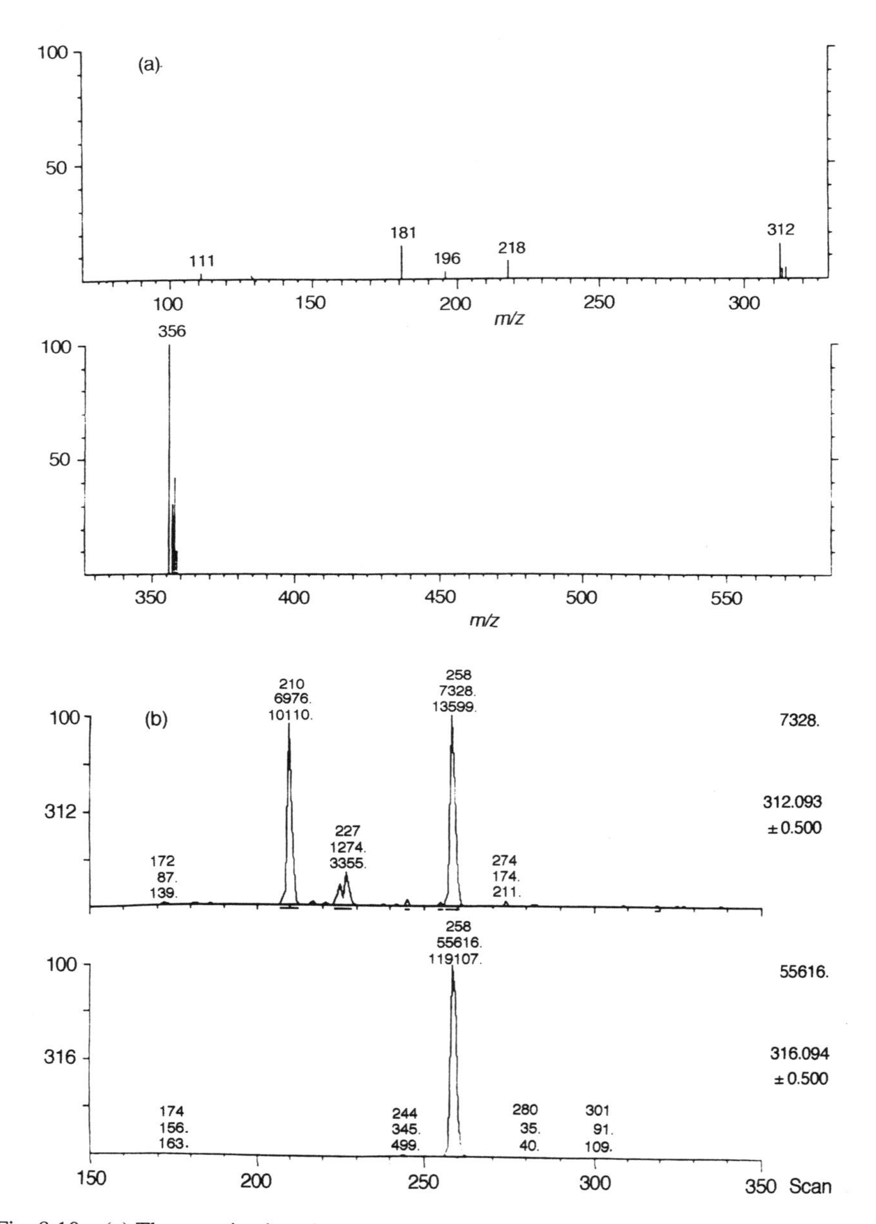

Fig. 8.10 (a) The negative ion electron capture spectrum of indomethacin pentafluorobenzyl ether and (b) chromatograms obtained from monitoring *m/z* 312 formed by collision induced dissociation of *m/z* 356 (indomethacin pentafluorobenzyl ether) and *m/z* 316 formed from *m/z* 360 (deuterated internal standard). Indomethacin was present at 0.1 ng cm^{-3} in the plasma sample analysed. (Reproduced with permission from ref. 88)

and negative CI [87–89] (Fig. 8.10), have been used. Most of these references demonstrate the selectivity and sensitivity of the method and, interestingly, ref. 28 shows that MS–MS assays on TQ instruments can also be robust since over 3000 plasmas were analysed over a two-year period in this case.

Several investigators [57,90,91] have compared selected reaction monitoring using an instrument of hybrid geometry with high resolution selected ion monitoring for the quantitative analysis of polychlorodibenzodioxins and dibenzofurans. Overall, the conclusion is that, while selected reaction monitoring is less sensitive than high resolution selected ion monitoring, it is, nevertheless, a more selective technique, even when compared with an instrumental resolution of 18 000 [90]. Gaskell and co-workers have used a hybrid instrument in conjunction with FAB ionization for the quantitative analysis of a number of compounds of biomedical importance, e.g. leukotrienes [92], phospholipids [93], steroid sulphates [94], and acylcarnitines [95]. These workers have also demonstrated that low energy collisions in the quadrupole collision region constitute a preferred method for sensitive and selective detection using selected reaction monitoring [95].

8.3.3 Use of other ionization methods: derivatization

An increasing number of low resolution selected ion monitoring determinations are now performed under chemical ionization conditions. In many cases this is because the compounds of interest do not provide either molecular or any other higher mass, high intensity ions suitable for monitoring under electron impact conditions. Although a standard positive ion chemical ionization reagent gas such as ammonia, methane or isobutane is most often used in these selected ion monitoring experiments, a reagent gas is sometimes chosen specifically as a means of making the determination a selective one. For example, $^{15}NH_3$ in ammonia may be determined by protonation using isobutane reagent gas when water, which is also present and which would otherwise form H_3O^+ with the same nominal mass as $^{15}NH_4^+$, is not protonated [96]. Other published examples are the selective ionization of unsaturated fatty acid esters in a mixture with saturated esters using benzene as a charge exchange reagent (section 3.1.1.2) [97] and the selective ionization of aliphatic ketones in a mixture with aliphatic and aromatic hydrocarbons using protonated acetone as a proton transfer reagent [98].

The best-known example of selective determination using chemical ionization is undoubtedly negative ion electron capture ionization where a significant increase in sensitivity as well as selectivity is apparent for compounds that have a high cross-section for resonance electron capture. Thus, Abdel-Baky and Giese [99] have demonstrated the detection of approximately 0.1 femtograms (10^{-16} g) of a pentafluorobenzyl derivative of uracil with a signal-to-noise ratio of 2:1 under electron capture conditions.

Negative ion CI has been applied widely to the GC–MS analysis of eicosanoids, biogenic amines and drugs and their metabolites, almost invariably after reaction to form a polyfluorinated ester derivative [88,100,101] which provides the necessary electron capture cross-section and chromatographic properties. Optically active fluorinated reagents have been used for the separation of enantiomeric forms [66,102].

Ions from the reagent gas itself can produce a relatively intense background under chemical ionization conditions. Ammonia is somewhat better than methane or isobutane in this respect since it provides the least intense background as well as often providing a slightly higher sample sensitivity under positive ion conditions [103,104]. Under negative ion electron capture chemical ionization conditions, however, all three gases contribute very little to the background. The low background, high sensitivity, and selectivity that can be achieved using electron capture ionization make it a particularly suitable method for quantitative analysis.

In some cases, e.g. where a number of related organophosphorus pesticides are to be determined, negative ion electron capture ionization may be insufficiently specific, although sensitivity for the determination of these compounds is, in general, much better than in positive ion chemical ionization or in the electron impact mode. Thus, chemical ionization of organophosphorus pesticides usually generates an abundant quasi-molecular ion in the positive mode and mainly abundant group-specific fragments in the negative mode. In this case, the optimum with regard to sensitivity and specificity is to use positive–negative switching and monitor three ions that include the base peaks in the positive and the negative spectra [105].

Most chemical ionization determinations have used gas chromatography as a separation technique, but for compounds that are less amenable to gas chromatography, appropriately purified extracts may be introduced by distillation from a direct insertion probe under conventional chemical ionization conditions [106–109]. Quantitative data can then be calculated from the areas under the evaporation profiles at selected sample masses relative to the area for a standard compound.

Since there is much less separation from other components of the sample than under chromatographic conditions, the use of the probe relies to a considerable extent on the production of simple chemical ionization spectra with an intense quasi-molecular ion offering good analyte sensitivity and minimum interference from other components. For this reason the method has been applied to the determination of components present at relatively high concentrations. Alternatively, high resolution could be used to provide the necessary freedom from interference. As with all non-chromatographic methods, the results may be strongly influenced by matrix effects [13,14].

An alternative to chemical ionization is to prepare a stable higher molecular weight derivative which is of suitable volatility for introduction from the direct insertion probe and which also gives an intense molecular or other high mass ion under EI conditions, e.g. bansyl [110] or dansyl-acetyl [111] derivatives of amines:

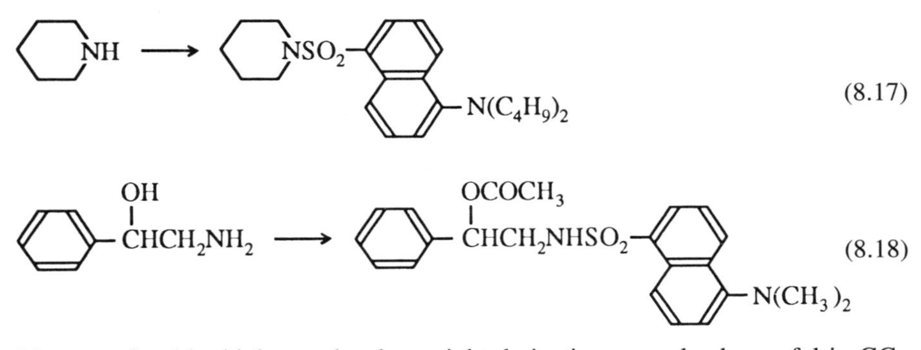

The use of stable, higher molecular weight derivatives can also be useful in GC–MS quantitation. Not only does the higher mass result in a cleaner MS background, but, under repetitive analysis conditions, the higher column temperatures needed for these derivatives contribute substantially to cleaner high resolution chromatography with less interference [112]. Some of these advantages have been incorporated into an analytical procedure for methamphetamine and amphetamine which uses 4-carbethoxy-hexafluorobutyryl chloride as a derivatizing reagent [113]:

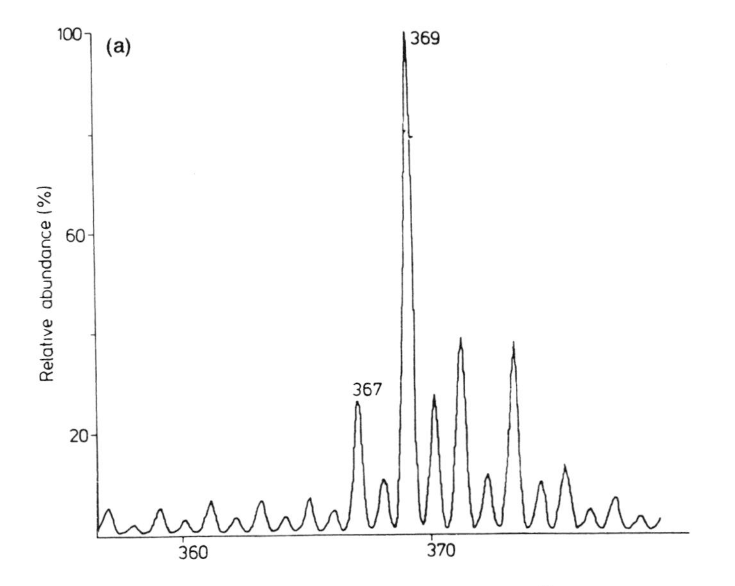

Fig. 8.11 Analysis of ($^2H_2$)dehydroepiandrosterone sulphate (($^2H_2$)DHAS) in serum using negative ion FAB ionization. (a) Narrow mass range scan of ion source derived ions; glycerol matrix (*m/z* 369 represents ($^2H_2$)DHAS). (b) As (a), thioglycerol matrix. (c) Narrow mass range scan of ions yielding fragment ions at *m/z* 97 following collisional activation (*m/z* 97 represents HSO_4^-). (Reproduced with permission from ref. 94)

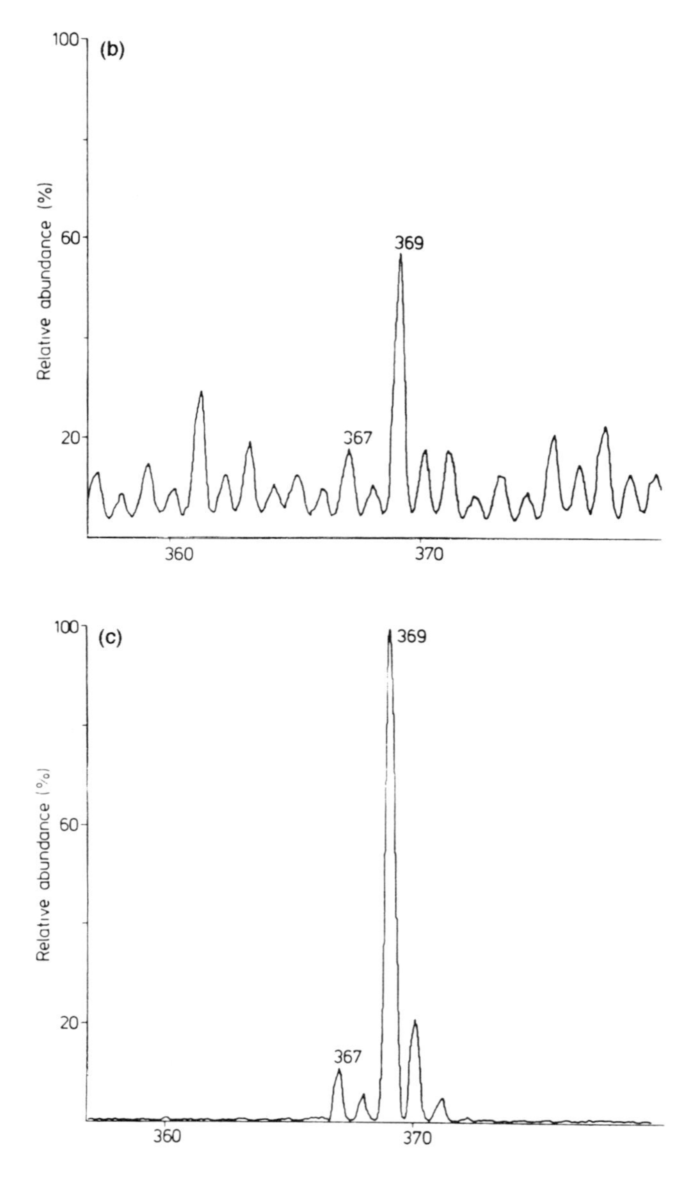

Fig. 8.11 (*continued*)

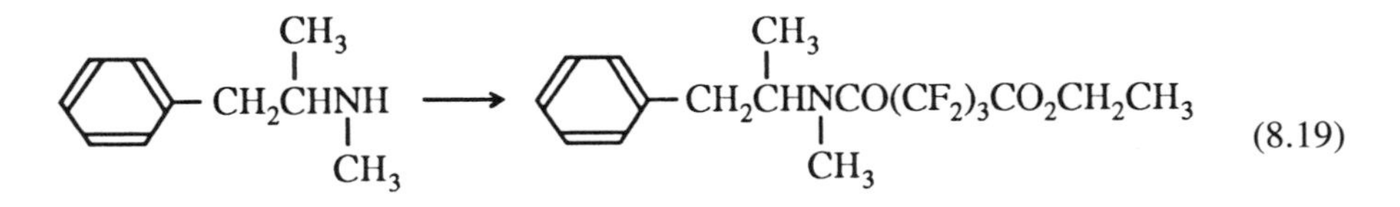

The use of ionization methods such as fast atom bombardment [71,83,92–95,114,115], thermospray [5,52,116,117], and, to a lesser extent, field desorption [118,119] permits the direct quantitative analysis of many more polar molecules that can otherwise only be measured after chemical degradation and/or derivatization. Many quantitative applications of FAB/LSIMS involve the determination of targeted compounds in biological matrices. A number of environmental applications have also been published, including a report of particularly high sensitivity for cationic surfactants [71]. The long-lasting, stable ion beams observed in FAB/LSIMS allow the use of scanning data acquisition for quantitation. Typically several restricted range scans are acquired over a period of a few minutes and then summed.

With a technique such as FAB/LSIMS where no chromatographic separation is involved, careful attention must be paid to the pre-purification of samples. Thus, impurities with higher surface activity may cause suppression of the analyte signal (section 5.3.2.2). Furthermore, background ions produced from the matrix or from the sample itself may require the use of higher mass resolution or MS-MS (e.g. Fig. 8.11) to provide the necessary specificity.

A particular problem in quantitation using thermospray ionization is signal instability. This can arise from instrumental changes or from competitive ionization of co-eluting compounds. Oxford and Lant [49] have suggested that these problems may be overcome by improving the cleanliness of the biological extract or, as in this case, by the use of a co-eluting internal standard. Bellar and Budde [5] have demonstrated that, for the quantitation of involatile organics in aqueous environmental samples, peak stability observed with buffer thermospray operation is significantly better than that observed with filament or discharge ionization, but not as good as capillary column GC–MS. They have also demonstrated that in this case an external calibration can be used instead of an isotopically labelled internal standard so long as the external calibration is repeated at frequent intervals.

References

1. Sweeley, C.C., Elliott, W.M., Fries, I., and Ryhage, R. (1966), *Anal. Chem.*, **38**, 1549.
2. Shoemaker, J.D., and Elliott, W.H. (1991), *J.Chromatogr.*, **562**, 125.
3. Budde, W.L., and Alford-Stevens, A.L. (1988), *ASTM Spec.Tech.Publ.*, **976**, 204.
4. Slivon, L.E., Gebhart, J.E., Hayes, T.L., Alford-Stevens, A., and Budde, W.L. (1985), *Anal. Chem.*, **57**, 2464.
5. Bellar, T.A., and Budde, W.L. (1988), *Anal. Chem.*, **60**, 2076.
6. Eichelberger, J.W., Kerns, E.H., Olynyk, P., and Budde, W.L. (1983), *Anal. Chem.*, **55**, 1471.

7. Chapman, J.R. (1978), *Computers in Mass Spectrometry*, Academic Press, London.
8. Moon, D.-C., and Kelley, J.A. (1988), *Biomed. Environ. Mass Spectrom.*, **17**, 229.
9. Li, Q.M., Dillen, L., and Claeys, M. (1992), *Biol. Mass Spectrom.*, **21**, 408.
10. Hurst, R.E., Settine, R.L., Fish, F., and Roberts, E.C. (1981), *Anal. Chem.*, **53**, 2175.
11. Stanley, J.S., Sack, T.M., Tondeur, Y., and Beckert, W.F. (1988), *Biomed. Environ. Mass Spectrom.*, **17**, 27.
12. Kondrat, R.W., and Cooks, R.G. (1978), *Anal. Chem.*, **50**, 81A.
13. Kallos, G.J., Caldecourt, V., and Tou, J.C. (1982), *Anal. Chem.*, **54**, 1313.
14. Weinkam, R.J., Gal, J., Callery, P., and Castagnoli, N., Jr (1976), *Anal. Chem.*, **48**, 203.
15. *Eight Peak Index* (4th Edition) (1992), Royal Society of Chemistry, Cambridge.
16. ACS Committee on Environmental Improvement (1980), *Anal. Chem.*, **52**, 2242.
17. Clement, R.E., and Tosine, H.M. (1988), *Mass Spectrom. Rev.*, **7**, 593.
18. Clayton, C.A., Hines, J.W., and Elkins, P.D. (1987), *Anal. Chem.*, **59**, 2506.
19. Delaney, M.F. (1988), *Chemometrics and Intelligent Laboratory Systems*, **3**, 45.
20. Long, G.L., and Winefordner, J.D. (1983), *Anal. Chem.*, **55**, 713A.
21. De Leenheer, A.P., and Cruyl, A.A. (1980), in *Biomedical Applications of Mass Spectrometry* (First Supplementary Volume) (Eds. G.R.Waller and O.C.Dermer), pp. 1170–1207, John Wiley, New York.
22. Bedding, N.D., McIntyre, A.E., Lester, J.N., and Perry, R. (1988), *J. Chromatogr. Sci.*, **26**, 597.
23. Bedding, N.D., McIntyre, A.E., Lester, J.N., and Perry, R., (1988), *J. Chromatogr. Sci.*, **26**, 606.
24. Hawthorne, S.B., Miller, D.J., and Aulich, T.R. (1989), *Fresenius' Z. Anal. Chem.*, **334**, 421.
25. Millard, B.J. (1978), *Quantitative Mass Spectrometry*, Ch. 6, Heyden, London.
26. Ellerbe, P., Meiselman, S., Sniegoski, L.T., Welch, M.J., and White, E.V (1989), *Anal. Chem.*, **61**, 1718.
27. Murphy, R.C., and Clay, K.L. (1990), in *Methods in Enzymology*, Vol. 193: *Mass Spectrometry* (Ed. J.A. McCloskey), Ch. 17, Academic Press, San Diego.
28. Dobson, R.L.M., Neal, D.M., De Mark, B.R., and Ward, S.R. (1991), *Anal. Chem.*, **62**, 1819.
29. Leis, H.J., Gleispach, H., Nitsche, V., and Malle, E. (1990), *Biomed. Environ. Mass Spectrom.*, **19**, 382.
30. Murphy, R.C., and Clay, K.L. (1979), *Biomed. Mass Spectrom.*, **6**, 309.
31. Self, R. (1979), *Biomed. Mass Spectrom.*, **6**, 315.
32. Haskins, N.J. (1982), *Biomed. Mass Spectrom.*, **9**, 269.
33. Millard, B.J., Tippet, P.A., Couch, M.W., and Williams, C.M. (1977), *Biomed. Mass Spectrom.*, **4**, 381.
34. Falkner, F.C., Fouda, H.G., and Mullins, F.G. (1984), *Biomed. Mass Spectrom.*, **11**, 482.
35. Schoeller, D.A. (1986), *J. Clin. Pharmacol.*, **26**, 396.
36. Falkner, F.C. (1981), *Biomed. Mass Spectrom.*, **8**, 43.
37. Colby, R.N., and McCaman, M.W. (1979), *Biomed. Mass Spectrom.*, **6**, 725.
38. Garland, W.A., and Powell M.L. (1981), *J. Chromatogr. Sci.*, **19**, 392.
39. Thorne, G.C., Gaskell, S.J., and Payne, P.A. (1984), *Biomed. Mass Spectrom.*, **11**, 415.
40. Jonckheere, J.A., DeLeenheer, A.P., and Steyarrt H.L. (1982), *Anal. Chem.*, **55**, 153..
41. Millard, B.J. (1978), *Quantitative Mass Spectrometry*, Ch. 4, Heyden, London.
42. Schoeller, D.A. (1976), *Biomed. Mass Spectrom.*, **3**, 265.
43. Tondeur, Y., Beckert, W.F., Billets, S., and Mitchum, R.K. (1989), *Chemosphere*, **18**, 119.

44. Federal Register, Vol. 56, No. 26, 7 Feb. 1991, 40 CFR Part 136, Water Programs; Guidelines Establishing Test Procedures for the Analysis of Pollutants .
45. Schoeller, D.A. (1980), *Biomed. Mass Spectrom.*, **7**, 457.
46. Eicemann, G.A., Gardea-Torresdey, J.L., O'Connor, G.A., and Urquhart, N.S. (1989), *J. Environ. Qual.*, **18**, 374.
47. Matthews, D.E., Denson, K.B., and Hayes, J.M. (1978), *Anal. Chem.*, **50**, 681.
48. Alexander, L.R., Maggio, V.L., Gill, J.B., Green V.E., Turner, W.E., Patterson D.G.Jr., Green, B.N., Gray, B.W., Guyan, S.A., Krolik, S.T., and Nicolaysen, L.C. (1989), *Chemosphere*, **19**, 241.
49. Oxford, J., and Lant, M.S. (1989), *J. Chromatogr.*, **496**, 137.
50. Garland, W.A., Crews, T., Brown, S.Y., and Fukuda, E.K. (1989), *J.Chromatogr.*, **472**, 250.
51. Thorne, G.C., and Gaskell, S.J. (1986), *Biomed. Environ. Mass Spectrom.*, **13**, 605.
52. Throck-Watson, J. (1990) in *Methods in Enzymology*, Vol. 193: *Mass Spectrometry* (Ed. J.A. McCloskey), Ch.4, Academic Press, San Diego.
53. Campana, J.E., Schoengold, D.M., and Butler, L.C. (1989), *Chemosphere*, **18**, 169.
54. Patterson, D.G., Jr, Turner, W.E., Alexander, L.R., Isaacs, S., and Needham, L.L. (1989), *Chemosphere*, **18**, 875.
55. Chapman, J.R. (1978), *Computers in Mass Spectrometry*, Ch. 8, Academic Press, London.
56. Tondeur, Y., Albro, P.W., Hass, D.J., Harvan, D.J., and Schroeder, J.L. (1984), *Anal. Chem.*, **56**, 1344.
57. Tondeur, Y., Niederhut, W.N., Campana, J.E., and Missler, S.E. (1987), *Biomed. Environ. Mass Spectrom.*, **14**, 449.
58. Harvan, D.J., Hass, J.R., Schroeder, J.L., and Corbett, B.J. (1981), *Anal. Chem.*, **53**, 1755.
59. Harvan, D.J., Hass, J.R., and Wood, D. (1982), *Anal. Chem.*, **54**, 332.
60. Chapman, J.R. (1982), *Int. J. Mass Spectrom. Ion Processes*, **45**, 207.
61. Grange, A.H., and Brumley, W.W. (1992), *Rapid Commun. Mass Spectrom.*, **6**, 68.
62. Finlay, E.M., and Gaskell, S.J. (1981), *Clin. Chem.*, **27**, 1165.
63. Millington, D.S., Buoy, M.E., Brooks, G., Harper, M.E., and Griffiths, K. (1975), *Biomed. Mass Spectrom.*, **2**, 219.
64. Gaskell, S.J., Finney, R.W., and Harper, M.E. (1979), *Biomed. Mass Spectrom.*, **6**, 113.
65. Davis, B.A., Durden, D.A., and Boulton, A.A. (1982), *J. Chromatogr.*, **230**, 219.
66. Miwa, B.J., Choma, N., Brown, S.Y., Keigher, N., Garland, W.A., and Fukuda, E.K. (1988), *J. Chromatogr.*, **431**, 343.
67. Fukuda, E.K., Choma, N., and Preston, P.P. (1989), *J. Chromatogr.*, **491**, 97.
68. Jones, D., Curvall, M., Abrahamsson, L., Kazemi-Vala, E., and Enzell, C. (1982), *Biomed. Mass Spectrom.*, **9**, 539.
69. Dahl, S.G., Johnsen, H., and Lee, C.R. (1982), *Biomed. Mass Spectrom.*, **9**, 534.
70. Webb, K.S., Gough, T.A., Carrick, A., and Hazelby, D. (1979), *Anal. Chem.*, **51**, 989.
71. Simms, J.R., Keough, T., Ward, S.R., Moore, B.L., and Bandurraga, M.M. (1988), *Anal. Chem.*, **60**, 2613.
72. Gallegos, E. (1981), *Anal. Chem.*, **53**, 187.
73. Voyksner, R.D., Hass, J.R., Sovocool, G.W., and Bursey, M.M. (1983), *Anal. Chem.*, **55**, 744.
74. Gaskell, S.J., and Millington, D.S. (1978), *Biomed. Mass Spectrom.*, **5**, 557.
75. McGhie, T.K., Holland, P.T., and Malcolm, C.P. (1990), *Biomed. Environ. Mass Spectrom.*, **19**, 267.
76. Harvey, D.J., Leuschner, J.T.A., and Paton, W.D.M. (1980), *J. Chromatogr.*, **202**, 83.
77. Durden, D.A. (1982), *Anal. Chem.*, **54**, 666.

78. Warburton, G.A., and Zumberge, J.E. (1983), *Anal. Chem.*, **55**, 123.
79. Patouraux, D., Lacave, C., and Prome, J.-C., (1981), *Biomed. Mass Spectrom.*, **8**, 118.
80. Kondrat, R.W., McClusky, G.A., and Cooks, R.G. (1978), *Anal. Chem.*, **50**, 2017.
81. Glish, G.L., Shaddock, V.M., Harmon, K., and Cooks, R.G. (1980), *Anal. Chem.*, **52**, 165.
82. Youssefi, M., Cooks, R.G., and McLaughlin, J.L. (1979), *J. Am. Chem. Soc.*, **101**, 3400.
83. Newton, R.P., Walton, T.J., Brenton, A.G., Kingston, E.E., and Harris, F.M. (1989), *Rapid Commun. Mass Spectrom.*, **3**, 178.
84. Schweer, H., Seyberth, H.W., and Schubert, R. (1986), *Biomed. Environ. Mass Spectrom.*, **13**, 611.
85. McCurvin D.M.A., Schellenberg, D.H., Clement, R.E., and Taguchi, V.Y. (1989), *Chemosphere*, **19**, 201.
86. McCurvin, D.M.A., Clement, R.E., Taguchi, V.Y., Reiner, E.J., Schellenberg, D.H., and Bobbie B.A. (1989), *Chemosphere*, **19**, 205.
87. Brumley, W.C., Canas, B.J., Perfetti, G.A., Mossoba, M.M., Sphon, J.A., and Corneliussen, P.E. (1988), *Anal. Chem.*, **60**, 975.
88. Dawson, M., Smith, M.D., and McGee, C.M. (1990), *Biomed. Environ. Mass Spectrom.*, **19**, 453.
89. Schweer, H., Meese, C.O., and Seyberth, H.W. (1990), *Anal. Biochem.*, **189**, 54.
90. Charles, M.J., Green, B., Tondeur, Y., and Hass, J.R. (1989), *Chemosphere*, **19**, 51.
91. DeJong, A.P.J.M., Liem, A.K.D., DenBoer, A.C., Van der Heft, E., Marsman, J.A., Van de Werken, G., and Wegman, R.C.C. (1989), *Chemosphere*, **19**, 59.
92. Raftery, M.J., Thorne, G.C., Orkiszewski, R.S., and Gaskell, S.J. (1990), *Biomed. Environ. Mass Spectrom.*, **19**, 465.
93. Haroldsen, P.E., and Gaskell, S.J. (1989), *Biomed. Environ. Mass Spectrom.*, **18**, 439.
94. Gaskell, S.J. (1988), *Biomed. Mass Spectrom.*, **15**, 99.
95. Gaskell, S.J., and Reilly, M.H. (1988), *Rapid Commun. Mass Spectrom.*, **2**, 139.
96. Lundeen, C.V., Viscomi, A.S., and Field, F.H. (1973), *Anal. Chem.*, **45**, 1288.
97. Subba Rao, S.C., and Fenselau, C. (1978), *Anal. Chem.*, **50**, 511.
98. Hatch, F., and Munson, B. (1977), *Anal. Chem.*, **49**, 731.
99. Abdel-Baky, S., and Giese, R.W. (1991), *Anal. Chem.*, **63**, 2986.
100. Girault, J., Gobin, P., and Fourtillan, J.B. (1990), *Biomed. Environ. Mass Spectrom.*, **19**, 90.
101. Weber, C., Höller, M., Beetens, J., deClerck, F., and Tegtmeier, F. (1991), *J. Chromatogr.*, **562**, 599.
102. Changchit, A., Gal, J., and Zirrolli, J.A. (1991), *Biomed. Environ. Mass Spectrom.*, **20**, 751.
103. Mauchamp, B., Lafont, R., Hardy, M., and Jourdain, D. (1979), *Biomed. Mass Spectrom.*, **6**, 276.
104. Jacobsson, S.-E., Jönsson, S., Lindberg, C., and Svensson, L.Å. (1980), *Biomed. Mass Spectrom.*, **7**, 265.
105. Stan, H.-J., and Kellner, G. (1987), *Biomed. Environ. Mass Spectrom.*, **18**, 645.
106. Issachar, D., and Yinon, J. (1980), in *Advances in Mass Spectrometry*, Vol.8B (Ed. A. Quayle), p.1321, Heyden & Son, London.
107. Howald, W.N., Busch, E.D., Trager, W.F., O'Reilley, R.A., and Motley, C.H. (1980), *Biomed. Mass Spectrom.*, **7**, 35.
108. Yergey, A.L., Kody, M.H., and Gershfeld, N.L. (1981), *Biomed. Mass Spectrom.*, **8**, 503.
109. Weinkam, R.J., Rowland, M., and Meffin, P.J. (1977), *Biomed. Mass Spectrom.*, **4**, 42.
110. Seiler, N., and Schneider, H.H. (1974), *Biomed. Mass Spectrom.*, **1**, 381.

111. Durden, D.A., Juorio, A.V., and Davies, B.A. (1980), *Anal. Chem.*, **52**, 1815.
112. Hornbeck, C.L., and Czarny, R.J. (1989), *J. An. Toxicol.*, **13**, 144.
113. Czarny, R.J., and Hornbeck, C.L. (1989), *J. An. Toxicol.*, **13**, 257.
114. Chen, S., Kirschner, G., and Benefenati, E. (1990), *Rapid Commun. Mass Spectrom.*, **4**, 214.
115. Wernery, J.D., and Peake, D.A. (1989), *Rapid Commun. Mass Spectrom.*, **3**, 396.
116. Yergey, A.L., Esteban, N.V., Vieira, N.E., and Vicchio, D. (1990), in *Mass Spectrometry of Biological Materials* (Eds. C.N. McEwen and B.S. Larsen), Marcel Dekker Inc.
117. Voysner, R.D. (1987), in *Applications of New Mass Spectrometry Techniques in Pesticide Chemistry* (Ed. J.D. Rosen), p.146, John Wiley, New York.
118. Lehmann, W.D., and Schulten, H.-R. (1978), *Fresenius' Z. Anal. Chem.*, **290**, 121.
119. Shiraishi, H., Otsuki, A., and Fuwa, K. (1982), *Bull. Chem. Soc. Jpn.*, **55**, 1410.
120. Benfenati, E., De Belli, G., Chen, S., Bettazolli, L., Fanelli, R., Tacconi, M.T., Kirschner, G., and Pege, G. (1989), *J. Lipid Res.*, **30**, 1983.

Appendix A

Introduction

The measurements made in mass spectrometry are those of ionic mass (more correctly, mass-to-charge ratio) and relative intensity. Mass is measured on the basis of a calibration with a reference compound which provides ions of known mass and which is used to establish a correlation between peak 'position' and mass.

A good reference compound has the following characteristics. It should cover the mass range of interest, providing regularly spaced peaks of moderate to high relative intensity. For gas phase ionization techniques, such as EI or CI, the reference compound should be relatively volatile and chemically inert so that its vapour may be introduced in a steady flow and then pumped out again rapidly without problems due to condensation, absorption or decomposition. The requirement for volatility is less relevant for ionization techniques such as fast atom bombardment, electrospray or laser desorption, but it is still important that the reference material does not contaminate the source region or interfere with sample ionization when used as an internal reference. A good reference compound should also be readily available in quantity consistently to provide a material of known composition and behaviour.

For most routine experiments in which only integer masses (e.g. m/z 200) are measured, an initial calibration of the mass scale with an external reference is usually sufficient. The same material may also be used for general source tuning. For the measurement of accurate masses (e.g. m/z 200.1234) to give elemental compositions, the reference compound is usually introduced and ionized concurrently with the sample so that it is desirable that reference peaks can be resolved from sample peaks [1]. This is most often accomplished by the use of a mass deficient reference compound and high instrumental resolution. The reference compound should also contain as small a range of heteroatoms and isotopes as possible to facilitate the assignment of reference masses and minimize the occurrence of unresolved multiplets within the reference spectrum itself.

Electron Impact Ionization

Many of the above requirements are met by the use of perfluorinated reference compounds. Perfluorinated reference materials used under electron impact conditions are listed in Table A1, and mass spectral data from the two most commonly used compounds, perfluorokerosene and heptacosafluorotributylamine (also referred to as PFTBA), are reproduced in Tables A2 and A3 respectively. Other, high mass, fluorinated reference compounds which provide useful ions up to m/z 2000–3000 are also available. The most common of these high mass materials are Fomblin oils (Ultramark-F series) [2–4], phosphazine mixtures (Ultramark 1621) [2,5], and various triazines [5–7](Table A1). As will be seen in the following sections, most of these materials can also be used as reference compounds for one or more alternative ionization techniques.

Chemical Ionization

Under EI conditions there is little difficulty in finding suitable reference compounds. Under conventional CI conditions, however, a number of reference compounds are either only weakly ionized or produce only quasi-molecular ions and very few fragment ions so that some change in operating conditions is necessary. For example, the reagent gas pressure can be reduced to somewhat below the normal operating value. Under appropriate conditions some fragmentation of reference compounds such as perfluorokerosene with methane reagent gas [8] and particularly Fomblin (Ultramark-F) is observed. Alternatively, an argon–water mixture reagent gas (section 3.1.1.2) ionizes PFK by charge transfer from Ar^+ and results in a spectrum that is very suitable for accurate mass measurement [9].

Conventional reference compounds may also be replaced with other materials that fragment under CI conditions, e.g. higher n-alkanes with isobutane reagent gas [10] or a mixture of poly(dimethylsiloxanes) with methane [11] or isobutane [12] reagent gas. A commercial phosphazine mixture (Ultramark 1621) has also been found to be a good calibrant up to m/z 2000 under CI conditions [13].

An alternative approach is to calibrate under EI conditions prior to CI operation. The assumption made in this approach that there is no shift in the mass scale on changing to CI conditions, cannot be justified if accurate rather than nominal mass assignment is required. An extension of this method, which may be used for accurate mass measurement, requires the presence of a very few (usually 2–3) reference ions under CI conditions to correct the preceding calibration [14].

Negative Ion Chemical Ionization

With most instrumentation there is only a small discrepancy between a mass scale obtained by calibration in the positive ion mode and masses recorded in the negative ion mode. Thus, an initial calibration in the positive ion mode using a standard

Table A1 EI reference compound data (for use with other ionization techniques, see text)

Reference compound	Formula	Composition of major ions in reference spectrum or in the spectrum of an individual component	Upper mass limit
Perfluorokerosene (perfluoroalkane mixture; PFK)	$CF_3(CF_2)_nCF_3$	See Table A2	800–900
Heptacosafluorotributylamine (perfluorotributylamine; 'Heptacosa'; PFTBA	$(C_4F_9)_3N$	See Table A3	671
Ultramark F-series (Fomblin diffusion pump fluid; perfluoro polyether	$CF_3O[CF(CF_3)CF_2O]_m(CF_2O)_nCF_3$	Mainly $(C_xF_{2x+1}O_y)$ with some $(C_xF_{2x-1}O_y)$ ions below m/z 1000	≥ 3000
Ultramark 1621(Fluoroalkoxy cyclotriphosphazine mixture)	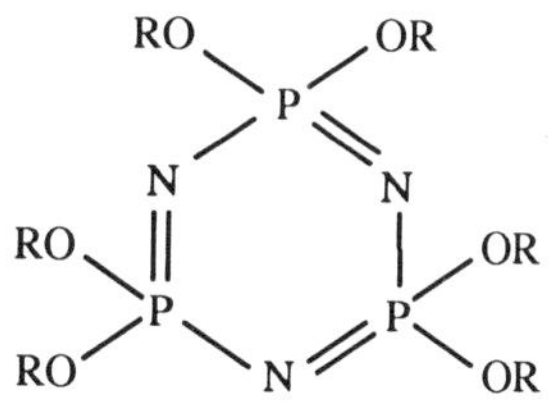 $R = CH_2(CF_2)_nH$	$C_6H_{18}(C_2F_4)_nO_6N_3P_3$ $C_5H_{17}CF_2(C_2F_4)_{n-1}O_6N_3P_3$	~ 2000
Tris(perfluoroalkyl)-*s*-triazine	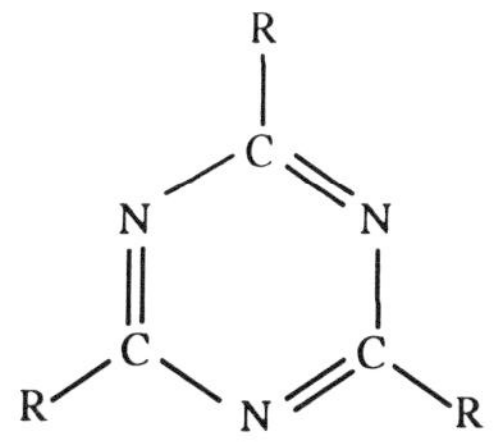 $R = (CF_2)_nCF_3$	$C_{25}F_{45}N_3(1185)$, $C_{24}F_{44}N_3(1166)$ $C_{22}F_{40}N_3(1066)$, $C_{20}F_{36}N_3(966)$ $C_{18}F_{32}N_3(866)$, $C_{16}F_{29}N_2(771)$ [n=6]	285, 435 585, 1185 1485, (n = 0, 1,2,6,8)

Table A2 Major reference masses in the spectrum of high boiling PFK

Accurate mass	Relative intensity	Formula	Accurate mass	Relative intensity	Formula
880.9441	0.23	$C_{18}F_{35}$	442.9278	1.70	$C_{10}F_{17}$
866.9473	0.41	$C_{20}F_{33}$	430.9278	3.79	C_9F_{17}
854.9473	0.38	$C_{19}F_{33}$	416.9760	1.06	$C_{11}F_{15}$
842.9473	0.34	$C_{18}F_{33}$	404.9760	2.04	$C_{10}F_{15}$
830.9473	0.30	$C_{17}F_{33}$	392.9760	2.72	C_9F_{15}
816.9505	0.52	$C_{19}F_{31}$	380.9760	4.54	C_8F_{15}
804.9505	0.47	$C_{18}F_{31}$	368.9760	1.08	C_7F_{15}
792.9505	0.41	$C_{17}F_{31}$	354.9792	1.79	C_9F_{13}
780.9505	0.31	$C_{16}F_{31}$	342.9792	4.26	C_8F_{13}
766.9537	0.64	$C_{18}F_{29}$	330.9792	5.67	C_7F_{13}
754.9537	0.61	$C_{17}F_{29}$	318.9792	2.81	C_6F_{13}
742.9537	0.51	$C_{16}F_{29}$	304.9824	2.26	C_8F_{11}
730.9537	0.47	$C_{15}F_{29}$	292.9824	5.88	C_7F_{11}
716.9569	0.78	$C_{17}F_{27}$	280.9824	8.51	C_6F_{11}
704.9569	0.72	$C_{16}F_{27}$	268.9824	5.74	C_5F_{11}
692.9569	0.60	$C_{15}F_{27}$	254.9856	3.57	C_7F_9
680.9569	0.56	$C_{14}F_{27}$	242.9856	8.22	C_6F_9
666.9601	0.82	$C_{16}F_{25}$	230.9856	11.5	C_5F_9
654.9601	0.94	$C_{15}F_{25}$	218.9856	10.9	C_4F_9
642.9601	0.75	$C_{14}F_{25}$	204.9888	3.28	C_6F_7
630.9601	0.71	$C_{13}F_{25}$	192.9888	5.70	C_5F_7
616.9633	0.92	$C_{15}F_{23}$	180.9888	21.9	C_4F_7
604.9633	1.08	$C_{14}F_{23}$	168.9888	21.5	C_3F_7
592.9633	1.06	$C_{13}F_{23}$	161.9904	4.79	C_4F_6
580.9633	0.94	$C_{12}F_{23}$	149.9904	1.63	C_3F_6
566.9664	0.98	$C_{14}F_{21}$	130.9920	33.3	C_3F_5
554.9664	1.06	$C_{13}F_{21}$	118.9920	28.1	C_2F_5
542.9664	1.11	$C_{12}F_{21}$	111.9936	0.19	C_3F_4
530.9664	1.19	$C_{11}F_{21}$	99.9936	6.60	C_2F_4
516.9696	0.85	$C_{13}F_{19}$	92.9952	0.92	C_3F_3
504.9696	1.32	$C_{12}F_{19}$	80.9952	0.88	C_2F_3
492.9696	1.49	$C_{11}F_{19}$	68.9952	100	CF_3
480.9696	1.79	$C_{10}F_{19}$	49.9968	0.21	CF_2
466.9728	0.75	$C_{12}F_{17}$	30.9984	0.23	CF
454.9728	1.57	$C_{11}F_{17}$			

Table A3 Major reference masses in the spectrum of heptacosafluorotributylamine

Accurate mass	Relative intensity	Formula	Accurate mass	Relative intensity	Formula
613.9647	2.6	$C_{12}F_{24}N$	180.9888	1.9	C_4F_7
575.9679	1.7	$C_{12}F_{22}N$	175.9935	1.0	C_4F_6N
537.9711	0.4	$C_{12}F_{20}N$	168.9888	3.6	C_3F_7
501.9711	8.6	$C_9F_{20}N$	163.9935	0.7	C_3F_6N
463.9743	3.8	$C_9F_{18}N$	161.9904	0.3	C_4F_6
425.9775	2.5	$C_9F_{16}N$	149.9904	2.1	C_3F_6
413.9775	5.1	$C_8F_{16}N$	130.9920	31	C_3F_5
375.9807	0.9	$C_8F_{14}N$	118.9920	8.3	C_2F_5
325.9839	0.4	$C_7F_{12}N$	113.9967	3.7	C_2F_4N
313.9839	0.4	$C_6F_{12}N$	111.9936	0.7	C_3F_4
263.9871	10	$C_5F_{10}N$	99.9936	12	C_2F_4
230.9856	0.9	C_5F_9	92.9952	1.1	C_3F_3
225.9903	0.6	C_5F_8N	68.9952	100	CF_3
218.9856	62	C_4F_9	49.9968	1.0	CF_2
213.9903	0.6	C_4F_8N	30.9984	2.1	CF

reference material such as perfluorokerosene or heptacosafluorotributylamine is often adequate for nominal mass operation in both the negative and positive ion modes.

Reference peaks from a spectrum recorded under negative ion conditions are, however, essential for accurate mass measurement and have been advocated in some circumstances for nominal mass calibration using quadrupole instruments [15]. Fortunately, fluorinated reference compounds are readily ionized under negative ion CI conditions to give spectra which, although containing some ions in common with the corresponding EI spectra, are generally considerably different [16].

Heptacosafluorotributylamine provides ions over only a limited mass range and no major ions are seen below m/z 169 in the spectrum of perfluorokerosene (Table A4) although I^- (m/z 127) and Cl^- (m/z 35, 37) are often present and isotope clusters due to $ReO_3{}^-$ and $ReO_4{}^-$ (m/z 233 to 251) are also seen when a rhenium filament is used [17]. One of the most useful fluorinated reference compounds is Fomblin oil which gives an excellent negative ion spectrum covering a wide mass range [15].

Fast Atom Bombardment/LSIMS

The most commonly used material for calibration and source tuning in fast atom bombardment is caesium iodide [18](Tables A5,A6) which provides both positively and negatively charged cluster ions up to high mass, although intensity falls off much more rapidly at higher mass in the negative ion mode. A mixture of CsI with NaI in the positive ion mode and with CsF in the negative ion mode has been used [19] to fill in the gaps between CsI peaks at lower masses (Table A6) and,

Table A4 Major ions in the negative ion CI spectrum of PFK (ammonia moderating gas)

Accurate mass	Relative intensity	Formula	Accurate mass	Relative intensity	Formula
799.9489	0.9	$C_{16}F_{32}$	516.9696	7.3	$C_{13}F_{19}$
787.9489	2.8	$C_{15}F_{32}$	504.9696	18.3	$C_{12}F_{19}$
773.9521	2.5	$C_{17}F_{30}$	492.9696	23.4	$C_{11}F_{19}$
761.9521	2.4	$C_{16}F_{30}$	480.9696	18.7	$C_{10}F_{19}$
750.9601	4.3	$C_{23}F_{25}$	473.9712	11.3	$C_{11}F_{18}$
735.9553	5.5	$C_{17}F_{28}$	454.9728	15.9	$C_{11}F_{17}$
723.9553	12.2	$C_{16}F_{28}$	442.9728	35.7	$C_{10}F_{17}$
711.9553	7.5	$C_{15}F_{28}$	430.9728	39.0	$C_{9}F_{17}$
704.9569	5.6	$C_{16}F_{27}$	423.9744	15.3	$C_{10}F_{16}$
699.9553	6.6	$C_{14}F_{28}$	411.9744	16.0	$C_{9}F_{16}$
685.9585	20.7	$C_{16}F_{26}$	392.9760	47.9	$C_{9}F_{15}$
673.9585	36.7	$C_{15}F_{26}$	380.9760	63.2	$C_{8}F_{15}$
661.9585	13.2	$C_{14}F_{26}$	361.9776	34.5	$C_{8}F_{14}$
654.9601	21.3	$C_{15}F_{25}$	342.9792	32.0	$C_{8}F_{13}$
635.9617	32.0	$C_{15}F_{24}$	330.9792	100	$C_{7}F_{13}$
623.9617	41.5	$C_{14}F_{24}$	311.9808	50.6	$C_{7}F_{12}$
611.9617	13.1	$C_{13}F_{24}$	292.9824	12.2	$C_{7}F_{11}$
604.9633	30.8	$C_{14}F_{23}$	280.9824	81.2	$C_{6}F_{11}$
585.9648	19.7	$C_{14}F_{22}$	261.9840	40.4	$C_{6}F_{10}$
573.9648	17.0	$C_{13}F_{22}$	249.9840	3.3	$C_{5}F_{10}$
554.9664	21.1	$C_{13}F_{21}$	230.9856	25.8	$C_{5}F_{9}$
542.9664	12.8	$C_{12}F_{21}$	218.9856	7.5	$C_{4}F_{9}$
535.9680	8.0	$C_{13}F_{20}$	211.9872	5.4	$C_{5}F_{8}$
530.9664	10.9	$C_{11}F_{21}$	199.9872	1.3	$C_{4}F_{8}$
523.9680	7.0	$C_{12}F_{20}$	168.9888	1.9	$C_{3}F_{7}$

more recently, CsF alone has been proposed [20] as a replacement for CsI in both positive and negative ion modes. Another recent publication [21] has demonstrated that a gold target provides relatively intense cluster ions (Table A6) up to very high mass (m/z 10,000) in the negative ion mode. The FAB matrix itself can provide a source of ions for calibration and tuning at lower masses, e.g. ions from glycerol (Table A6) can be used up to m/z 1000.

For accurate mass measurement in FAB, a number of techniques have been devised to overcome the problem of suppression of sample ions following the addition of an internal reference. These techniques include successive analysis of reference and sample [22,23] and the use of a split target for reference and sample on variously designed direct insertion probes [24–26]. For the best accuracy of mass measurement, however, the reference material should be mixed with the sample as an internal calibrant. Polyethylene glycol is suitable as an internal standard up to m/z 750 in the positive ion mode, while Siegel and co-workers [27] have demonstrated the use of a number of related non-ionic surfactants as internal

Table A5 Major ions in the positive and negative* ion FAB/LSIMS spectra of dry CsI

Accurate mass	Formula	Accurate mass	Formula
132.9054	Cs	6368.3430	$(CsI)_{24}Cs$
265.8109	Cs_2	6628.1529	$(CsI)_{25}Cs$
392.7153	(CsI)Cs	6887.9628	$(CsI)_{26}Cs$
652.5252	$(CsI)_2Cs$	7147.7727	$(CsI)_{27}Cs$
912.3351	$(CsI)_3Cs$	7407.5826	$(CsI)_{28}Cs$
1172.1450	$(CsI)_4Cs$	7667.3925	$(CsI)_{29}Cs$
1431.9549	$(CsI)_5Cs$	7927.2024	$(CsI)_{30}Cs$
1691.7648	$(CsI)_6Cs$	8187.0123	$(CsI)_{31}Cs$
1951.5747	$(CsI)_7Cs$	8446.8222	$(CsI)_{32}Cs$
2211.3846	$(CsI)_8Cs$	8706.6321	$(CsI)_{33}Cs$
2471.1945	$(CsI)_9Cs$	8966.4420	$(CsI)_{34}Cs$
2731.0044	$(CsI)_{10}Cs$	9226.2519	$(CsI)_{35}Cs$
2990.8143	$(CsI)_{11}Cs$	9486.0618	$(CsI)_{36}Cs$
3250.6242	$(CsI)_{12}Cs$	9745.8717	$(CsI)_{37}Cs$
3510.4341	$(CsI)_{13}Cs$	126.9045*	I
3770.2440	$(CsI)_{14}Cs$	386.7144*	(CsI)I
4030.0539	$(CsI)_{15}Cs$	646.5243*	$(CsI)_2I$
4289.8638	$(CsI)_{16}Cs$	906.3342*	$(CsI)_3I$
4549.6737	$(CsI)_{17}Cs$	1166.1441*	$(CsI)_4I$
4809.4836	$(CsI)_{18}Cs$	1425.9540*	$(CsI)_5I$
5069.2935	$(CsI)_{19}Cs$	1685.7639*	$(CsI)_6I$
5329.1034	$(CsI)_{20}Cs$	1945.5738*	$(CsI)_7I$
5588.9133	$(CsI)_{21}Cs$	2205.3837*	$(CsI)_8I$
5848.7232	$(CsI)_{22}Cs$	2465.1936*	$(CsI)_9I$
6108.5331	$(CsI)_{23}Cs$	2725.0035*	$(CsI)_{10}I$
		2984.8134*	$(CsI)_{11}I$
		3244.6233*	$(CsI)_{12}I$

standards up to m/z 2100. Ultramark 1621 has been used as an internal reference covering the mass range m/z 700–1900 in both the negative and positive ion modes, while polyethylene glycol sulphates and related esters have been used as reference materials in negative ion FAB [28].

Field Desorption

Polystyrene and polypropylene glycol mixtures (Table A6) provide ions that are suitable for calibration over a wide mass range in field desorption [29]. Alternatively, the use of a combined FD/FAB source offers a very simple means of high mass calibration using CsI in the FAB mode. Accurate mass measurements in field desorption are ideally obtained by peak matching against a reference material which is ionized together with the sample and which gives an ion of known composition close in mass to that from the unknown [30]. These criteria are usually

Table A6 Reference compounds for use with alternative ionization techniques (for further details, see text)

Reference compound	Composition of major ions in reference spectrum [ionization mode]	Upper mass limit
Caesium iodide	See Table A5	$> 25\,000$
Caesium iodide and sodium iodide	$[(CsI)_nCs]^+$, $[(CsI)_nNa]^+$, $[(NaI)_nNa]^+$ [FAB/LSIMS]	≥ 3500
Caesium iodide and caesium fluoride	$[(CsI)_nI]^-$, $[(CsI)_nF]^-$, $[(CsF)_nF]^-$ [FAB/LSIMS]	≥ 3000
Gold	Au_n^- [FAB/LSIMS]	$\geq 10\,000$
Glycerol (G)	$(G_n + H)^+$, $(G_n - H)^-$ [FAB/LSIMS]	~ 1000
Polystyrene	$[CH_3(CH_2)_3(CH_2CHPh)_nH]^+$ [FD]	10 000 for average m.wt. 8500
Polypropylene glycol	$[HO(C_3H_6O)_nH + Cat]^+$: Cat = H,Na,NH_4 [ES,FD,TS]	Up to 2000 depending on average m.wt.
Polypropylene glycol sulphate	$[(C_3H_6O)_n + HSO_4]^-$, $[(C_3H_6O)_n + S_2O_7]^{2-}$ [ES]	1500 for average m.wt. 1000
Polyethylene glycol	$[HO(C_2H_4O)_n + Cat]^+$: Cat = H,Na,NH_4 [FAB/LSIMS,TS]	Up to 2000 depending on average m.wt.
Ammonium acetate	$[(CH_3CO_2H)_m(NH_3)_n(H_2O)_pNH_4]^+$ [TS]	~ 800

met by the field ionization of a volatile reference material such as a phosphazine or triazine [5].

Thermospray Ionization

Polyethylene glycol and polypropylene glycol, which both give series of adducted molecular ions with the composition $(M + NH_4)^+$ (Table A6) under thermospray conditions, were proposed as calibrants for thermospray operation quite soon after the introduction of the technique [31]. In addition, post-column introduction of these materials has been used as the basis of accurate mass measurement of thermospray ions [32]. However, a number of authors [33,34] have noted that rapid contamination of the ion source can occur with these compounds. Furthermore, a definite identification of individual oligomer ions from one of these materials is difficult on the basis of relative intensity alone.

As alternative low resolution calibration and tuning methods, Niessen *et al.* [34] and more recently Saar and Haimberg [33] have described experimental thermospray conditions under which cluster ions are formed from ammonium acetate together with acetic acid, and from ammonium acetate alone respectively (Table A6). These methods allow calibration up to m/z 800 and require a minimum of change from normal LC-MS analytical conditions. The use of perfluorinated straight chain acids together with their ammonium salts has been presented as a method of calibrating to high mass (m/z 2000) in both positive and negative ion modes in thermospray [35].

Electrospray Ionization

Since the object of most electrospray experiments is the assignment of molecular weight to larger molecules, a procedure which demands an accuracy of $\pm 0.01\%$, these experiments require a calibration in the same ionization mode, preferably recorded close in time to the molecular weight determination experiment. Both CsI [36,37] and polypropylene glycols [37,38] have been suggested as ES calibrants. Unfortunately, CsI contaminates the ion source at the concentration levels required [37]. With polypropylene glycol mixtures, it can, as previously mentioned, prove difficult to identify individual oligomer ions in the spectra, particularly since, depending on experimental conditions, $(M + H)^+$ and $(M + Na)^+$ [37] or $(M + NH_4)^+$ [38] ions (Table A6) and multiply charged ions may be present. Despite these difficulties, polypropylene glycol has been used to provide accurate mass assignments in the positive ion mode [37,38] and polypropylene glycol sulphate (Table A6) has been used for determinations in the negative ion mode [37].

An alternative approach to mass calibration is to use the range of multiply charged molecular ions available from the electrospray ionization of well-characterized, homogeneous protein molecules. This method of mass assignment for higher molecular weight molecules has typically used calibrants such as myoglobin or cytochrome *c*, perhaps admixed with a lower molecular weight peptide [39]. In this instance, it is important that both the origin (e.g. 'equine skeletal muscle myoglobin', 'horse heart cytochrome *c*') and the amino acid composition (cf. ref. 40) of these protein materials should be precisely defined. A new method for ES calibration under LC–MS conditions uses clusters generated over the mass range m/z 100–1500 from aqueous solutions of acetonitrile or methanol which also contain a small percentage of acid [36].

Matrix Assisted Laser Desorption

Mass assignment in MALDI presents a particular problem which is partly a result of the nature of the ionization process and partly related to the nature of the mass analyser used [41]. Thus, while the accuracy of mass assignment with pure samples may be acceptable, this is generally not the case with contaminated samples. For example, the presence of inorganic impurities often leads to a requirement for higher laser power which in turn enhances adduct ion formation. These adduct ions are not resolved at high mass with a time-of-flight analyser and consequently lead to systematic errors in mass assignment. One procedure has been relatively successful is the use of a well characterized, co-ionized internal standard which is close in molecular weight and chemically similar to the sample. For example, Beavis and Chait [42] have used the singly and doubly protonated molecular ions from equine skeletal muscle myoglobin as internal calibration points to achieve routine accuracies of $\pm 0.01\%$. Other internal standards such as bovine insulin for lower molecular weight samples and bovine serum albumin [41] have been used in a similar manner.

References

1. Chapman, J.R. (1978), *Computers in Mass Spectrometry*, Academic Press, London.
2. Morris, H.R., Dell, A., and McDowell, R.A. (1981), *Biomed. Mass Spectrom.*, **8**, 463.
3. Warburton G.A., McDowell, R.A., Taylor, K.T., and Chapman, J.R. (1980), in *Advances in Mass Spectrometry*, Vol. 8B (Ed. A. Quayle), p.1952, Heyden and Son, London.
4. Ligon, W.V. Jr. (1978), *Anal. Chem.*, **50**, 1228.
5. Olson, K.L., Rinehart, K.L. Jr., and Cook, J. C., Jr. (1977), *Biomed Mass Spectrom.*, **4**, 284.
6. Aczel, T. (1968), *Anal. Chem.*, **40**, 1917.
7. Wallick, R.H., Peele, G.L., and Hynes, J.B. (1969), *Anal. Chem.*, **41**, 388.
8. Dzidic, I., Desiderio, D.M., Wilson, M.S., Crain, P.F., and McCloskey, J.A. (1971), *Anal. Chem.*, **43**, 1877.
9. Hunt, D.F., and Ryan, J.F. III (1972), *Anal. Chem.*, **44**, 1306.
10. Maeder, H., and Gunzelmann, K.H. (1988), *Rapid Commun. Mass Spectrom.*, **2**, 199.
11. Bertrand, M.J., Maltais, L., and Evans, M.J. (1987), *Anal. Chem.*, **59**, 194.
12. McCamish, M., Allan, A.R., and Roboz, J. (1987), *Rapid Commun. Mass Spectrom.*, **1**, 124.
13. Jiang, L., and Moini, M. (1992), *J. Am. Soc. Mass Spectrom.*, **3**, 842.
14. Haddon, W.F., and Lukens, H.C. (1974) *Proceedings of the 22nd ASMS Conference on Mass Spectrometry and Allied Topics*, Philadelphia, Paper U2, p.436.

15. Bruins, A.P. (1980), *Biomed. Mass Spectrom.*, **7**, 454.
16. Dougherty, R.C., Whitaker, M.J., Smith, L.M., Stalling, D.L., and Kuehl, D.W. (1980), *EHP, Environ. Health Perspect.*, **36**, 103.
17. Budzikiewicz, H. (1986), *Mass Spectrom. Rev.*, **5**, 345.
18. Barlak, T.M., Wyatt, J.R., Colton, R.J., DeCorpo, J.J., and Campana, J.E. (1982), *J. Am. Chem. Soc.*, **104**, 1212.
19. Vekey, K. (1989), *Org. Mass Spectrom.*, **24**, 183.
20. Rubino, F.M. (1991), *Org. Mass Spectrom.*, **26**, 718.
21. Sim, P.G., and Boyd, R.K. (1991), *Rapid Commun. Mass Spectrom.*, **5**, 538.
22. Morgan, R.P., and Reed, M.L. (1982), *Org. Mass Spectrom.*, **17**, 537.
23. Chang, T.T., Tsou, H.-R., and Siegel, M.M. (1987), *Anal. Chem.*, **59**, 614.
24. Gilliam, J.M., Landis, P.W., and Occolowitz, J.L. (1983), *Anal. Chem.*, **55**, 1531.
25. Heller, D., Hansen, G., Yergey, J., Cotter, R.J., and Fenselau, C. (1982), *Proceedings of the 30th ASMS Conference on Mass Spectrometry and Allied Topics*, Honolulu, p.560.
26. Rinehart, K.L. Jr. (1985), in *Mass Spectrometry in the Health and Life Sciences* (Eds. A.L. Burlingame and N. Castagnoli, Jr.), p. 119, Elsevier, Amsterdam.
27. Siegel, M.M., Tsao, R., Oppenheimer, S., and Chang, T.T. (1990), *Anal. Chem.* **62**, 322.
28. Hemling, M.E. (1990), in *Proceedings of the 38th ASMS Conference on Mass Spectrometry and Allied Topics*, Tucson, Arizona, p.174.
29. Matsuo, T., Matsuda, H., and Katakuse, I. (1979), *Anal. Chem.*, **51**, 1329.
30. Rinehart, K.L. Jr., Cook, J.C. Jr., Meng, H., Olson, K.L., and Pandey, R.C. (1977), *Nature*, **269**, 832.
31. Yang, L., and Fergusson, G.J. (1985), *33rd ASMS Conference on Mass Spectrometry and Allied Topics*, San Diego, California, p.775.
32. Davidson, W.C., Dinallo, R.M., and Hansen G.E. (1991), *Biol. Mass Spectrom.*, **20**, 389.
33. Saar, J., and Haimberg, M. (1991), *Org. Mass Spectrom.*, **26**, 660.
34. Heeremans, C.E.M., van der Hoeven, R.A.M., Niessen, W.M.A., and van der Greef, J. (1989), *Org. Mass Spectrom.*, **24**, 109.
35. Stout, S.J., and DaCunha, A.R. (1990), *Org. Mass Spectrom.*, **25**, 187.
36. Anacleto, J.F., Boyd R.K., Pleasance, S., Sim, P.G., and Thibault, P. (1991), in *Proceedings of the 39th ASMS Conference on Mass Spectrometry and Allied Topics*, Nashville, Tennessee, p.1366.
37. Cody, R.B., Tamura, J., and Musselman, B.D. (1992), *Anal. Chem.*, **64**, 1561.
38. Feng, R., Konishi, Y., and Bell, A.W. (1991), *J. Am. Soc. Mass Spectrom.*, **2**, 387.
39. Chowdhury, S.K., Katta, V., and Chait, B.T. (1990), *Rapid. Commun. Mass Spectrom.*, **4**, 81.
40. Zaia, J., Annan, R.S., and Biemann, K. (1992), *Rapid Commun. Mass Spectrom.*, **6**, 32.
41. Lewis, S., Korsmeyer, K.K., and Correia, M.A. (1993), *Rapid Commun. Mass Spectrom.*, **7**, 16.
42. Beavis, R.C., and Chait, B.T. (1990), *Anal. Chem.*, **62**, 1836.

Appendix B

Introduction

The main purpose of this book is to introduce the reader, via the basic principles of instrumental mass spectrometry, to the different analytical techniques offered by modern mass spectrometric instrumentation. As a result of this emphasis on methodology, little space has been allocated to any discussion of the problems associated with interpretation of data produced by the use of these techniques.

The data compilations in this appendix attempt to make some amends in this respect, although for more detailed information, the reader is directed to the use of more specialised texts [1–3]. The tables, which deal with the interpretation of electron impact ionization spectra, are, hopefully, self-explanatory and should provide some useful starting points in any consideration of EI spectra. It should be emphasised, however, that these lists of masses and neutral losses are not exhaustive and that the compositions and structural inferences suggested are only those most commonly encountered.

Table B1 Masses and possible compositions of common fragment ions

Mass	Groups Commonly Associated with the Mass	Possible Inference
15	CH_3^+	—
18	$H_2O^{+\cdot}$	—
26	$C_2H_2^{+\cdot}$	—
27	$C_2H_3^+$	—
28	$C_2H_4^{+\cdot}$,$CO^{+\cdot}$	—
28	$N_2^{+\cdot}$	Background air
29	$C_2H_5^+$,CHO^+	—
30	CH_2=NH_2^+	Amine
30	NO^+	Nitro compound
31	CH_2=OH^+	Primary alcohol
32	$O_2^{+\cdot}$	Background air
36 & 38 (3:1)	$HCl^{+\cdot}$	Chloro compound
39	$C_3H_3^+$	—
40	$C_3H_4^{+\cdot}$	—
40	$Ar^{+\cdot}$	Background air
41	$C_3H_5^+$	—
41	$C_2H_3N^{+\cdot}$	Aliphatic nitrile
42	$C_3H_6^{+\cdot}$	—
43	$C_3H_7^+$	C_3H_7–, hydrocarbon chain
43	CH_3CO^+	CH_3CO–
44	$C_2H_6N^+$	Aliphatic amine
44	O=C=NH_2^+	Primary amide
44	$CO_2^{+\cdot}$	Background air
45	CH_2=OCH_3^+,CH_3CH=OH^+	Ether, alcohol
47	CH_2=SH^+, CH_3S^+	Thio compound
49 & 51 (3:1)	CH_2Cl^+	Chloro compound
50	$C_4H_2^{+\cdot}$	Aromatic compound
51	$C_4H_3^+$	C_6H_5–
55	$C_4H_7^+$	—
56	$C_4H_8^{+\cdot}$	—
57	$C_4H_9^+$	C_4H–$_9$
57	$C_2H_5CO^{+\cdot}$	Ethyl ketone, propionate ester
58	$C_3H_8N^+$	Aliphatic amine
58	CH_2=$C(CH_3)OH^{+\cdot}$	Some aliphatic ketones
59	CH_2=$OC_2H_5^+$, C_2H_5CH=OH^+	Ether, alcohol
59	$CO_2CH_3^+$	Methyl ester
59	CH_2=$C(OH)NH_2^{+\cdot}$	Primary amide
60	CH_2=$C(OH)OH^{+\cdot}$	Carboxylic acid
61	$C_2H_5S^+$	Thio compound
61	$CH_3CO_2H_2^+$	Alkyl acetate

Mass	Groups Commonly Associated with the Mass	Possible Inference
65	$C_5H_5^+$	Aromatic compound
69	$C_5H_9^+$	—
69	CF_3^+	Fluorocarbon
71	$C_5H_{11}^+$	C_5H_{11}–
71	$C_3H_7CO^+$	Propyl ketone, butyrate ester
72	$C_4H_{10}N^+$	Aliphatic amine
72	$CH_2{=}C(C_2H_5)OH^{+\cdot}$	Some aliphatic ketones
73	$C_4H_9O^+$	Ester, alcohol
73	$(CH_3)_3Si^+$	Trimethylsilyl derivative
74	$C_3H_6O_2^{+\cdot}$	Methyl ester, carboxylic acid
75	$(CH_3)_2Si{=}OH^+$	Trimethylsilyl derivative
75	$C_6H_3^+$	Disubstituted benzene
76	$C_6H_4^{+\cdot}$	Benzene derivative
77	$C_6H_5^+$	Monosubstituted benzene
78	$C_6H_6^{+\cdot}$	Monosubstituted benzene
79 & 81 (1:1)	Br^+	Bromo compound
80 & 82 (1:1)	$HBr^{+\cdot}$	Bromo compound
80	$C_5H_6N^+$	Some pyridines, pyrroles, anilines
81	$C_5H_5O^+$	Some furans
83 & 85 & 87 (9:6:1)	$CHCl_2^+$	*gem*-dichloro compounds
85	$C_5H_9O^+$	butyl ketone,
85	$C_4H_5O_2^+$	
86	$C_5H_{12}N^+$	Aliphatic amine
87	$C_4H_7O_2^+$	Methyl ester
91	$C_7H_7^+$	Aromatic (possibly benzylic) compound
91 & 93 (3:1)	$C_4H_8Cl^+$	Alkyl chloride
92	$C_7H_8^{+\cdot}$	Benzylic compound
93 & 95 (1:1)	CH_2Br^+	Bromo compound
94	$C_6H_6O^{+\cdot}$	C_6H_5O–
95	$C_5H_3O_2^+$	Furyl–CO–
95	$C_6H_7O^+$	Some methyl furans
97	$C_5H_5S^+$	Some thiophenes
99	$C_5H_7O_2^+$	Ethylene ketal
105	$C_6H_5CO^+$	Benzoyl compound
105	$C_8H_9^+$	Benzene derivative
106	$C_7H_8N^+$	Some pyridines or anilines
107	$C_7H_7O^+$	Some phenols

Mass	Groups Commonly Associated with the Mass	Possible Inference
121	$C_8H_9O^+$	Some phenols and anisoles
122	$C_6H_5CO_2H^{+\cdot}$	Alkyl benzoate
123	$C_6H_5CO_2H_2^+$	Alkyl benzoate
127	I^+	Iodo compound
130	$C_9H_8N^+$	Some indoles
135 & 137 (1:1)	$C_4H_8Br^+$	Alkyl bromide
147	$(CH_3)_2Si{=}OSi(CH_3)_3^+$	Trimethylsilyl derivative
149	$C_8H_5O_3^+$	Dialkyl phthalate (plasticizer)
207	$C_5H_{15}Si_3O_3^+$	Methyl silicone (silicone grease or GC column bleed)
446	$C_{30}H_{22}O_4^{+\cdot}$	Polyphenylether (diffusion pump oil)

Table B2 Some common neutral losses from molecular ions

Mass	Groups Commonly Associated with the Mass Lost	Possible Inference
M–1	H	—
M–2	H_2	—
M–14	—	Homologue
M–15	CH_3	—
M–16	O	Nitro compound, *N*-oxide, sulphoxide
M–16	NH_2	$ArCONH_2$, $ArSO_2NH_2$
M–17	OH	$ArCO_2H$
M–18	H_2O	Alcohol, ketone, ether, etc.
M–19	F	Fluoro compound
M–20	HF	Fluoro compound
M–26	C_2H_2	Aromatic hydrocarbon, ArF
M–27	HCN	Aromatic amine or nitrile, nitrogen heterocycle
M–28	C_2H_4	Ethyl ester, aromatic ethyl ether, *n*-propyl ketone
M–28	CO	Quinone, phenol, oxygen heterocycle
M–29	C_2H_5	Ethyl group
M–29	CHO	Aromatic aldehyde, phenol
M–30	CH_2O	Aromatic methyl ether
M–30	NO	Aromatic nitro compound
M–31	CH_3O	Methyl ester, methyl ether
M–32	CH_3OH	Methyl ether, some methyl esters
M–33	SH	Thiol
M–33	H_2O+CH_3	—
M–34	H_2S	Thiol
M–35 & M–37	Cl	Chloro compound
M–36 & M–38	HCl	Chloro compound
M–41	C_3H_5	Propyl ester
M–42	CH_2CO	Acetamides, aromatic acetates
M–42	C_3H_6	Butyl ketone, aromatic propyl ether
M–43	C_3H_7	Propyl ketone
M–43	CH_3CO	Methyl ketone
M–44	CO_2	Anhydride, unsaturated ester
M–45	CO_2H	Carboxylic acid
M–45	C_2H_5O	Ethyl ester, ethyl ether
M–46	C_2H_5OH	Ethyl ether, some ethyl esters
M–46	NO_2	$ArNO_2$
M–48	CH_3SH	Methyl thioether
M–55	C_4H_7	Butyl ester
M–56	C_4H_8	Aromatic butyl ether, amyl ketone

Table B2 Some common neutral losses from molecular ions

Mass	Groups Commonly Associated with the Mass Lost	Possible Inference
M–57	C_4H_9	Butyl ketone
M–57	C_2H_5CO	Ethyl ketone
M–60	CH_3CO_2H	Acetate

References

1. McLafferty, F.W., and Turecek, P. (1993), *Interpretation of Mass Spectra* (Fourth Edition), University Science Press, Mill Valley, CA.
2. Davis, R., and Frearson, M. (1987), *Mass Spectrometry (Open Learning)*, John Wiley, Chichester.
3. Chapman, J.R. (1992), *Interpretation of Mass Spectra. An ACS Audio-Cassette-Workbook*, American Chemical Society, Washington, DC.

Appendix C

Some acronyms and other abbreviations used in this text and in mass spectrometry generally:

APCI	Atmospheric pressure chemical ionization.
API	Atmospheric pressure ionization.
APSI	Atmospheric pressure (thermo)spray ionization.
B	Magnetic sector – also used in instrument configurations, e.g. BEEB, and in linked scanning modes, e.g. B/E.
CAD	Collision-activated decomposition.
CDD	Chlorinated dibenzo-*p*-dioxins.
CDF	Chlorinated dibenzofurans.
CE-MS	Capillary electrophoresis-mass spectrometry.
CF-FAB	Continuous flow – fast atom bombardment.
CI	Chemical ionization.
CID	Collision-induced dissociation.
DCI	Desorption chemical ionization.
DEI	Desorption electron ionization.
DHB	2,5-Dihydroxybenzoic acid (gentisic acid).
DLI	Direct liquid introduction.
E	Electrostatic sector – also used in instrument configurations, e.g. BEEB, and in linked scanning modes, e.g. B/E.
e^-	Electron.
e^-th	Electron with thermal energy.
EI	Electron impact (ionization), more correctly electron ionization.
EPA	Environmental Protection Agency.
ES	Electrospray (ionization).
FAB	Fast atom bombardment.
FD	Field desorption.
FFT	Fast Fourier Transform.
FI	Field ionization.
FTMS	Fourier transform mass spectrometry.
GC-MS	Gas chromatography-mass spectrometry.
GLT	Glass-lined tubing.
IP	Ionization potential.
IS	Ion-spray.
ITMS	Ion trap mass spectrometry.

LC-MS	Liquid chromatography-mass spectrometry.
LDMS	Laser desorption mass spectrometry.
LOD	Limit of detection.
LOQ	Limit of quantitation.
LSIMS	Liquid secondary-ion mass spectrometry.
$M^{+\cdot}$	Molecular ion(-radical – positively charged.
$(M + H)^+$	Protonated molecular ion – positively charged.
$(M + NH_4)^+$	Ammoniated molecular ion – positively charged.
$(M + Na)^+$	Sodiated molecular ion – positively charged.
$(M - h)^+$	Molecular ion after loss of hydrogen radical – positively charged.
$M^{-\cdot}$	Molecular ion(-radical) – negatively charged.
$(M + Cl)^-$	Chloride addition to molecular ion – negatively charged.
$(M - H)^-$	Molecular ion after loss of hydrogen radical – negatively charged.
MALDI	Matrix-assisted laser desorption-ionization.
MIKES	Mass-analysed ion kinetic energy scan.
MS-MS	Mass spectrometry-mass spectrometry or tandem mass spectrometry.
$(MS)^2$	As MS-MS.
$(MS)^n$	A multiple-stage MS-MS experiment.
m-NBA	*m*-Nitrobenzyl alcohol.
PA	Proton affinity.
PAD	Post-acceleration detector.
PAH	Polycyclic aromatic hydrocarbon.
PCB	Polychlorinated biphenyl.
PDMS	Plasma desorption mass spectrometry.
PFK	Perfluorokerosene.
PFTBA	Perfluorotri-*n*-butylamine.
ppm	Parts per million.
Q	Quadrupole mass filter – also used in instrument configurations, e.g. BEqQ, and in linked scanning modes, e.g. B^2/Q.
q	Quadrupole mass filter used as a collision cell – see preceding definition for an example.
RE	Recombination energy.
REMPI	Resonance-enhanced multiphoton ionization.
RIC	(Data system) reconstructed ion current.
RP	Resolving power.
R2PI	Resonant two-photon ionization.
SFC-MS	Supercritical fluid chromatography-mass spectrometry.
SID	Surface-induced decomposition.
SIM	Selected ion monitoring.
SIMS	Secondary-ion mass spectrometry.
S/N	Signal-to-noise ratio.
SRM	Selected reaction monitoring.
TCDD	Tetrachlorodibenzo-*p*-dioxin.
TIC	Total ion current.
TLC	Thin layer chromatography.
TOF	Time-of-flight (instrument).
TQ	Triple quadrupole (instrument).
TS	Thermospray (ionization).
TZ	Separation number (Trennungszahl).

Index

Compound entries that are not otherwise qualified refer to the analysis of that compound